Sensor Systems for Environmental Monitoring

Volume One: Sensor Technologies

Sensor Systems for Environmental Monitoring

Volume One: Sensor Technologies

Edited by

M. Campbell
Department of Physical Sciences
Glasgow Caledonian University
Glasgow
UK

BLACKIE ACADEMIC & PROFESSIONAL
An Imprint of Chapman & Hall
London · Weinheim · New York · Tokyo · Melbourne · Madras

Published by Blackie Academic & Professional, an imprint of Chapman & Hall, 2–6 Boundary Row, London SE1 8HN, UK

Chapman & Hall, 2–6 Boundary Row, London SE1 8HN, UK

Chapman & Hall GmbH, Pappelallee 3, 69469 Weinheim, Germany

Chapman & Hall, 115 Fifth Avenue, New York NY 10003, USA

Chapman & Hall Japan, ITP-Japan, Kyowa Building, 3F, 2-2-1 Hirakawacho, Chiyoda-ku, Tokyo 102, Japan

DA Book (Aust.) Pty Ltd, 648 Whitehorse Road, Mitcham 3132, Victoria, Australia

Chapman & Hall India, R. Seshadri, 32 Second Main Road, CIT East, Madras 600 035, India

First edition 1997

© 1997 Chapman & Hall

Typeset in 10/12 pt Times by AFS Image Setters (Glasgow) Ltd
Printed in Great Britain by Alden, Oxford, Didcot and Northampton

Cover photography by Jackie McWilliams and John Giles, Audio-visual Dept, Glasgow Caledonian University

ISBN   0 7514 0267 2 (Two volume set)
           0 7514 0418 7 (Volume One)
           0 7514 0419 5 (Volume Two)

A catalogue record for this book is available from the British Library

Library of Congress Catalog Card Number: 96–83011

To my wife, Imelda Anne,
for her encouragement and
support during the compilation
of this book.

Contents

3   Laser-based sensors   65
K.W.D. Ledingham and M. Campbell

4   Electrochemical sensors   100
R.O. Ansell and A. McNaughtan

Contents

Volume 2

5   Industrial methods of spectrophotometric measurements in process control   236

A.D. McIntyre

Contributors

R.O. Ansell
Department of Physical Sciences
Glasgow Caledonian University
Glasgow, UK

J. Bates
Department of Electrical and Electronic
Engineering and Physics
University of Northumbria
Newcastle-upon-Tyne, UK

G. Bishop
SGS Environment
Liverpool, UK

B.I. Brookes
Department of the Regional Chemist
Glasgow, UK

J.N. Cape
Institute of Terrestrial Ecology
Edinburgh Research Station
Penicuik, UK

M. Campbell
Department of Physical Sciences
Glasgow Caledonian University
Glasgow, UK

M. Cardosi
Department of Biology
University of Paisley
Paisley, UK

J.M. Crowther
Department of Physical Sciences
Glasgow Caledonian University
Glasgow, UK

J.L.E. Flack
SGS Environment
Colwyn Bay
Clwyd, UK

J. Gemmill
Department of the Regional Chemist
Glasgow, UK

B. Haggett
The Research Centre
University of Luton,
Luton UK

M.J. Hepher
Department of Energy and
Environmental Technology
Glasgow Caledonian University
Glasgow, UK

E.A. Knight
Department of Physical Sciences
Glasgow Caledonian University
Glasgow, UK

K.W.D. Ledingham
Department of Physics and Astronomy
University of Glasgow
Glasgow, UK

A.D. McIntyre
Zeneca Fine Chemicals
Grangemouth
Stirlingshire, UK

A. McNaughtan
Department of Physical Sciences
Glasgow Caledonian University
Glasgow, UK

J.V. Magill
Department of Electronics and
Electrical Engineering
University of Glasgow
Glasgow, UK

J.R. Pugh
Department of Physical Sciences
Glasgow Caledonian University
Glasgow, UK

D. Reilly
Department of Physical Sciences
Glasgow Caledonian University
Glasgow, UK

G. Stewart
Department of Electronic and Electrical
Engineering
University of Strathclyde
Glasgow, UK

P.A. Strachan
Department of Mechanical Engineering
University of Strathclyde
Glasgow, UK

1 Fibre optic sensors

G. STEWART

1.1 Introduction to fibre optics

Fibre optics is best known today for its widespread and ever growing application in communication networks around the world, linking continents by undersea fibre cables and forming the backbone of the major communication routes within major nations. This trend is set to continue with further growth of 'information superhighways' and the increased deployment of fibre closer to the office and home in the local area network. These developments have been fuelled by the inherent capacity of fibre optic systems to carry a huge information volume as well as the significant advances since the early 1970s in fibre optics, components and laser technology.

Particularly from the early 1980s however, the application of fibre optics in measurement and sensor systems began to emerge, and although not nearly so huge as the communications market, the area of fibre sensors has grown steadily to become of major importance. A key stimulus has been the availability of high-quality fibres and optoelectronic components as a spin-off from the communications field; in more recent times, the growth and demand within the sensor area has warranted research and development of components in their own right for sensor application. Today sensors have been demonstrated for a wide range of physical and chemical parameters including temperature, pressure, strain, vibration, acceleration, electric/magnetic fields, current, etc., and several fibre systems have reached the production stage, the most notable being the fibre optic gyroscope.

Within the chemical sensing area, there have been several factors that have helped to accelerate the interest in fibre sensors. Increasing concerns over environmental pollution means that environmental protection is receiving national and global attention; there is continual need for safe and accurate monitoring of industrial processes; in the medical field, new trends in medical diagnostics and preventive medicine have led to demands for new chemical/biochemical sensor strategies.

1.1.1 The need for environmental monitoring and new technology

Recently, the European Union (EU) has launched the new European Environmental Protection Agency and plans to adopt a new approach to pollution reduction, namely, the principle of integrated pollution control

Table 1.1 Pollutants targeted by IPC directive of the EU

Discharges to air	Discharges to water
Sulphur dioxide and other sulphur compounds	Organohalogen compounds and substances that may form these in the aquatic environment
Oxides of nitrogen and other nitrogen compounds	Organophosphorus compounds
Carbon monoxide and carbon dioxide	Organo-tin compounds
Volatile organic compounds	Mercury and its compounds
Heavy metals and their compounds	Cadmium and its compounds
Dust, asbestos, glass and mineral fibres	Persistent mineral oils and petroleum-derived hydrocarbons
Chlorine and its compounds	Persistent synthetic substances that may float, remain in suspension or sink and interfere with any use of water
Fluorine and its compounds	
Ammonia	
Arsenic and its compounds	Nutrients, e.g. nitrates and phosphates
Hydrogen fluoride	Substances proved to have carcinogenic properties in or via the aquatic environment
Hydrogen cyanide	
Nitric acid	Metals: Zn, Cu, Ni, Cr, Pb, Se, As, Sb, Mo, Ti, Sn, Ba, Be, B, U, V, Co, Tl, Te and Ag
Known airborne carcinogens	

(IPC) (Bogue, 1994). The idea is to ensure overall pollution reduction rather than concentrating on the individual zones of land, air and water. The IPC programme will incorporate many of the earlier environmental laws and has been designated a priority field of action by the European Commission. Consequently, in member states throughout the EU, a very large number of companies, including those involved with energy, metals, minerals, chemicals, waste management, leather tanneries, etc., will need to comply with legislation governing discharge limits of pollutants. While companies will seek to avoid or minimise their involvement with hazardous materials, this will never be totally possible and some form of monitoring will be necessary to ensure compliance with regulations. This could take the form of sampling and analysis of discharges or spot checks with portable instruments, but the most desirable technique is continuous on-line monitoring, which would also provide evidence of compliance with regulations. Fibre optic sensors are ideally suited to this type of *in situ* real time monitoring. The list of substances implicated in the EU directive is quite large as shown in Table 1.1 and for many substances convenient and reliable sensor technologies have yet to be developed. The total market in Western Europe for process gas and water analysers is estimated to grow to over $700 million by 1997 and hence there exists a strong stimulus for research and development of new and improved on-line gas and chemical analysers. Fibre optic systems are likely to play a key role here.

A similar situation prevails in other continents. In the USA, for example, the US Environmental Protection Agency has listed over 180 substances under pollution control directives. It is estimated that there are more than

28 000 uncontrolled or abandoned hazardous waste sites and monitoring is required for many of the landfill and existing municipal sites. There are about $1\frac{1}{2}$ million underground fuel storage tanks of which about one third leak into ground water and hence into drinking water which may be obtained from these supplies. There is clearly a need for new chemical sensors covering a wide range of substances and environments for both identifying contamination and its sources (diagnostic function) and for continuous monitoring of known pollutants. Fibre optic sensors are seen as primarily suited to the latter task where numerous repetitive measurements are required at a low cost (Klainer, 1991).

1.1.2 Advantages and disadvantages of fibre optic sensors

Perhaps the most outstanding advantage of optical fibres is their ability to transmit light over large distances, typically tens to hundreds of metres, with low loss allowing a sensor head to be remotely located from the instrumentation. This feature is particularly useful for sensing in harsh environments where hazardous chemicals may be present or extremes in temperature occur. Additionally the flexibility and small cross-sectional size of fibres allows access to areas that would be otherwise difficult to reach.

Provided the optical power densities are within certain limits, fibre optic chemical sensors are much safer in explosive environments compared with sensors involving electrical signals, where a spark may trigger a gas explosion. Optical signals are immune to electrical or magnetic interference from, for example, power lines and electrical machinery.

The amount of information that can be carried by an optical fibre is very much greater than that carried by electrical wires and this can often be utilised in sensor systems. The intensity, wavelength, phase and polarisation of light can all be used as measurement parameters, and several wavelengths launched in the same optical fibre in either direction form independent signals. This gives the possibility of monitoring several chemicals with the same fibre sensor or simultaneously monitoring, for example, the temperature. Multiplexing of fibre optic systems is relatively easy, allowing expensive source or analysis instrumentation to be shared among a number of sites.

On the negative side, fibre sensors are often sensitive to parameters other than the desired measurand, for example, temperature or disturbance of the fibre. Cross-sensitivity to other chemical species also needs to be considered. There are, in addition, a number of disadvantages specific to certain types of fibre sensor. The optical fibre itself has a background absorption, fluorescence and Raman scatter spectrum and these need to be taken into account, especially in fibre Raman systems where signals are generally very weak. Many types of fibre sensor are based on indicator dyes or other chemically sensitive reagents, and problems can arise through the dye leaching out or through photobleaching effects. Although reference electrodes are not required as in conventional potentiometric sensors, optical fibre sensors generally require optical referencing, usually at a second (non-absorbing) wavelength.

1.2   Physical principles of fibre optic chemical sensors

Optical methods for the identification of and for concentration measurements of chemical species are well known and have been widely used long before the advent of modern fibre optics. Apart from a few exceptions, what is new about fibre optic chemical sensors is not so much the fundamentals but more the way in which the optics are assembled, involving fibre or optical waveguides. In this section, we review some of the more important physical principles behind fibre sensors. The most common phenomena used are absorption, fluorescence, Raman scattering and refractive index effects. The principles behind evanescent field devices, where the fibre itself is directly involved in the sensing, are also discussed.

1.2.1   Absorption sensors

The various discrete energy levels associated with atoms and molecules means that, in general, absorption of electromagnetic radiation by a chemical species takes place at well-defined wavelengths (λ) or frequencies (f) defined by:

$$E_2 - E_1 = \frac{hc}{\lambda} = hf \tag{1.1}$$

where $E_2 - E_1$ is the difference in energy levels of an electronic or molecular transition, hf is the photon energy required to induce the transition, h is Planck's constant and c is the free space velocity of light.

Since the energy levels associated with an atom or molecule are unique, the absorption spectrum serves as a 'fingerprint' identification of the chemical. The actual wavelengths at which absorption takes place depends on the transitions involved. Electron transitions within atoms and molecules occur at short wavelengths (high photon energy) corresponding to the UV–visible region of the spectrum; nitrogen dioxide gas, for example, has an electronic absorption band around a wavelength of 405 nm. Transitions in vibrational or rotational states of molecules occur at energies corresponding to the near- and mid-IR band of the spectrum. Methane gas (CH_4) has a fundamental molecular vibration giving rise to strong absorption lines around a wavelength of 3.3 μm. Typically, a vibrational absorption line has a degree of fine structure superimposed on

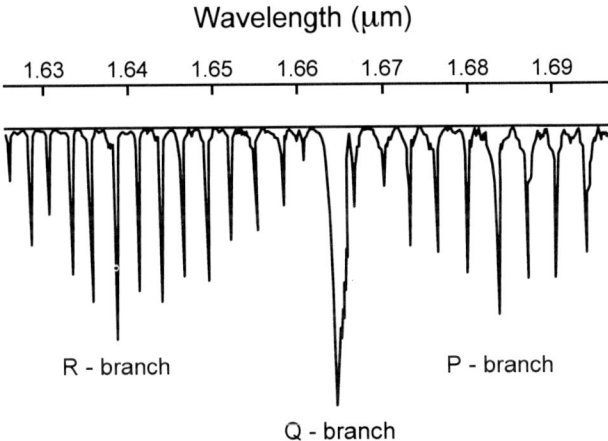

Figure 1.1 Methane absorption spectrum.

it corresponding to quantised rotational energy levels, so rather than being a single line the absorption spectra consists of a series of closely spaced lines. Also, absorption can occur at frequencies that are harmonics or a combination of harmonics of the fundamental. Again, with methane as an example, there is a series of weak absorption lines at 1.66 µm (i.e. at twice the fundamental absorption frequency) and at 1.33 µm (combination band). Figure 1.1 illustrates the methane absorption spectrum in the range 1.63–1.69 µm and Table 1.2 lists the absorption lines of some common gases in the near-IR, within the transmission window of optical fibres (Cooper and Martinelli, 1992).

The chemical concentration may be related to the measured optical absorption through the Beer–Lambert law. Figure 1.2 shows light of intensity I_0 incident on a cell of length l containing a chemical species which has an absorption line or band at $k\lambda_0$. The output intensity I is related to the

Table 1.2 Near-IR absorption lines for some common gases

Gas		Wavelength (μm)
Oxygen	O_2	0.761
Carbon dioxide	CO_2	1.573
Carbon monoxide	CO	1.567
Water vapour	H_2O	1.365
Nitrogen dioxide	NO_2	0.8
Methane	CH_4	1.665
Acetylene	C_2H_2	1.53
Ammonia	NH_3	1.544
Hydrogen sulphide	H_2S	1.578

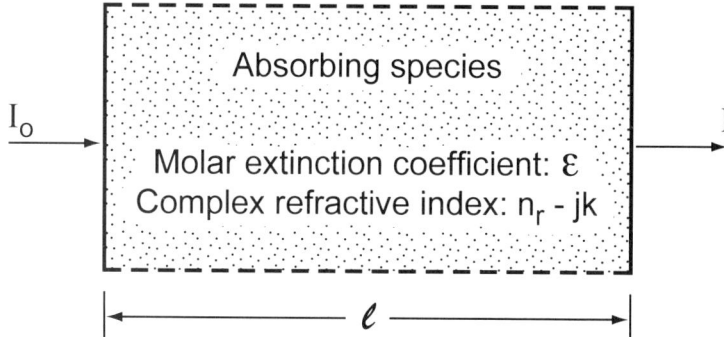

Figure 1.2 The output intensity I of incident light I_0 is related to the concentration of the chemical in the cell C (see equation 1.2).

chemical concentration C through the expression:

$$I = I_0 \cdot 10^{-\varepsilon l C} \tag{1.2}$$

where ε is the (molar) extinction coefficient.

Alternatively equation 1.2 may be written as:

$$I = I_0 \cdot \exp[-\alpha_m l C] \tag{1.3}$$

where α_m is the (molar) absorption coefficient.

Some typical values for α_m are as follows. For pure methane Tai *et al.*, (1987) quote a value for α_m of $8.2 \times 10^{-5}\,\mathrm{Pa^{-1}\,cm^{-1}}$ ($8.3\,\mathrm{atm^{-1}\,cm^{-1}}$) at 3.39 µm, while, from the data given by Chan and Inaba (1985), values of $1.4 \times 10^{-6}\,\mathrm{Pa^{-1}\,cm^{-1}}$ ($0.14\,\mathrm{atm^{-1}\,cm^{-1}}$) and $3 \times 10^{-7}\,\mathrm{Pa^{-1}\,cm^{-1}}$ ($0.03\,\mathrm{atm^{-1}\,cm^{-1}}$) may be calculated for the Q-branch of the absorption lines at 1.665 and 1.331 µm, respectively, where l in equation 1.3 is in centimetres and C is the gas pressure in Pascals or atmospheres. From the data given by Inn and Tanaka (1953) and Griggs (1968), ozone has a peak absorption coefficient in the UV (the Hartley band) of $\sim 1 \times 10^{-21}\,\mathrm{m^2}$ at 250 nm and in the visible (Chappius band) of $\sim 5 \times 10^{-25}\,\mathrm{m^2}$ at 600 nm where l in equation 1.3 is in metres and C is in molecules $\mathrm{m^{-3}}$. Data for other gases may be found from Erley and Blake (1964; 1965).

In the design of optical fibre sensors, it is often convenient to use a complex refractive index to describe the chemical absorption rather than the absorption coefficient. These quantities may be related as follows. The electric field E of the plane electromagnetic wave propagating through the absorption cell in Figure 1.2 is described by:

$$E = E_0 \exp j[\omega t - \beta z] \tag{1.4}$$

where β is the propagation constant given by:

$$\beta = k_0 n = \frac{2\pi}{\lambda_0}(n_r - jk) \tag{1.5}$$

and n_r, k are the real and imaginary parts of the complex index.

Substitution of equation 1.5 into 1.4 gives:

$$E = E_0 \exp(-k_0 kz) \exp j(\omega t - k_0 n_r z) \tag{1.6}$$

Since the intensity is proportional to the square of the amplitude, then, after traversing a cell of length l, we have:

$$I = I_0 \exp(-2k_0 kl) \tag{1.7}$$

Comparing equations 1.7 and 1.3 the relation between absorption coefficient and the imaginary part of the complex index is:

$$k = \frac{\alpha_m C \lambda_0}{4\pi} \tag{1.8}$$

For gases, $n_r \sim 1$ and values for k typically lie in the range 10^{-3} to 10^{-6} so that $k \ll n_r$. The complex refractive index can be used in theoretical and computer models of optical fibres or in waveguides to predict the sensitivity of an absorption-based sensor.

1.2.2   Sensing through indicator dyes

In a number of cases, the direct absorption properties of the chemical may not be amenable to the design of a fibre optic chemical sensor. This may be because the absorption is too weak or the wavelength of absorption is not compatible with the transmission window of silica fibres. Additionally, availability and cost of light sources and detectors at a given wavelength determine what is realistically possible. In these cases, it is sometimes possible to introduce an intermediate compound to mediate the sensing. The intermediate compound, typically an organic dye, undergoes an optical change through a chemical reaction or other process on exposure to the chemical of interest and it is the optical properties of the dye which are monitored in the sensor.

The most common example of this technique is that of pH measurements based on the colour change of an indicator dye, as in litmus paper. There are a large number of pH-sensitive dyes, such as cresol red, bromocresol green, bromothymol blue, etc., which have different pH ranges and colour changes. The principles of operation are as follows. The indicator dye, which is denoted by HI, is itself a weak acid (or base) which ionises according to the equation:

$$HI \rightleftharpoons H^+ + I^- \tag{1.9}$$

The acid and base forms of the dye (i.e. HI and I^-) have different colours or absorption coefficients; for example with the indicator phenol red, HI is red and I^- is yellow.

Equation 1.9 is governed by a thermodynamic equilibrium constant K_a;

$$K_a = \frac{\{H^+\}\{I^-\}}{\{HI\}} \tag{1.10}$$

where { } represents the activity of the species.

For dilute solutions, the activity is approximately equal to the concentration and with the definitions: $pH = -\log[H^+]$ and $pK_a = -\log K_a$ we obtain:

$$pH = pK_a - \log\frac{[HI]}{[I^-]} \tag{1.11}$$

where [ ] represents the concentration of the species.

For non-ideal solutions where the activity is not equal to the concentration, the thermodynamic equilibrium constant in equation 1.11 may be replaced by an effective value K_e given by Edmonds et al., (1988):

$$pK_e = pK_a + \log\frac{\gamma_{HI}}{\gamma_I} \tag{1.12}$$

where each γ is an activity coefficient.

It is clear from equations 1.9 and 1.10 that an increase in the H^+ concentration (i.e. a reduction in the solution pH) will shift the reaction in equation 1.9 to the left, increasing the concentration of HI and reducing the concentration of I^-. This produces the colour change because of the different absorption characteristics of HI and I^-. The pH can be determined by measuring the absorption at a suitable wavelength. The absorbance (or optical density) at a wavelength λ_0 is given by:

$$A = \varepsilon_{HI}[HI]l + \varepsilon_I[I^-]l \tag{1.13}$$

where each ε is an extinction coefficient at λ_0 and l is the interaction length.

The total indicator concentration may be assumed constant so that $[HI]+[I]=\text{constant}=C$ and we obtain from equation 1.13:

$$A = A_{max} - b[HI]$$
$$A = b[I^-] + A_{min} \tag{1.14}$$

where:

$$A_{max} = \varepsilon_I Cl$$
$$A_{min} = \varepsilon_{HI} Cl \tag{1.15}$$
$$b = (\varepsilon_I - \varepsilon_{HI})l$$

Hence equation 1.11 can be written in terms of the measured absorbance, A, as:

$$\text{pH} = \text{p}K_a - \log \frac{A_{\max} - A}{A - A_{\min}} \qquad (1.16)$$

On re-arranging, the absorbance as a function of pH has the form:

$$A = A_{\max} \cdot \frac{1 + r10^{-d}}{1 + 10^{-d}} \qquad (1.17)$$

where $d = \text{pH} - \text{p}K_a$ and r is the ratio of the extinction coefficients, $r = \varepsilon_{\text{HI}}/\varepsilon_{\text{I}}$.

Figure 1.3 illustrates the typical sigmoid curve obtained from equation 1.17 at a wavelength where the absorption of the undissociated species HI is negligible (i.e. $r = 0$) and for $\text{p}K_a = 7$. Figure 1.4 illustrates the typical absorbance of a pH-sensitive dye in solution as a function of wavelength for different pH values. Note that at one wavelength, the *isosbestic wavelength* λ_i, the absorption is unaffected by the pH value. This wavelength is particularly useful in fibre optic pH sensors as a reference to remove the effects of other optical losses, i.e. the output signal is taken as the ratio of absorption at a pH-sensitive wavelength and the isosbestic wavelength.

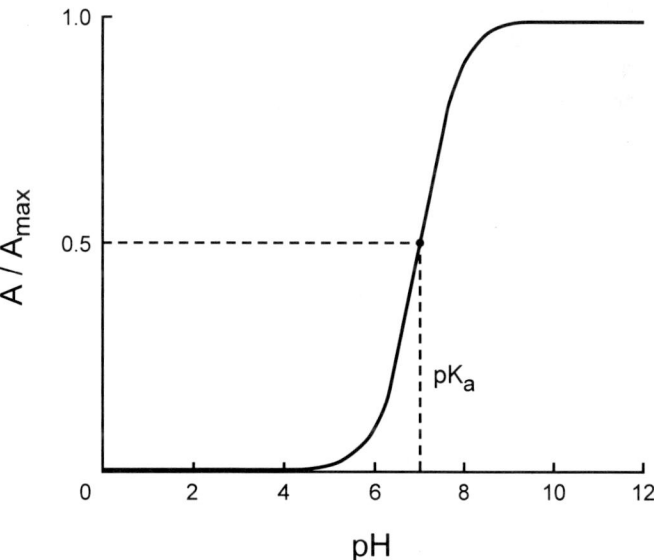

Figure 1.3 Absorbance as a function of pH (from equation 1.17).

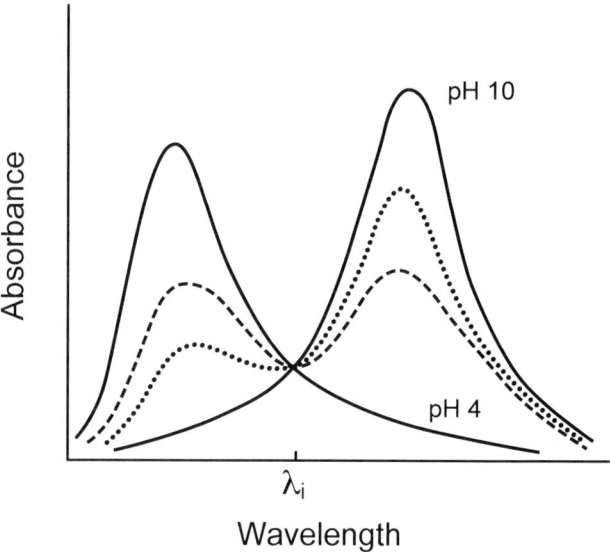

Figure 1.4 Absorbance of a pH-sensitive dye in solution as a function of wavelength at varying pH values.

So far we have been discussing pH sensors, but the same principles are employed for the optical detection of acidic or basic vapours and gases such as carbon dioxide and ammonia (Wolfbeis, 1991).

For carbon dioxide in aqueous solution, hydration and dissociation occur as:

$$CO_2 + H_2O \leftrightharpoons H_2CO_3$$
$$H_2CO_3 \leftrightharpoons H^+ + HCO_3^-$$
$$HCO_3^- \leftrightharpoons H^+ + CO_3^{-2}$$

(1.18)

Hence the carbon dioxide concentration can be determined from the pH changes induced in an aqueous solution.

Similarly Chernyak *et al.*, (1990) demonstrated a sensor for atmospheric ammonia based on colour change in the dye oxazine-170 (Ox). The sensor operates through the reaction:

$$OxH^+ + NH_3 \leftrightharpoons Ox + NH_4^+$$

(1.19)

with the equilibrium constant for the two forms of oxazine as:

$$K = \frac{[Ox][H^+]}{[OxH^+]}$$

(1.20)

Therefore, the colour change occurs as a result of the ammonia altering the relative concentrations of Ox and OxH$^+$, similar to pH sensing.

1.2.3 Fluorescence sensors

The concentration of certain chemicals may be measured through the fluorescence emitted from the chemical itself or from an intermediate fluorescent dye. Fluorescence occurs when a species in an electronically excited state releases its energy through light emission (Stokes' transition). Excitation is by absorption of photons from incident radiation at a wavelength λ_{ex} and the emitted fluorescence is at a longer wavelength λ_{fl}, where $(\lambda_{fl} - \lambda_{ex})$ is the Stokes' shift, as shown in Figure 1.5. In fluorescence, the species remains in the excited state for a very short time, with typical lifetimes of 1–20 ns (cf. phosphorescence where the lifetimes are much longer — typically 1 µs to 10 s).

Fibre optic sensors based on fluorescence make use of changes in either the intensity or lifetime of the fluorescence emission to monitor the chemical concentration. This type of sensor is generally more selective than absorption-based sensors because the sensing is performed at a different wavelength from the excitation and it is unlikely that other fluorescent species will be present at the same emission wavelength to cause errors. In practice, a large Stokes' shift is desirable to separate the excitation from the fluorescence at the fibre output with inexpensive wavelength filters and, ideally, the excitation wavelength should be in the visible or near IR region of the spectrum where light-emitting diodes (LED) or laser sources are cheap and readily available.

The simplest type of fluorescence sensor involves measuring the fluorescence intensity at a single wavelength, such as in pH sensors using the dye fluorescein. Here the fluorescence from the base form of the dye is monitored and the intensity increases when a pH change

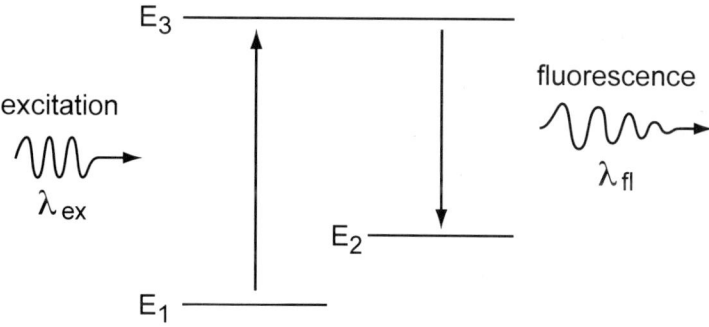

Figure 1.5 The Stokes' shift in fluorescence.

increases the concentration of the base form according to the chemical equilibrium equations 1.9 and 1.10. For weak absorption by the fluorescent species, the fluorescence intensity is given by Parker's law (Parker, 1968):

$$I_{fl} = 2.3kI_{ex}\xi\varepsilon l[D] \tag{1.21}$$

where I_{ex} is the excitation intensity, ξ is the fluorescence quantum efficiency, ε is the extinction coefficient at λ_{ex}, l is the path length, [D] is the concentration of the fluorescent species and k is a factor to take into account that only a fraction of the total emission is measured (depending on the sensor configuration).

As another example, sensors for metal ions in solution have been made using ligands which are only weakly fluorescent by themselves but form fluorescent complexes with ions such as Al^{3+}, Be^{2+}, Mg^{2+}, Zn^{2+}, Cd^{2+} (Seitz, 1984).

A second form of fluorescence sensor involves 'dynamic quenching' by the chemical of interest. Here the interaction is with the *excited* state of the fluorescent species, resulting in a *decrease* in the fluorescence intensity upon association with the chemical to be monitored. In this case there is also a reduction in the fluorescence lifetime. The emission is described by the Stern–Volmer equation (Lakowicz 1983, Schulman 1988):

$$\frac{I_0}{I} = \frac{\tau_0}{\tau} = 1 + k_q\tau_0[Q] \tag{1.22}$$

where I_0, I, τ_0, τ are the intensities and lifetimes in the absence and presence of the quencher, respectively, [Q] is the quencher concentration and k_q is the bimolecular quenching constant.

Although more sophisticated instrumentation is required, there are a number of important advantages in sensing through the change in τ (lifetime-based sensing) rather than through the change in intensity I in equation 1.22. Unlike intensity measurements, lifetime measurements are not affected by changes in the dye concentration through leaching or photobleaching. Similarly, variation in the intensity of the excitation source or in the photodetector sensitivity will not affect lifetime measurements.

There are two methods that can be used for fluorescence lifetime measurement: pulse decay time and the phase-shift method. In the first method, excitation is by a short pulse (nanoseconds or less) and the resultant fluorescence decays as shown in Figure 1.6(a):

$$I_{fl}(t) = I\exp(-t/\tau) \tag{1.23}$$

Changes in the lifetime as a function of chemical concentration can then be observed. In the second method (Lakowicz, 1992), the excitation light is modulated sinusoidally at frequency f (angular frequency $\omega = 2\pi f$) and

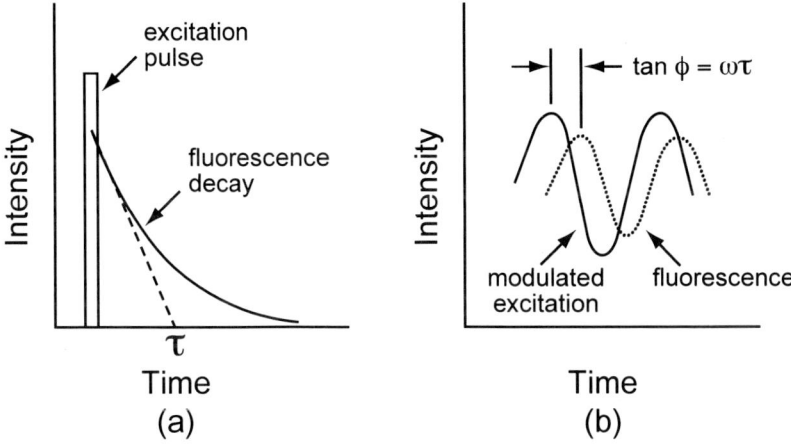

Figure 1.6 Fluorescence lifetime measurements. (a) Pulse decay time method. (b) Phase-shift method.

the phase shift of the fluorescence relative to the excitation is measured as shown in Figure 1.6(b). The phase shift ϕ is related to the lifetime τ through the expression:

$$\tan \phi = \omega\tau \qquad (1.24)$$

Since typical fluorescence lifetimes are in the 1–20 ns range, modulation frequencies in the 1–200 MHz band are required for measurable changes in ϕ.

Typical gases which are good quenchers are oxygen, nitrogen and nitrogen dioxide. Oxygen sensors have been demonstrated based on fluorescence quenching in ruthenium complexes (Klimant and Leiner, 1992; MacCraith *et al.*, 1994; O'Keeffe *et al.*, 1994). Lifetime probes have also been identified for Ca^{2+}, Mg^{2+}, K^+, Cl^- and pH.

1.2.4 Raman sensors

Chemical analysis can be performed using the Raman scattered light from a sample excited by a visible or near-IR laser source (Tobias, 1967; Barnwell, 1983). Raman scattered light differs from fluorescence in two important respects: (i) the wavelength of excitation must *not* correspond to an electronic absorption line or band; and (ii) the intensity of the Raman scattered light is very weak, typically 4–6 orders of magnitude less than fluorescence. The Raman spectrum reveals information about the vibrational states of a molecule and a necessary condition for Raman scattering is that the molecule must undergo a change in its polarisability during vibration.

The origin of Raman scattered light can be viewed from two standpoints: (i) the photon picture, and (ii) the electromagnetic field description. In the photon model, the incident photons of the excitation light collide with molecules of the sample and two types of collision may occur, elastic or inelastic. In elastic collisions, photons neither gain nor lose energy and the scattered light has the same frequency as the incident light — this is called 'Rayleigh scattering'. With an inelastic collision, a molecule may either absorb energy from the photon and move into a higher vibrational state or a molecule in an upper state may lose energy to a photon, making a downward transition. Here, the scattered photon will either have a lower or a higher energy than the incident photon and the change in energy will correspond to the difference in energy levels between two vibrational states of the molecule, as illustrated in Figure 1.7. The lower energy photons correspond to Raman lines of lower frequency than the incident radiation and are called 'Stokes' radiation', while the higher energy photons give rise to the 'anti-Stokes' radiation'. The characteristic Raman line pattern may be used for identification of the molecule. Note that lower energy vibrational states will normally have the higher population, so the anti-Stokes' radiation will be weaker than the Stokes'.

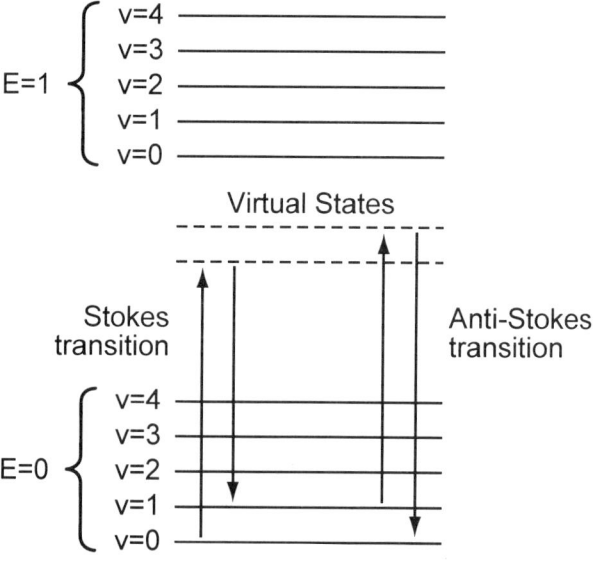

Figure 1.7 The photon model for inelastic collisions in Raman scattering of light.

The wave description of the Raman effect demonstrates the need for a change in the polarisability of the molecule during vibration. The excitation light subjects a molecule to an oscillating field of angular frequency ω_i, inducing a classical oscillating dipole as a result of the electron displacements under the influence of the field. The induced dipole moment **M** is given by:

$$\mathbf{M} = p[\mathbf{E}\cos(\omega_i t)] \tag{1.25}$$

where p is the polarisability tensor relating the dipole moment vector **M** to the electric field vector **E** (Tobias, 1967).

For simplicity, consider the case where **M** and **E** are parallel and p is a scalar. For a diatomic molecule, p is a function of the bond length and if the molecule is vibrating at frequency ω_v the bond length will be varying so we can write (for small vibration amplitude):

$$p = p_0 + \Delta p \cos(\omega_v t) \tag{1.26}$$

Substituting equation 1.26 into 1.25 yields:

$$\mathbf{M} = [p_0 \cos \omega_i t + \frac{\Delta p}{2}\cos(\omega_i + \omega_v)t + \frac{\Delta p}{2}\cos(\omega_i - \omega_v)t]\mathbf{E} \tag{1.27}$$

The scattered light is represented by the radiation from the dipole and equation 1.27 shows that there are three components, namely Rayleigh scattering at ω_i and the two components of Raman scattering at $\omega_i - \omega_v$ (Stokes) and $\omega_i + \omega_v$ (anti-Stokes). Note that the existence and strength of the Raman scatter is dependent on Δp, the polarisability change on molecular vibration.

Chemical sensors using the Raman effect have, until recent years, required expensive instrumentation because of the weakness of the Raman signal. Traditionally, a powerful visible source such as a krypton- or argon-ion laser is used for excitation with a high-resolution monochromator and sensitive photomultiplier detector at the system output. Such sources are associated with problems such as fluorescence from the sample itself or from traces of impurities, which can easily overwhelm the Raman signal and obscure the entire spectrum. Recently, however, there has been a number of important innovations in instrumentation to deal with this and other problems, making the technique useful for a wider range of chemical analysis. By using similar techniques to the Fourier Transform IR (FTIR) spectrometer (which employs a Michelson interferometer), FT–Raman spectrometers have been developed in combination with 1.06 μm Nd:YAG laser sources and germanium detectors (Fleischli and Walder, 1992). The longer wavelength of the source virtually eliminates the fluorescence problem and although the scattering is weaker at longer wavelengths (because of the $1/\lambda^4$ factor), this is compensated by the multiplex advantage and the large entrance aperture of FT spectrometers. In an alternative approach, multichannel Raman systems (Niemczyk *et al.*, 1993) have been designed using diode lasers in the near IR (with powers of 100 mW or more), holographic edge filters to provide rejection of the Rayleigh light (allowing collection of Raman–shifts to within $50 \, \text{cm}^{-1}$ of the laser line) and CCD detectors giving high signal-to-noise ratio and short measurement time (1–10 s) for sample analysis.

The introduction of fibre optics into Raman systems has also had a significant impact, particularly in carrying light to and from samples and in the design of multiplexed systems to share the expensive instrumentation between a number of sites (Leugers and McLachlan, 1988; Dao and Jouan, 1993; Schoen, 1994; Newbery *et al.*, 1994; Angel and Cooney, 1994). Since the excitation light for Raman is in the visible or near-IR, standard silica fibres can be used. Raman scattering from the silica fibre itself may interfere with measurements in some cases and needs to be filtered out. Recently, the technique of surface enhanced Raman spectroscopy (SERS) has been applied to fibre optic probes (Vo-Dinh *et al.*, 1990; Mullen and Carron, 1991; Mullen *et al.*, 1992; Vo-Dinh, 1994). In SERS, the intensity of the Raman scattered light can be enhanced by a factor of up to 10^8 by use of a roughened metal surface or metal-coated microspheres.

Since the intensity of Raman lines is proportional to the number of scattering sites, the technique can be used both for detecting the presence of a specific chemical from its characteristic Raman signature and for concentration measurement. Raman techniques are particularly suitable for analysis of chemical species in water, such as metal complexes. This is because the background Raman scatter from water is very weak since the polarisability of the water molecule changes little during its vibrations. Highly polarisable bonds such as $C=C$ or $S-S$ produce strong Raman lines. Raman scattering has been applied to the detection of contaminants in soil, and for detection of species such as carbon tetrachloride, benzene, ethanol and other hydrocarbons (Schoen 1994; Angel and Cooney, 1994).

1.2.5   *Refractive index and optical path length sensors*

Information about concentration and composition of solutions and solvent mixtures can be gained through refractive index measurements; sensors for glucose and for water/organic solvent mixtures have been demonstrated using this principle. These sensors, however, are not particularly specific and may be subject to error from other sources of index change, such as temperature. Coatings whose refractive index change on exposure to a certain chemical have been investigated. Coatings of heteropolysiloxanes (HPS) formed by sol–gel type methods have indices close to that of silica fibres (1.44–1.46) and undergo an index change on exposure to certain chemicals (Archenault *et al.*, 1992). Depending on the type of HPS used, the HPS coating can be made sensitive to different chemicals; for example, glycidoxyl propyl siloxane has been shown to undergo an index change on exposure to hydrocarbons, especially toluene. As discussed later, various fibre optic configurations can be used to monitor the changes in refractive index, the simplest being an unclad multimode fibre whose transmission is modified by the surrounding index. Other examples of fibre optic refractive index-based sensors include sensors for detecting oil in water and for trichloroethylene (Wolfbeis, 1991).

Sensors based on optical path length changes have also been investigated and interferometric methods can be used to monitor the change, such as a fibre Michelson interferometer or a Fabry–Perot type system. Hydrogen sensors have been demonstrated using a film of palladium which swells on exposure to hydrogen. Polymer films of tetrafluoroethylene (C_2F_4) swell on exposure to hydrocarbon vapours such as trichloroethylene, ethanol, benzene, pentane and hexane. Fibre sensors have been made using plasma-polymerised C_2F_4 films of a few micron thickness deposited on the end of a fibre (Butler and Buss, 1992; Gauglitz, 1992) forming a Fabry–Perot type interferometer (see section 1.3.5 and Figure 1.14(b)).

1.2.6   Evanescent field sensors

In evanescent field sensors, the fibre itself is intimately involved in the sensing process. The evanescent field interacts either directly with the chemical itself (monitoring absorption or index change) or with an intermediate dye, involving absorption or fluorescence as described in the previous sections.

An evanescent field is produced when a light ray incident on a boundary undergoes total internal reflection (TIR), as illustrated in Figure 1.8(a). For TIR to occur, $n_2 > n_1$ and $\theta > \theta_c$ where θ_c is the critical angle, $\theta_c = \sin^{-1}(n_1/n_2)$. For light incident on a planar boundary, the evanescent field has an exponential decay, given by:

$$E = E_0 \exp\left(-\frac{x}{d_p}\right) \tag{1.28}$$

where d_p is the penetration depth ($1/e$ depth) given by:

$$d_p = \frac{1}{k_0\sqrt{n_2^2 \sin^2\theta - n_1^2}} \tag{1.29}$$

and $k_0 = 2\pi/\lambda_0$.

From a ray optic's perspective, we can picture light in an optical waveguide as undergoing a series of total internal reflections, as shown in Figure 1.8(b). If an absorbing chemical or dye is present within the

(a)

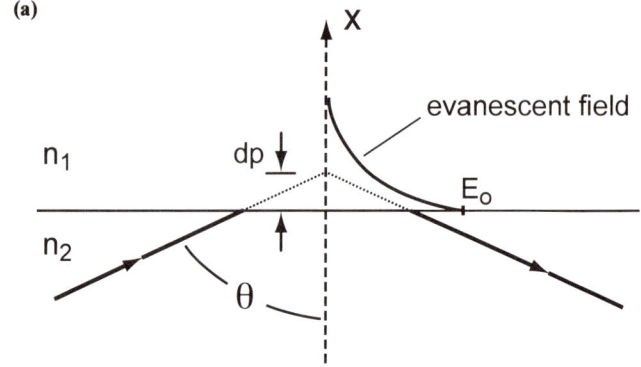

(b)

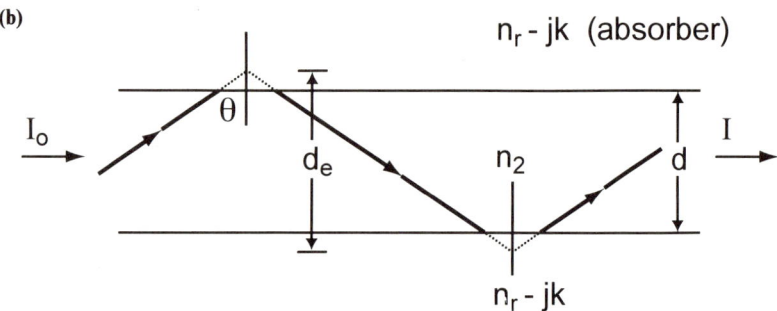

Figure 1.8 Evanescent field. (a) Production by total internal reflection. (b) A ray optics perspective.

evanescent field region, the reflection coefficient will be less than unity and the light will be attenuated as it travels along the guide. The attenuation experienced by the guided light will depend on the fraction of the total power that is carried in the evanescent field region.

Referring back to Figure 1.2, equation 1.3 describes the attenuation in a direct absorption cell. In a similar way, an effective attenuation coefficient γ for an optical waveguide can be defined through the relation (Ruddy, 1990):

$$I = I_0 \cdot \exp\left[-(r_f \alpha_m) l C\right] = I_0 \cdot \exp\left[-\gamma l C\right] \tag{1.30}$$

where $\gamma = r_f \alpha_m$ and r_f represents the reduction factor ($r_f < 1$) in attenuation with an optical waveguide as compared with a direct absorption cell for the same absorber.

In optical waveguides, it is often convenient to measure the attenuation in decibels per unit length (dB/unit length) and from equation 1.30 this attenuation (A_g) can be written as:

$$A_g = -10 \log_{10} \left(\frac{I}{I_0} \right)$$

$$= 4.343 r_f \alpha_m C$$

(1.31)

Hence we can write:

$$A_g = r_f \cdot A_d$$

(1.32)

where A_d is the attenuation in decibels per unit length for the direct absorption cell.

The evanescent wave sensor is much less sensitive per unit length than the direct absorption cell because r_f is small, typically < 0.1, but this can be partially compensated by the long interaction lengths possible with an optical fibre sensor. Nevertheless, it is important in the design of evanescent wave sensors to maximise the value of r_f.

In general terms, r_f can be written for any optical waveguide as (Stewart, 1994):

$$r_f = \frac{n_r}{n_e} \cdot f$$

(1.33)

where n_r is the index (real part) of the absorber, n_e is the effective index of the guided wave ($n_e = n_2 \sin \theta$) and f is the fraction of the total optical power that interacts with the absorber.

If the field distribution in the waveguide or fibre is known (either analytically or numerically through, for example, finite difference methods), the power fraction f may be calculated (Stewart and Culshaw, 1994). Alternatively, r_f can be determined for a simple planar guide by calculating the loss per reflection and multiplying by the number of reflections per unit length (Stewart et al., 1991). For the planar guide of Figure 1.8(b) which serves as a reasonable model for meridional rays in a multimode optical fibre, r_f is given by (Stewart et al., 1994):

$$r_f = \frac{2n_r}{k_0 n_e d_e \sqrt{n_e^2 - n_r^2}} \frac{n_2^2 - n_e^2}{n_2^2 - n_r^2}$$

(1.34)

where d_e is the effective thickness, $d_e = d + 2d_p$.

Note from equation 1.34 that the value of r_f is dependent on n_e. In an optical waveguide or fibre, only a discrete set of values of θ (and hence n_e) are allowed, as determined by transverse standing wave conditions in the fibre. The number of guided modes which are allowed depends on the wavelength, core diameter and index of core and cladding. (The fibre is single-moded if $v < 2.4$ or highly multimode if v is large, where the

v-number is given by:

$$v = k_0 a \sqrt{n_2^2 - n_1^2} \qquad (1.35)$$

and a is the core radius).

For multimode fibres with large v-numbers the fibre supports many thousands of modes so there is a virtual continuum in n_e values between the core and cladding index. Figure 1.9(a) shows r_f plotted against n_e for a multimode fibre with an aqueous absorber ($n_r = 1.33$) at a wavelength of 0.6328 μm. The fibre core index is 1.45 and core diameter 200 μm. It is clear from the figure that the overall sensitivity of the sensor will depend on the mode distribution launched into the fibre (Gupta et al., 1994) and will vary if external disturbances to the fibre result in intermode coupling.

More predictable performance can be achieved with a single mode guide such as a D-fibre (see Figure 1.12(c)) and Figure 1.9(b) shows the relative sensitivity r_f of D-fibre as a function of the core diameter computed using numerical analysis (Stewart and Culshaw, 1994) for a gaseous absorber ($n_r = 1$) at 1.66 μm wavelength. The core and cladding indices of the fibre are 1.45 and 1.42, respectively. Note that r_f is very small since the evanescent field decays rapidly into the gas. Because of the asymmetry of the structure, with the absorber on one side only, the core diameter must be chosen carefully to maximise the sensitivity, as indicated in Figure 1.9(b).

So far we have been discussing evanescent wave absorption sensors, but the same methods can be applied to refractive index and fluorescence sensors. If the presence of a chemical results in a change in the real part of the index (Δn_r) in the evanescent field region, then the effective index change (Δn_e) of the guided mode is:

$$\Delta n_e = r_f \cdot \Delta n_r \qquad (1.36)$$

where r_f is the same as that defined earlier.

The change in optical path length will be $\Delta n_e l$ where l is the interaction length. Interferometric methods (e.g. a fibre Mach–Zehnder or polarimetric interferometer; see section 1.3.5) may be used to monitor this change. (As noted earlier, an alternative simple scheme uses changes in transmission of a multimode fibre with index change. An increase in the index within the evanescent field region results in the escape of higher order modes to radiation modes when the index exceeds their effective index, i.e. total internal reflection no longer occurs.)

In fluorescence-based sensors, the evanescent wave can be used to excite the dye, as already described, and a small fraction of the fluorescence emission is coupled into guided modes of the fibre. For this to occur, the fluorescent dye must be within the evanescent field region associated with the fluorescence wavelength. A detailed theoretical analysis of the efficiency of coupling cladding-generated fluorescence to guided modes has been

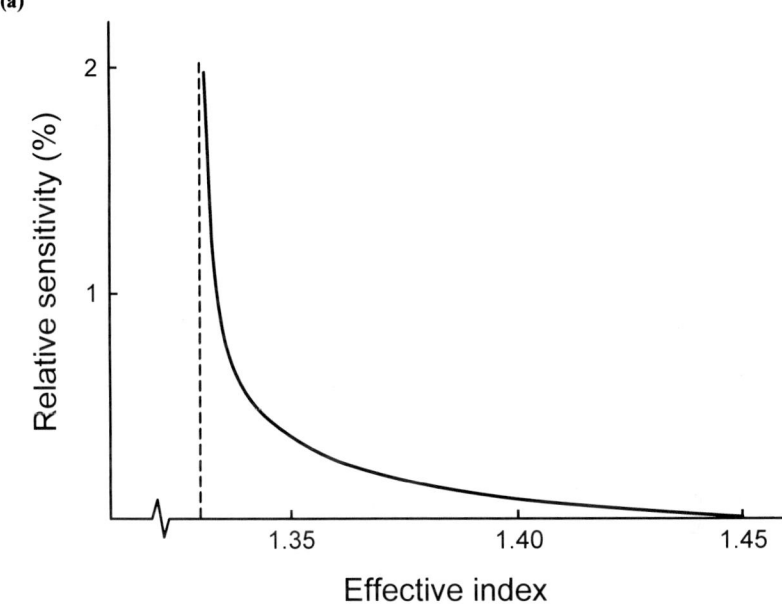

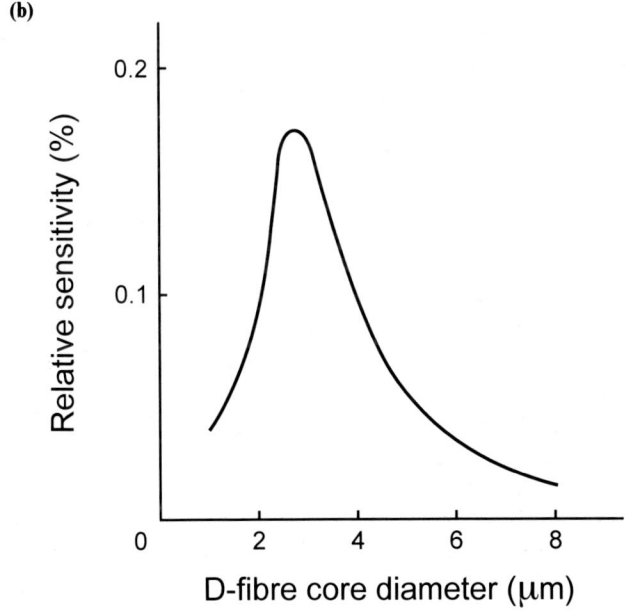

Figure 1.9 (a) Multimode fibre with an aqueous absorber. (b) Single mode fibre.

carried out by Marcuse (1988). The efficiency increases with fibre v-number and typically has a value of 0.01–0.02% for v-numbers around 300–400 (Marcuse, 1988; Lieberman *et al.*,1990).

1.3   Sensor construction and basic types

As far as the sensor construction is concerned, fibre optic chemical sensors can broadly be divided into three categories:

1. Plain fibre sensors, where the fibre performs a purely passive role in transferring light to and from an absorption or fluorescence cell. The fibre provides a convenient way to access the cell in remote or distributed systems but plays no part in chemical sensing.
2. Fibre optrodes (or optodes), where the working sensor chemistry (involving, for example, absorbing or fluorescent dyes) is formed directly at or on the fibre tip.
3. Evanescent wave sensors, where interaction with the chemical occurs within the evanescent field region of the optical fibre. In this case, the fibre plays the most direct role in the sensing process.

There are, of course, a number of other configurations for sensor construction which do not quite fall into the above categories and some of these will also be briefly discussed.

1.3.1   Fibre optics with absorption cells

Figure 1.10 illustrates the typical construction of plain fibre sensors using an absorption cell. This form of design is particularly suitable for gas sensors where gas concentration is measured through absorption measurements in the near-IR (within the transmission window of silica fibres), as described in section 1.2.1. In Figure 1.10(a), a lens is used to collimate the output light from the fibre and, after passage through the cell, a second lens focuses the light into the output fibre. The cell may typically be 10–50 cm in length and is designed to allow the gas to penetrate into the cell volume. Using this type of construction Tai *et al.*, (1992) demonstrated a sensor for both methane and acetylene with a sensitivity of 5 and 3 ppm, respectively. The experimental system used a cell of length 10 cm coupled with optical fibres of 4 km in length and two distributed feedback (DFB) laser sources operating at 1.66 μm and 1.53 μm wavelengths to monitor near-IR absorption lines of methane and acetylene.

Figure 1.10(b) shows how the path length can be doubled by placing the cell at the end of the fibre and using a reflective coating to return the light back along the same path. A fibre coupler may be used to separate the output light but this results in a reduction in throughput power as

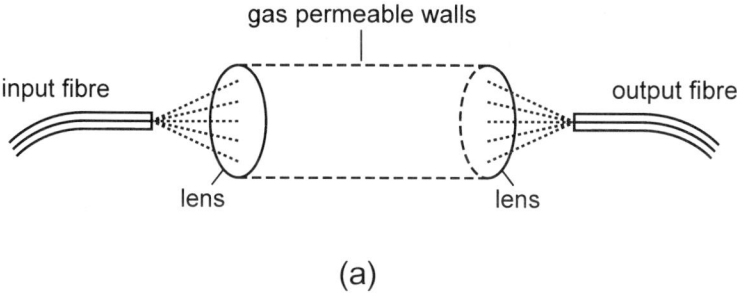

(a)

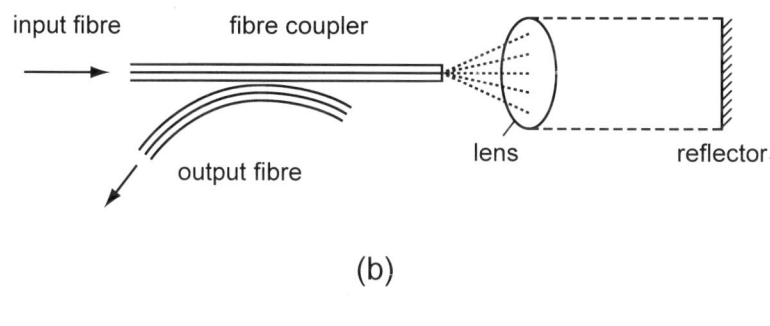

(b)

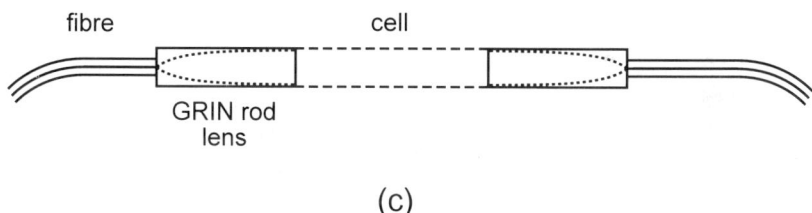

(c)

Figure 1.10 Plain fibre sensors using an absorption cell (see the text for details).

compared with the transmission cell. (If the coupling ratio is r, then the output power with no cell losses is: $r(1 - r) \times$ input power, which is maximised at 25% of input power when $r = 0.5$.)

The combination of fibre optics and bulk-optics in the cells as described is often undesirable. The cells are relatively bulky and care must be taken to ensure that misalignments do not occur during system operation which would affect light transmission. Effects of temperature change must also be properly controlled. Improvement can be gained through the use of micro-optic cells, as illustrated in Figure 1.10(c). In the micro-optic cell, the bulk lenses are replaced by miniature gradient-index rod lenses (GRIN lens) and the cell is made compatible with fibre optic connector technology. Micro-optic cells in both multimode and single-mode fibres are possible, and this is a subject of ongoing research.

1.3.2   Fibre opt(r)odes

Fibre optrodes (Figure 1.11) are particularly suitable for sensors based on indicator dye systems (either absorption or fluorescence, as discussed in

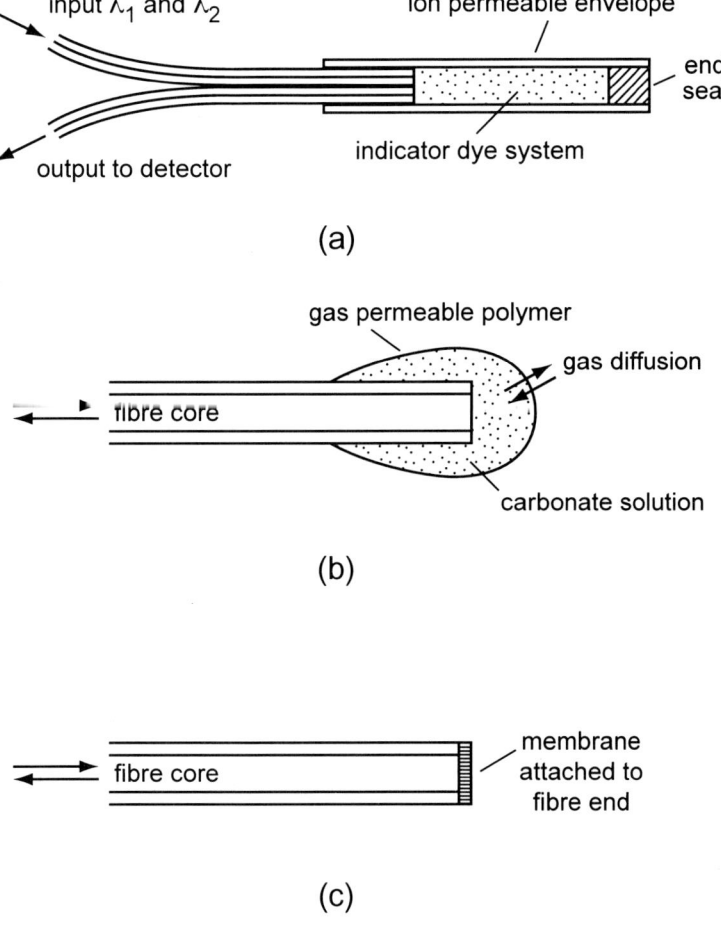

Figure 1.11 Fibre optrodes.

sections 1.2.2 and 1.2.3) and optrodes have been developed for pH, carbon dioxide, oxygen, etc. Peterson *et al.*, (1980) first developed a fibre optrode for pH measurement using the construction illustrated in Figure 1.11(a), consisting of two fibres, one for the input illumination and the other to collect the output signal. The indicator dye, phenol red, is immobilised in polyacrylamide microspheres and enclosed in the cellulosic tube at the end of the fibre pair. The cellulose membrane allows hydrogen ions from the external solution to migrate into the dye system and the change in dye absorption with pH is measured at 560 nm wavelength (green). Light-scattering microspheres are included to enhance the return signal. An additional source at 600 nm (red, where dye absorption is not affected by pH) is used to compensate for optical variations in the sensor by taking the output signal as the ratio of green to red intensity. A similar type of sensor was also demonstrated for oxygen (Peterson *et al.*, 1984) using fluorescence quenching by oxygen in the indicator perylene dibutyrate adsorbed on polystyrene beads.

Figure 1.11(b) shows a typical construction for an optrode to sense acidic or basic gases such as carbon dioxide. A gas-permeable membrane is used to contain the working chemistry at the end of the fibre, which consists of a pH-sensitive dye in a bicarbonate solution. The concentration of carbon dioxide affects the pH of the solution which is monitored through the indicator dye (see equation 1.18). Another form of fibre optrode is illustrated in Figure 1.11(c), where the indicator dye is covalently immobilised in a polymer film, porous glass bead or ion-exchange membrane that is attached to the end of the fibre. This form is commonly used for pH sensors.

1.3.3 Evanescent field sensor types

Typical examples of fibre construction for evanescent field sensors are shown in Figure 1.12. The simplest and cheapest is the plastic-clad silica (PCS) multimode fibre, Figure 1.12(a), with core diameter $\sim 200\,\mu m$. The plastic cladding can easily be removed over a suitable fibre length exposing the core and allowing access to the evanescent field. PCS fibres are commonly used in the construction of sensors based on indicator dyes, such as pH sensors (both absorption- and fluorescence-based systems). The exposed region, typically several centimetres in length, is usually coated with a polymer or sol–gel film containing the indicator dye, as discussed in section 1.3.4. However, as was explained in section 1.2.6, the disadvantage of the multimode fibre is that the sensitivity is dependent on the mode distribution and hence on launching conditions and external disturbances.

Single mode fibres have core diameters of only a few microns surrounded by a relatively thick cladding of $\sim 100\,\mu m$, and so removal of the cladding (or tapering of the fibre) to gain access to the evanescent field results in a

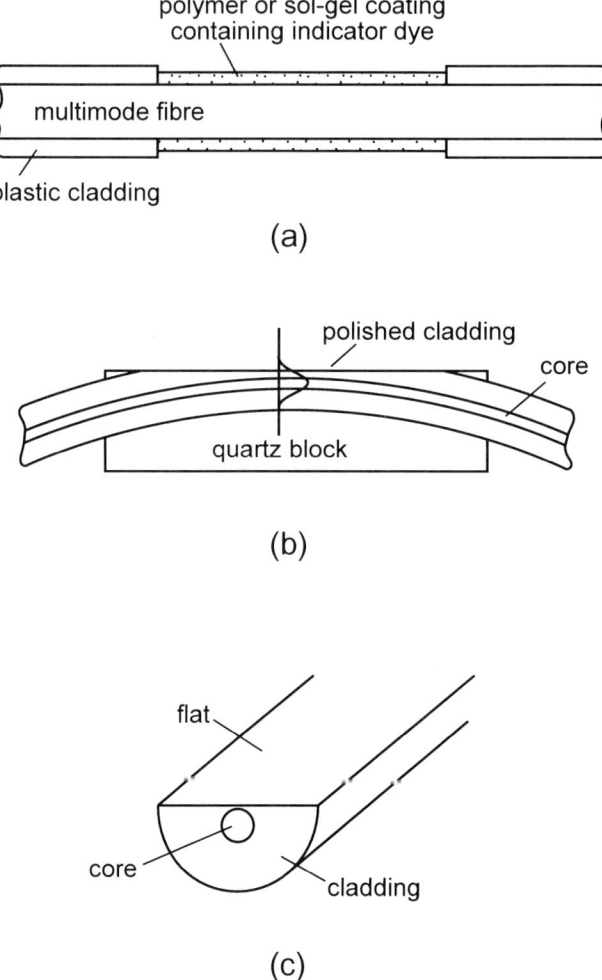

Figure 1.12 Evanescent field sensors.

very fragile filament. Two solutions are possible, as shown in Figure 1.12(b) and (c). In (b), the single mode fibre is first mounted in a curved slot cut in a quartz block and the top surface is then polished to remove the cladding region on one side of the fibre. The disadvantages here are (i) the process is time consuming (hence expensive), and (ii) there is only a relatively short length (< 1 cm) with the evanescent field exposed. In Figure 1.12(c), the fibre is made from a conventional preform but with half the cladding region removed; the fibres pulled from the preform have a D-shaped cross-section, allowing continuous access to the evanescent field along the whole length of the fibre. Again the disadvantages are the limited availability of D-fibre commercially and the more difficult problems associated with splicing and connecting these fibres into systems.

(Integrated optic waveguides are also used for evanescent wave sensors and are discussed in section 2.3.1.)

1.3.4   Sol–gel coatings

For evanescent field sensors using indicator dyes as described above, it is usually necessary to immobilise the indicator in a suitable host matrix that is then coated onto the exposed evanescent field region of the fibre. Sol–gel films (Brinker and Scherer, 1990) are particularly suitable for this purpose. Thin glassy films can be made with refractive indices between about 1.4 and 2.0, thicknesses from $\sim 0.1\,\mu m$ and with various degrees of porosity to allow molecules to diffuse into the films (McCulloch *et al.*, 1994). The sol–gel process is simple, inexpensive and low temperature, allowing organic dyes to be incorporated in the films during the processing stage. The starting compounds for film fabrication are alkoxides, commonly silicon tetraethoxide (TEOS), $Si(OC_2H_5)_4$, for silica sol–gels (or both silicon and titanium ethoxides for Si–Ti films with indices > 1.45). The key steps in the process are hydrolysis, condensation and polymerisation, which are stimulated by including a catalyst such as an acid. The hydrolysis step is described by the reaction:

$$Si(OC_2H_5)_4 + H_2O \rightarrow HO-Si(OC_2H_5)_3 + C_2H_5OH \qquad (1.37)$$

Condensation of the hydrolysis products above proceeds simultaneously according to:

$$Si(OC_2H_5)_4 + HO-Si(OC_2H_5)_3 \rightarrow (OC_2H_5)_3Si-O-Si(OC_2H_5)_3 + C_2H_5OH$$
$$(1.38)$$

The reaction of equation 1.38 continues to build up larger and larger molecules through polymerisation.

Typically, in the fabrication process, the TEOS is dissolved in ethanol and water is added. The indicator dye can also be included at this stage by dissolving it in ethanol. Thin films can be formed on a fibre surface from the sol–gel mixture by dip-coating at a coating speed of $\sim 10\,cm\,min^{-1}$. After drying and low-temperature annealing, thin glassy films on the fibre are obtained containing the trapped indicator dye. Single coats produce films of around $0.1\,\mu m$ thickness; thicker layers can be built up through multiple coats.

1.3.5   Other types of sensor construction

Figure 1.13 shows the construction of fibre optic Raman probes. Figure 1.13(a) shows the simplest form, consisting of a fibre bundle where the central fibre is used for excitation of the sample, and the outer ring of fibres is used for collection of the weak Raman signal. An improved version is shown in Figure 13(b), where two fibres are used at 90° to each other with GRIN lenses focused on a capillary tube containing the Raman sample

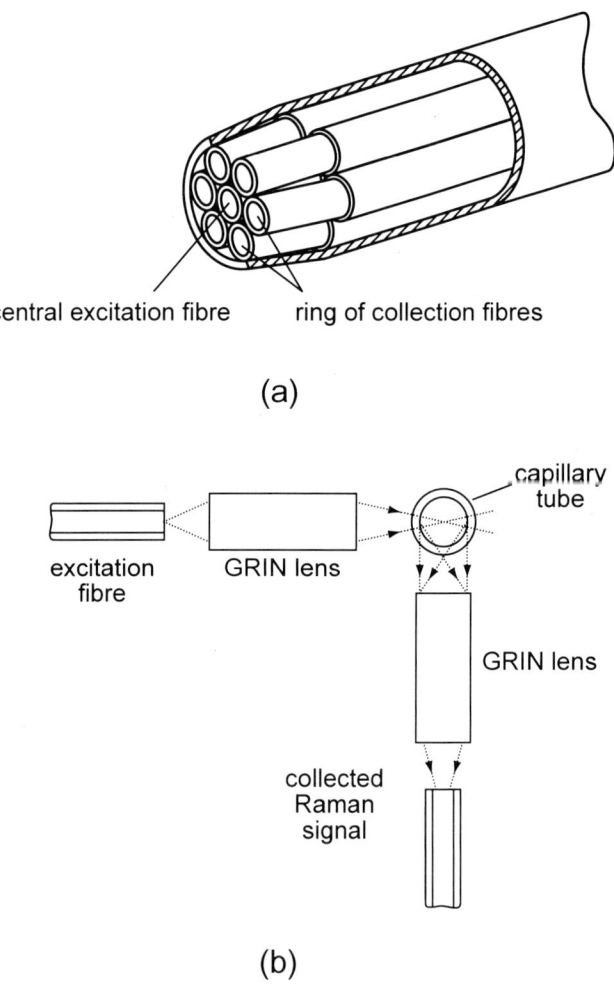

central excitation fibre   ring of collection fibres

(a)

capillary
tube

excitation
fibre   GRIN lens

GRIN lens

collected
Raman
signal

(b)

Figure 1.13 Fibre optic Raman probes.

(Dao and Jouan, 1993). This arrangement is very efficient for suppressing the Raman signal from the input fibre itself, which may interfere with measurements.

Figure 1.14 shows two forms of fibre optic interferometers for measuring changes in refractive index or optical path length caused by the presence of a chemical species. In Figure 1.14(a), a fibre optic Mach–Zehnder is formed using an input fibre coupler to split the light into a reference and measurement arm with a second coupler at the output to recombine the light. The fibres must be single mode throughout. A change in the optical path length in one arm through, for example, evanescent wave interaction causes the output fringe pattern to cycle through a series of maxima and minima. The system is very sensitive to changes in optical path length, but this can be a source of problems as it may also respond to other

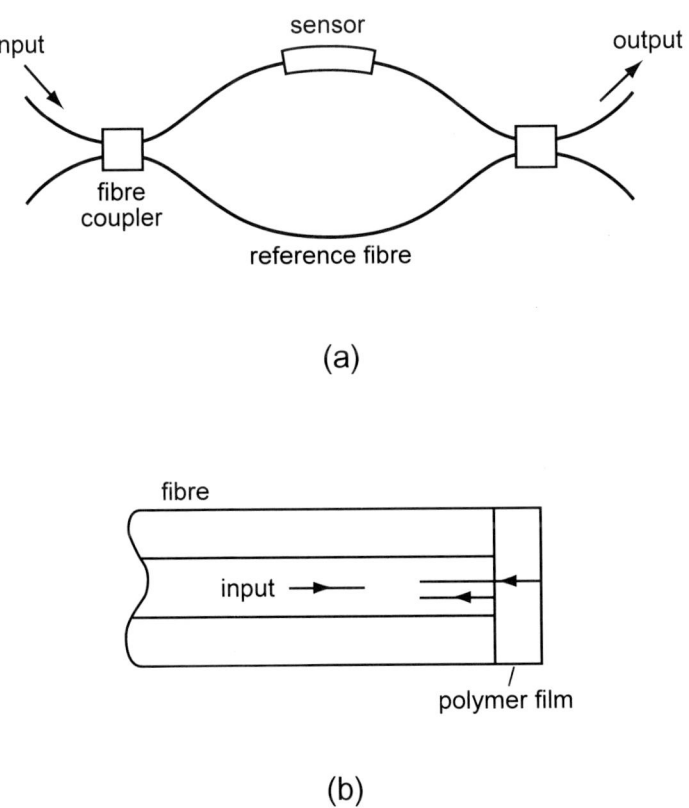

(a)

(b)

Figure 1.14 Fibre optic interferometers.

disturbances and temperature changes. In Figure 1.14(b), a thin film is formed on a fibre end and interference occurs between light reflected from the two surfaces. If the film swells or changes in refractive index on exposure to a gas, this can be monitored from the interference pattern (e.g. palladium films and certain polymer films swell on exposure to hydrogen and hydrocarbons, respectively, as discussed in section 1.2.5). In some cases, it is necessary to first coat the fibre end with a metal film before coating with the chemically sensitive film to improve reflectivity and fringe contrast.

1.3.6   Source and detector considerations

The most convenient light sources for optical fibre chemical sensors are light emitting diodes (LEDs) and diode lasers because they are compact and the output light can be modulated to improve signal-to-noise ratio in detection. The cheapest and most widely available are those which have been developed for communication applications of optical fibres, namely GaAs/AlGaAs LEDs and lasers at $\sim 0.8\,\mu m$ and InGaAsP/InP lasers at 1.3 and $1.55\,\mu m$ wavelengths. However, a much wider range of source wavelengths is required for sensor applications, and in the late 1980s and early 1990s, many new LED and diode laser sources have become available although often at much greater cost. Sources currently available in the visible spectrum range include orange, red, yellow, green and blue (at lower power levels) LEDs as well as visible (red) diode lasers. These developments are particularly relevant for indicator dye sensors, which often require excitation in the shorter-wavelength–visible or UV regions of the spectrum. For gas sensors based on absorption line measurements, sources in the IR region are required. Lead salt diode lasers are available for $> 3\,\mu m$ wavelengths covering the fundamental absorption lines of gases, but the need for cryogenic cooling of both laser and associated detector means they are expensive and cumbersome in use. Additionally, the use of silica fibres is precluded at these wavelengths and IR fibres are still relatively immature and expensive. For these reasons, attention has been given to LED and laser diode sources in the near IR region $(1-2\,\mu m)$ compatible with the transmission window of silica fibres, where many gases possess overtone or combination absorption lines (albeit much weaker than fundamental lines). A range of distributed feedback lasers (DFBs) based on InGaAsP/InP have been developed (Cooper and Martinelli, 1992) suitable for spectroscopy on gases such as methane $(1.65\,\mu m)$, carbon dioxide $(1.573\,\mu m)$, carbon monoxide $(1.567\,\mu m)$, ammonia $(1.544\,\mu m)$, hydrogen sulphide $(1.578\,\mu m)$, etc. The DFB lasers have very narrow linewidth $(\sim 50\,MHz)$, less than the width of a single gas absorption line, and can be tuned in wavelength over a few nano-

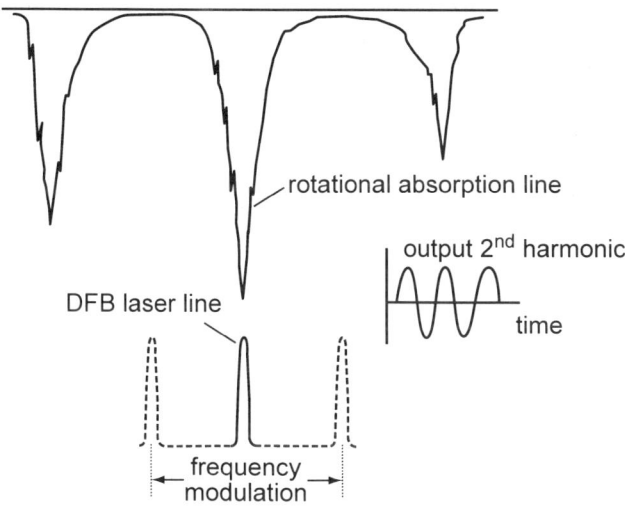

Figure 1.15 Frequency modulation in distributed feedback lasers.

metres by injection-current modulation. This means that frequency-modulation techniques, as illustrated in Figure 1.15, can be applied to give very high sensitivities. Here the laser is modulated at frequency f, generating a second harmonic output signal at $2f$ that is proportional to the gas concentration. The principle has been demonstrated by Uehara and Tai (1992) for methane detection using a 50 cm absorption cell (without fibres) giving sensitivities down to 0.3 ppm metre. Alternatively, two wavelength switching of the DFB can be used to reference the output signal to the intensity at a non-absorbing wavelength (Shimose *et al.*, 1991). At present, the main disadvantage of DFB lasers is their high cost, but this may come down in the future with an increase in sales volume.

The most convenient light detectors for fibre optic chemical sensors are junction photodiodes. The most common is the silicon photodiode for detection in the visible and near-IR over about 0.3–1 μm wavelength. Photodiodes of InGaAs are used for the 0.8–1.7 μm range. For longer wavelengths, 1–5 μm, lead salt photoconductive detectors may be used, such as PbS and PbSe, but these require thermoelectric cooling. Photomultiplier tubes are used when high sensitivity is required in the visible spectral region, such as in collecting a weak fluorescence signal. Of importance also are multichannel detector systems such as photodiode arrays and charge coupled devices (CCDs) which allow virtual instantaneous collection of data from spectrometer systems rather than scanning individual wavelengths across a single photodiode.

1.4   Examples of fibre optic sensors for environmental applications

In this section, we consider some examples of environmentally important chemicals or pollutants that can be monitored through fibre optic chemical sensor schemes. It is important to realise that this is very much an area of ongoing research and current developments range from practical systems to laboratory demonstrations and simply identification of potential strategies. Due care must be taken that cross-sensitivity of sensors to other gases or chemicals does not interfere with the desired measurements, and this is often an area requiring further investigation and test.

1.4.1   Air pollutants

Some of the more important air pollutants for environmental monitoring include carbon monoxide, nitrogen oxides and hydrocarbons from car exhaust emissions, as well as from other sources, sulphur dioxide from coal-fired power stations, resulting in acid rain, and hydrogen sulphide commonly found associated with subterranean methane repositories and produced as a side product in petrochemical processing. Ozone is an important constituent of low-level air pollution, arising from partially burnt hydrocarbons emitted from vehicle exhausts reacting in sunlight and from certain industrial processes such as welding shops.

Carbon monoxide is a highly toxic gas and extremely important for environmental monitoring. Optical sensing can be performed through direct absorption measurements using the IR absorption line at $4.66\,\mu m$ (not suitable, however, for silica fibres) or the near IR bands around $2.2\,\mu m$ and $1.567\,\mu m$, although this region has not been widely exploited as yet for fibre sensors. Fibre optrodes have also been reported using immobilised carbon monoxide-sensitive reagents (Stuart and Samson, 1988; Goswami *et al.*, 1994) giving sensitivities around 20 ppm or less.

Nitrogen dioxide (NO_2) and nitric oxide (NO) can be monitored through direct absorption measurements at particular wavelengths. Nitrogen dioxide has a broad absorption band in the visible, centred around $0.4\,\mu m$, and in the near-IR at $0.8\,\mu m$. Nitric oxide has no appreciable absorption in the visible and its fundamental absorption line around $5.3\,\mu m$ is unsuitable for silica fibres. There is a second harmonic around $2.69\,\mu m$ that may be suitable for fibre sensor application.

As already noted, methane and related aliphatic hydrocarbons possess a strong fundamental C−H absorption band at $3.3\,\mu m$. Although this is not suitable for silica fibre sensors, combination and overtone lines occur in the near-IR around $1.3\,\mu m$ and $1.6\,\mu m$, which may be used for direct absorption or even evanescent wave fibre sensors. Much effort has been concentrated on methane (lower explosive limit (LEL) 5% (V/V) in air), where fibre optic systems offer inherently safe detection in mines, land-fill sites and other environmental situations (Tai *et al.*, 1987; 1992; Chan *et al.*, 1985, Stewart *et al.*, 1992, Uehara and Tai, 1992). A complete multipoint system for methane monitoring in mines, with associated hardware and software, has already undergone trials in Australia (Stuart, 1992). The system employed several 0.5 m absorption cells coupled to optical fibres (wavelength of 1.66 μm from LED source) and gave a sensitivity of about 5% LEL. Development of DFB lasers that can be tuned over several nanometres in the near-IR region has allowed the demonstration of systems that can simultaneously detect two gases, such as methane and acetylene (Tai *et al.*, 1992) and methane and carbon dioxide (Weldon *et al.*, 1993).

Optical detection of hydrogen sulphide gas has been investigated using reflectance measurements through optical fibres on paper impregnated with lead acetate (Narayanaswamy and Sevilla, 1988). The paper darkens on exposure to the hydrogen sulphide, but the reaction is non-reversible and the paper has to be slowly moved for continuous sensing. Other reversible hydrogen sulphide-sensitive reagents are currently under investigation. Recently, the possibility of direct absorption sensing has been demonstrated using the near-IR line of hydrogen sulphide around 1.57 μm and a DFB laser (non-fibre at this stage) giving a sensitivity below the safe exposure level of 10 ppm (Weldon *et al.*, 1994).

Fibre optrodes for sulphur dioxide have been demonstrated based on dynamic fluorescence quenching by sulphur dioxide in certain polycyclic aromatic hydrocarbons such as benzo(b)fluoranthene (Wolfbeis and Sharma, 1988), giving a sensitivity of around 70 ppm. The fluorophor is immobilised in a silicone polymer positioned at the tip of a bifurcated optical fibre. More sensitive detection, down to 10 ppm, was demonstrated using a two-fluorophor system consisting of pyrene and perylene dissolved in the silicone polymer (Sharma and Wolfbeis, 1989). Here, the fluorescence quenching is enhanced through the action of sulphur dioxide in inhibiting electronic energy transfer between the two components. Other possibilities for fibre sensors include those based on pH optrodes (sulphur dioxide is a highly acidic gas and a primary cause of acid rain) and on the use of direct absorption at an appropriate wavelength (the strong 101 band of sulphur dioxide is at 4.0 μm and is used in conventional gas cells).

Ozone possesses a strong UV absorption band around 250 nm, which is traditionally used for optical sensors. This wavelength is not suited to fibre optic sensors but there is a weak absorption band in the visible around 600 nm that may be useful for measurement of high concentrations (Fowles and Wayne, 1981). There has also been some investigation into fluorescence-based sensors employing thin films of porphyrins and phthalocyanines as ozone-sensitive reagents (ozone is an electron acceptor gas).

Finally, carbon dioxide, important as a greenhouse gas in global warming, has its strong IR absorption in the 4.2–4.4 μm region. There is a weak absorption line at 1.573 μm and a detection limit of 100 ppm has been demonstrated here with a DFB laser although, as yet, in a non-fibre system (Weldon *et al.*, 1994).

1.4.2 Seawater monitoring

Three of the most important parameters in seawater and oceanic monitoring are pH, dissolved oxygen and carbon dioxide levels. Many oceanic processes such as mineral solubility depend on pH, while dissolved oxygen is necessary to support life in the marine environment. Oceans absorb approximately half of the carbon dioxide released into the atmosphere and the carbon dioxide enters the biological cycle when fixed by plankton. Hence, measurement of dissolved carbon dioxide is important in understanding the ocean's contribution to the global carbon dioxide cycle and possible global warming effects. For each parameter, a continuous, real time, *in situ* monitoring capability is desirable and fibre optic sensors are ideally suited for this type of measurement. The principles of pH, oxygen and carbon dioxide optrodes based on absorbing or fluorescent indicator dyes have already been discussed in sections 1.2.2, 1.2.3 and 1.3.2, but there are a number of unique and difficult problems in oceanic monitoring. Seawater has a high ionic strength resulting in a shift in the pK_a of the dye (see equation 1.12) and a reduction in its quantum efficiency. Concentrations of carbon dioxide are extremely low, requiring resolution of several parts per million, and instrument drift and instability must be avoided. These problems have been addressed using several dyes to compress changes in the fluorescence intensity into a narrow pH range (Walt, 1992).

1.4.3 Ground and drinking water contamination

Contamination of ground water with chloroform and other volatile solvents from hazardous waste sites is a major problem in the USA and other countries. A fibre optic sensor system for remote long-term monitoring of subsurface contamination has been developed by the Lawrence Livermore National Laboratory and field trials have been conducted giving sensitivities in the parts per billion range (by weight in water) for chlorinated solvents such as chloroform and trichloroethylene (TCE) (Milanovich *et al.*, 1994). The sensor is based on pyridine, which absorbs light in the green region (530–570 nm) in the presence of TCE. Optical fibres connect to the sensor cell and a cone penetrometer is used to drive the sensor head to various depths (tens of metres) below the surface for contamination monitoring.

Another source of contamination for ground and drinking water is gasoline leakage from surface and underground storage tanks. FiberChem

(Saini *et al.*, 1994) have developed a fibre optic system with associated process electronics providing real time *in situ* monitoring of hydrocarbons in water or vapour form. The sensor is a refractive index-based system using an index-matched coating on the fibre core. The coating has a selective affinity for certain hydrocarbons, increasing in index in the presence of gasoline, which results in escape of light from the fibre core. Simulated field tests on the system have shown excellent response to xylene, synthetic gasoline and diesel.

Control of the concentration of metal ions such as Al^{3+}, Hg^{2+}, Cd^{2+}, etc. in drinking water is extremely important. Fibre optic sensors have been demonstrated using the absorption or fluorescent properties of ligands such as morin and porphyrin that form fluorescent complexes with metal ions (Seitz, 1984; Reichert *et al.*, 1992), as discussed in section 1.2.3.

The level of nitrates in drinking water is the subject of legislation in the EU which will require levels below $30\,\mathrm{mg\,l^{-1}}$. Optical sensors capable of measurements in this range have been demonstrated, although a number of improvements are necessary in terms of long-term stability and cross-sensitivity to other anions (Reichert *et al.*, 1992; Mohr *et al.*, 1994). The sensing mechanism is based on having a nitrate carrier in a PVC membrane which has the property of extracting hydrogen ions along with a nitrate ion from the solution (co-extraction). A pH indicator dye is included in the membrane which undergoes a change in absorption or fluorescence, following the hydrogen ion concentration. The pH of the sample has to be kept constant or independently measured, but alternative techniques to avoid this are under investigation (Mohr *et al.*, 1994).

1.4.4 Soil contamination

As a result of collaboration between several groups, including the Naval Command Control and Ocean Surveillance Center (NCCOSC, San Diego, CA), a rapid optical screening tool (ROST) system has been developed for detection of contamination in soil and groundwater by petroleum hydrocarbons (Hobbs, 1994; Lieberman *et al.*, 1994). A cone penetrometer is used to drive the ROST probe into the ground and profiles of contamination versus depth can be obtained, thus defining the level and size of a contaminated sector or plume. Compared with the traditional method where samples are brought to the surface for analysis, the *in situ* system is fast, continuous and the cost per hole is much less. The probe is based on using a tunable dye laser, pumped by a Nd:YAG laser and frequency doubled into the UV, to induce fluorescence in the aromatic hydrocarbons. Silica fibres of typically $50\,\mathrm{m}$ length with $600\,\mu\mathrm{m}$ core diameter are used to transmit light to and from a sapphire window in the probe, which acts as the window to the external environment. The signature of the induced fluorescence is used to identify the type and level of contamination; for example, fluorescence from one- and two-ring aromatic hydrocarbons (as found in jet fuel) give fluorescence in the 320–370 nm band (290 nm excitation), whereas diesel fuel with a higher proportion of three- and four-ring hydrocarbons give fluorescence in the 350–400 nm range. Research is currently underway to extend the capabilities of the probe to other spectroscopic techniques including fibre optic Raman sensing and a fibre optic probe for metals based on laser-induced breakdown spectroscopy (Lieberman *et al.*, 1994). Investigation of fibre optic Raman systems for waste and contamination monitoring is also under investigation elsewhere (Angel and Cooney, 1994).

1.5 Conclusion

1.5.1 Summary

From the above discussion, it is apparent that fibre optic sensors offer considerable scope for environmental monitoring applications. A range of viable techniques are available to cover a large number of pollutants and this list is likely to grow as research advances and new regulations come into force governing acceptable pollution limits. As already demonstrated in a number of specific cases, fibre optic sensors have the capability to meet the sensitivity requirements of environmental monitoring and they offer the distinct advantage of being able to perform continuous *in situ* measurements. Cost is likely to be competitive with other sensor technologies. There is always perhaps a slight reluctance to accept new technology, but at least with fibre optic chemical sensors the basic spectroscopic techniques are well established and have been used for a long time, and the introduction of fibre optics adds a new versatility to the sensor.

1.5.2 Future trends

There are a number of areas where advancements are being made or are needed which are likely to have significant impact on future sensors. Further development of laser sources such as DFB sources for near-IR gas sensing as well as fibre laser and fibre amplifiers operating at convenient wavelengths for sensor applications will be very important. Currently the cost of suitable DFB sources is very high. The ongoing development of fibres which can transmit in the mid-IR region along with corresponding IR sources and detectors is also significant, for this would allow the fundamental absorption lines of gases to be exploited in fibre sensors, giving high-sensitivity measurements.

Development of new indicator dyes and other chemically sensitive reagents is a key area of development. Most existing dyes have their absorption spectrum in the UV or shorter wavelength–visible spectral region, and so there is much interest in developing near-IR absorbing dyes where inexpensive laser and LED sources are readily available. Generally, there is need for improvements in immobilisation techniques and in dealing with the problems of long-term stability and cross-sensitivity to other parameters.

Finally, fibre optic sensors offer unique possibilities in distributed and quasi-distributed systems, and this is likely to receive increased attention in the future. For example, evanescent wave systems in combination with optical time domain reflectometry (OTDR) offer the possibility of obtaining concentration profiles over extended lengths in a environmental situation. Advanced quasi-distributed systems are also possible, containing arrays of sensors in fibre networks; some may be configured to monitor several different species simultaneously. Fibre optic sensors are, therefore, likely to have an important role in the future in addressing the concerns of environmental pollution.

References

Angel, S.M. and Cooney, T. (1994) Chemical waste measurements using Raman fibre optic sensing. *Proceedings 2nd European Conference on Optical Chemical Sensors and Biosensors*, Florence, Italy, April 19–21, p. 74.

Archenault, M., Ronot, C., Gagnaire, H., Goure, J.P. and Jaffrezic-Renault, N. (1992) Detection of chemical vapours with a specifically coated optical fibre sensor. *Proceedings 1st European Conference on Optical Chemical Sensors and Biosensors*, Graz, Austria, April 12–15, p. 97.

Barnwell, C.N. (1983) *Fundamentals of Molecular Spectroscopy*. McGraw-Hill, London.

Bogue, R. (1994) Briefing: Integrated pollution control in the EU. *Environmental Sensors*, May, 6-10, IOP Publishing, Bristol, U.K.

Brinker, C.J. and Scherer, C.W. (1990) *Sol–Gel Science*. Academic Press, Boston, MA.

Butler, M.A. and Buss, R.J. (1992) Kinetics of the micromirror chemical sensor. *Proceedings 1st European Conference on Optical Chemical Sensors and Biosensors*, Graz, Austria, April 12–15, p. 49.

Chan, K., Ito, H. and Inaba H. (1985) 10 km long fibre optic remote sensing of CH_4 gas by near infrared absorption. *Applied Physics B*, **38**, 11–15.

Chernyak, V., Reisfeld, R., Gvishi, R. and Venezky, D. (1990) Oxazine-170 in sol–gel glass PMMA films as a reversible optical waveguide sensor for ammonia and acids. *Sensors and Materials*, **2** (2), 117–126.

Cooper, D.E. and Martinelli, R.U. (1992) Near-infrared diode lasers monitor molecular species. *Laser Focus World*, **28** (11), 133–146.

Dao, N.Q. and Jouan, M. (1993) The Raman laser fiber optics (RLFO) method and its applications. *Sensors and Actuators B*, **11**, 147–160.

Edmonds, T.E., Flatters, N.J., Jones, C.F. and Miller, J.N. (1988) Determination of pH with acid-base indicators: implications for optical fibre probes. *Talanta*, **35** (2), 103–107.

Erley, D.S. and Blake, B.H. (1964) *Infrared Spectra of Gases and Vapors*, The Dow Chemical Company, Michigan.

Erley, D.S. and Blake, B.H. (1965) *Infrared spectra of gases and vapors*, Vol. II, The Dow Chemical Company, Michigan.

Fowles, M. and Wayne, R.P. (1981) Ozone monitor using an LED source. *Journal of Physics E: Science Instruments*, **14**, 1143–1145.

Fleischli, M.A. and Walder F.T. (1992) Dedicated spectrometer aids Raman spectroscopic analysis. *Laser Focus World*, **28** (11), 149–153.

Gauglitz, G. (1992) Chemical and biochemical sensors based on interferometry at thin layers. *Proceedings 1st European Conference on Optical Chemical Sensors and Biosensors*, Graz, Austria, April 12–15, p. 12.

Griggs, M. (1968) Absorption coefficients of ozone in the ultraviolet and visible regions. *Journal of Chemical Physics*, **49** (2), 857–859.

Goswami, K., Ejiofor, C., Saini, D.P. and Klainer, S.M. (1994) Detection of carbon monoxide: the fiber optic way. *Proceedings 2nd European Conference on Optical Chemical Sensors and Biosensors*, Florence, Italy, April 19–21, p. 31.

Gupta, B.D., Singh, C.D. and Sharma, A. (1994) Fiber optic evanescent field absorption sensor: effect of launching condition and the geometry of the sensing region. *Optical Engineering*, **33** (6), 1864–1868.

Hobbs, J.R. (1994) Dye-based laser system probes soil contamination. *Laser Focus World*, **30** (11), 34–38.

Inn, E.C.Y. and Tanaka, Y. (1953) Absorption coefficient of ozone in the ultraviolet and visible regions. *Journal of the Optical Society of America*, **43** (10), 870–873.

Klainer, S.M., Goswami, K., Dandge, D.K., Simon, S.J., Herron, N.R., Eastwood, D.L. and Eccles, L.A. (1991) Environmental monitoring applications of fibre optic chemical sensors. *Fiber Optic Chemical Sensors*, Vol. II (ed. O.S. Wolfbeis), pp. 83–122, CRC Press, Boca Raton, FL.

Klimant, I. and Leiner, M.J.P. (1992) Recent investigations in oxygen sensing. *Proceedings 1st European Conference on Optical Chemical Sensors and Biosensors*, Graz, Austria, April 12–15, p. 131.

Lakowicz, J. R. (1983) *Principles of Fluorescence Spectroscopy*, Plenum Press, New York.

Lakowicz, J. R. (1992) Fluorescence lifetime sensing generates cellular images. *Laser Focus World*, **28** (5), 60–80.

Leugers, M.A. and McLachlan, R.D. (1988) Remote analysis by fiber optic Raman spectroscopy. *Chemical, Biochemical and Environmental Applications of Fibers*, The Society of Photo-Optical Instrumentation Engineers, **990**, 89–94.

Lieberman, R.A., Blyler, L.L. and Cohen, L.G. (1990) A distributed fiber optic sensor based on cladding fluorescence. *Journal of Lightwave Technology*, **18** (2), 212–220.

Lieberman, S.H., Theriault, G.A., Wu, K. and Davey, M. (1994) Remote fiber optic spectroscopy for *in-situ* monitoring of chemical contamination in soils. *Proceedings 2nd European Conference on Optical Chemical Sensors and Biosensors*, Florence, Italy, April 19–21, p. 34.

MacCraith, B.D., O'Keeffe, G., McDonagh, C. and McEvoy, A.K. (1994) LED-based fibre optic oxygen sensor using sol–gel coating. *Electronics Letters*, **30** (11), 888–889.

McCulloch, S., Stewart, G., Guppy, R.M. and Norris, J.O.W. (1994) Characterisation of TiO_2–SiO_2 sol–gel films for optical chemical sensor applications. *International Journal of Optoelectronics*, **9** (3), 235–241.

Marcuse, D. (1988) Launching light into fiber cores from sources located in the cladding. *Journal of Lightwave Technology*, **6** (8), 1273–1279.

Milanovich, F.P., Brown, S.B., Colston, B.W. and Daley, P.F. (1994) A fibre optic sensor system for remote long term monitoring of soil and groundwater contamination. *Proceedings 10th Optical Fibre Sensors Conference (OFS 10)*, The Society of Photo-Optical Instrumentation Engineers, **2360**, Glasgow, Scotland, Oct 11–13, pp. 98–100.

Mohr, G.J., Kovacs, B. and Wolfbeis, O.S. (1994) Solid state nitrate sensor based on potential-sensitive fluorescent dyes. *Proceedings 2nd European Conference on Optical Chemical Sensors and Biosensors*, Florence, Italy, April 19-21, p. 36.

Mullen K.I. and Carron K.T. (1991) Surface enhanced Raman spectroscopy with abrasively modified fiber optic probes. *Analytical Chemistry*, **63** (19), 2196–2199.

Mullen, K.I., Wang, D.X., Crane, L.G. and Carron, K.T. (1992) Determination of pH with surface-enhanced Raman fibre optic probes. *Analytical Chemistry*, **64** (8), 930–935.

Narayanaswamy, R. and Sevilla, F. (1988) Optosensing of hydrogen sulphide through paper impregnated with lead acetate. *Fresenius' Zeitschrift für Analytische Chemie*, **329**, 789–792.

Newbery R., Mosier-Boss, P. and Lieberman, S.H. (1994) Raman spectroscopy for remote fibre optic sensing. *Proceedings 2nd European Conference on Optical Chemical Sensors and Biosensors*, Florence, Italy, April 19–21, p. 103.

Niemczyk, T.M., Delgado-Lopez, M. and Newman C.D. (1993) Multichannel Raman spectroscopy tackles industrial problems. *Laser Focus World*, **29** (3), 85–98.

O'Keeffe, G., MacCraith, B.D., McEvoy, A.K. and McDonagh, C.M. (1994) Development of an LED-based phase fluorimetric oxygen sensor using evanescent wave excitation of a sol–gel immobilised dye, *Proceedings 2nd European Conference on Optical Chemical Sensors and Biosensors*, Florence, Italy, April 19–21, p. 149.

Parker, C.A. (1968) *Photoluminescence of Solutions*, Elsevier, Amsterdam

Peterson, J.I., Goldstein, S.R., Fitzgerald, R.V. and Buckhold, D.K. (1980) Fibre optic pH probe for physiological use. *Analytical Chemistry*, **52**, 864–869.

Peterson, J.I., Fitzgerald, R.V. and Buckhold, D.K. (1984) Fibre optic probe for in-vivo measurement of oxygen partial pressure. *Analytical Chemistry*, **56**, 62.

Reichert, J., Czolk, R., Morales-Bahnik, A., Sellien, W. and Ache, H.J. (1992) Optical chemical sensors for environmental analysis: ammonium, nitrate and heavy metal ion sensors. *Proceedings 1st European Conference on Optical Chemical Sensors and Biosensors*, Graz, Austria, April 12–15, p. 40.

Ruddy, V. (1990) An effective attenuation coefficient for evanescent wave spectroscopy using multimode fibre. *Fiber and Integrated Optics*, **9**, 142–150.

Saini, D.P., Leclerc, R., Klainer, S.M., Himka, R.L., Arman, H., Dandge, D.K., Wolfbeis, O.S. and Kovács, B. (1994) Petrosense CMS 5000: a fibre optic sensors sensing system for the continuous monitoring of hydrocarbons. *Proceedings 2nd European Conference on Optical Chemical Sensors and Biosensors*, Florence, Italy, April 19–21, p. 39.

Schoen, C.L. (1994) Fiber probes permit remote Raman spectroscopy. *Laser Focus World*, **30** (5), 113–120.

Schulman, S. G. (ed.) (1988) *Molecular Luminescence Spectroscopy: Methods and Applications*, John Wiley, New York.

Seitz, W.R. (1984) Chemical sensors based on fiber optics. *Analytical Chemistry*, **56** (1), 16–34.

Sharma, A. and Wolbeis, O.S. (1989) Fibre optic fluorosensor for sulphur dioxide based on energy transfer and exciplex quenching. *Proceedings of The Society of Photo-Optical Instrumentation Engineers*, **990**, 116.

Shimose, Y., Okamoto, T., Maruyama, A., Aizawa, M. and Nagai, H. (1991) Remote sensing of methane gas by differential absorption measurement using a wavelength tunable DFB LD. *IEEE Photonics Technology Letters*, **3** (1), 86–87.

Stewart, G. and Culshaw, B. (1994) Optical waveguide modelling and design for evanescent field chemical sensors. *Optical and Quantum Electronics*, **26**, S249–S259.

Stewart, G., Norris, J., Clark, D.F. and Culshaw, B. (1991) Evanescent wave chemical sensors — a theoretical evaluation. *International Journal of Optoelectronics*, **6** (3), 227–238.

Stewart, G., Culshaw, B., Muhammad, F., Murray, S., Pinchbeck, D., Norris, J., Cassidy, S., Wilkinson, M., Williams, D., Crisp, I., Van Ewyk, R. and McGhee, A. (1992) Evanescent wave methane detection using optical fibres. *Electronics Letters*, **28** (4), 2232–2234.

Stuart, A. D. (1992) Some applications of infrared optical sensing. *Proceedings 1st European Conference on Optical Chemical Sensors and Biosensors*, Graz, Austria, April 12–15, p. 25.

Stuart, A.D. and Samson, P.J. (1988) Optrode sensors for carbon monoxide and relative humidity. *Proceedings 13th Australian Conference on Optical Fibre Technology*, Hobart, Australia, Dec 4–7, p. 117.

Tai, H., Tanaka, H. and Yoshino, T. (1987) Fibre optic evanescent wave methane gas sensor using optical absorption for the 3.392 μm line of a He–Ne laser. *Optics Letters*, **12** (6), 437–439.

Tai, H., Yamamoto, K., Uchida, M., Osawa, S. and Uehara, K. (1992) Long distance simultaneous detection of methane and acetylene by using diode lasers coupled with optical fibers. *IEEE Photonics Technology Letters*, **4** (7), 804–807.

Tobias, R. S. (1967) Raman spectroscopy in inorganic chemistry: I theory and II applications. *Journal of Chemical Education*, **44**, 2–8; 70–79.

Uehara, K. and Tai, H. (1992) Remote detection of methane with a 1.66 μm diode laser. *Applied Optics*, **31** (6), 809–814.

Vo-Dinh, T. (1994) SERS chemical sensors and biosensors: new tools for environmental and biomedical analysis. *Proceedings 2nd European Conference on Optical Chemical Sensors and Biosensors*, Florence, Italy, April 19–21, p. 101.

Vo-Dinh, T., Stokes, D.L., Li, Y.S. and Miller, G.H. (1990) Fiber optic sensor probe for in-situ surface-enhanced Raman monitoring. *Chemical, Biochemical and Environmental Fiber Sensors II*, The Society of Photo-Optical Instrumentation Engineers, **1368**, 203–209.

Walt, D.R. (1992) A fibre optic sensor for measuring CO_2 in seawater. *Proceedings 1st European Conference on Optical Chemical Sensors and Biosensors*, Graz, Austria, April 12–15, p. 71.

Weldon, V., Phelan, P. and Hegarty, J. (1993) Methane and carbon dioxide sensing using a DFB laser diode operating at 1.64 μm. *Electronics Letters*, **29** (6), 560–561.

Weldon, V., Phelan, P., Hegarty, J. and Tanbun-Ek, T. (1994) H_2S and CO_2 gas sensing using a 1.57 μm DFB laser diode. *Proceedings 2nd European Conference on Optical Chemical Sensors and Biosensors*, Florence, Italy, April 19–21, p. 26.

Wolfbeis, O.S. (ed.) (1991) *Fiber Optic Chemical Sensors*, Vol. I and II. CRC Press, Boca Raton, FL.

Wolfbeis, O.S. and Sharma, A. (1988) Fibre optic fluorosensor for sulphur dioxide. *Analytica Chimica Acta*, **208**, 53.

2 Integrated optic sensors

J.V. MAGILL

2.1 Introduction to integrated optics

Integrated optical devices use optical signals for their operation. While electrical currents carry the signals in electronic circuits, beams of light have this function in optical circuits. Thus, electrical connections such as wires are replaced by optical waveguides and semiconductor integrated circuits are replaced by optical circuits. Figure 2.1 shows a schematic representation of an integrated optical sensor, which illustrates the principal components of such devices and also some of the advantages of optical circuitry. Many volumes provide a detailed description of integrated optic theory and technology, and in this section only a brief overview of the most important aspects will be given (Hunsperger, 1991; Tamir, 1979).

The optical waveguide is the basic element used to connect the parts of an optical integrated circuit; it is analogous to a metal wire or strip in an electrical or electronic circuit. Light is guided along and controlled by thin dielectric films or strips. The wavelengths of operation generally lie between 0.1 and 10 μm (10^3–10^5 Å), which cover the visible and parts of the UV and IR regions of the spectrum. These regions are used principally because the

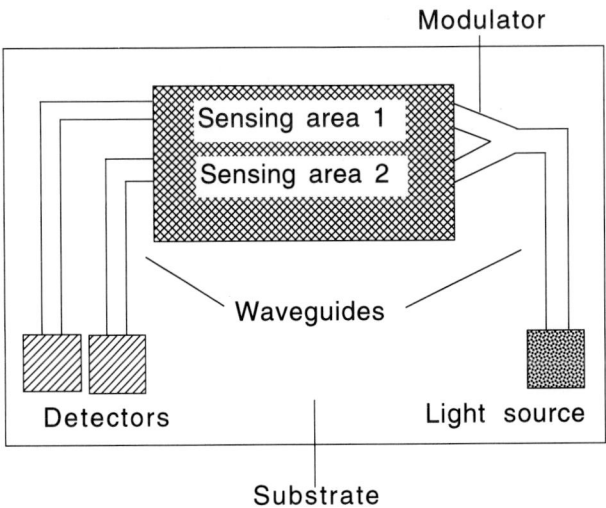

Figure 2.1 Schematic representation of an integrated optical sensor.

materials available for waveguide fabrication and light sources operate within this range. Outside this range, other techniques are preferable (e.g. metallic waveguides for microwave technology) and losses are also higher as a result of scattering. Light enters and leaves the waveguiding area through connectors, which in integrated optical devices are called couplers. These can take a number of forms, including prisms, gratings or fibres. Alterations to the light signal in time or space are made by means of optical modulators and switches. These components are used extensively in optical communications but, to date, less so for optical sensor applications. The options for light sources in integrated optics include LEDs and laser diodes, which are small in size and can be relatively easily integrated with other components. However, there are few examples of sensor devices using fully integrated light sources and most use a distinct light source with access into the sensor device by fibre, prism or grating coupling. Integrated optical detectors are generally photodiodes, which offer the high sensitivity, rapid response and low power consumption required for the devices.

Integrated optics sensors offer many features that can result in significant advantages in fabrication and performance over other sensing techniques. These include low fabrication cost, high sensitivity, high selectivity, rapid sensing, safe and simple (e.g. one-step) operation, ruggedness, miniaturisation, multiple sensor operation and resistance to electromagnetic interference.

In this area, a distinction has always been made between integrated optical devices and optical fibre devices. This is largely historical since the two areas have much in common and the distinction between the two is not clear at the boundaries. In addition, many devices benefit from the use of both technologies together. However, this chapter will concentrate primarily on integrated optical devices.

2.1.1 Waveguides

Waveguides are used to connect together components in integrated optics circuits. Whilst an optical fibre is an example of an optical waveguide of circular cross-section, those used in integrated optics are usually thin planar films or strips. The simplest waveguide consists of a thin film sandwiched between a substrate and a superstrate, each with lower refractive indices, as illustrated in Figure 2.2. A typical planar waveguide may have a superstrate which is air and a typical film thickness of 1 μm. The operation of waveguides may be described in terms of a simple ray optics approach, but a more complete description is given by electromagnetic theory (Tamir, 1975). The ray optics approach shows that light is confined by total internal reflection at two interfaces, i.e. the film substrate and film superstrate interfaces (Figure 2.3). Total internal reflection will only occur when the incident angle at each

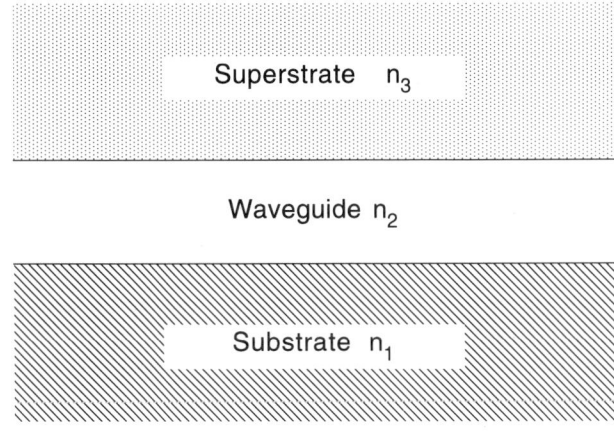

Refractive index $n_1 > n_2 < n_3$

Figure 2.2 Schematic structure of a simple planar waveguide.

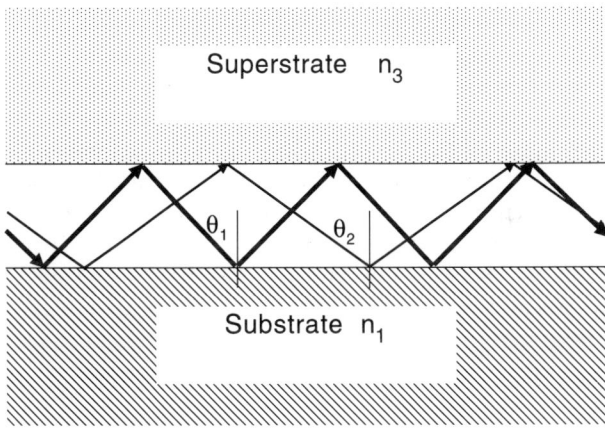

Figure 2.3 Propagation of light in a waveguide structure—the ray optics approach.

interface is greater than the critical angle. Below the critical angle, there is only partial reflection and also refraction. Above the critical angle, where total internal reflection occurs, there is a phase shift of the reflected light depending on whether or not the polarisation is TE and TM. In addition, not all angles of incidence are allowed, and theory shows a discrete set of angles corresponding to the guided modes for a particular waveguide structure. Thus, the ray optics approach describes a series of zig-zag waves propagating along the waveguide, each with its own angle of incidence at the interfaces.

A particular property of waveguides used extensively in integrated optic sensors and also providing the mechanism for coupling is the evanescent field, as described by electromagnetic theory or by the Goos–Hanchen shift in the ray optics approach. These theories show that the light energy extends beyond and decays exponentially away from the waveguide interface. The number of waveguide modes propagating in any structure depends on the thickness of the waveguiding layer, the frequency of the propagating light and the refractive indices of the waveguide, substrate and superstrate. The wavelength to be used is often a fixed parameter and, therefore, the guide thickness and refractive indices must be chosen to permit the waveguiding required.

The efficiency of an optical waveguide is determined by losses that occur as light travels through the waveguide. These losses are caused by scattering, absorption or radiation of the light, and the types of loss in any particular waveguide structure are generally determined by the materials and refraction processes used. The primary origin of scattering losses is roughness at the waveguide interfaces. Losses will occur at each reflection as the light beam follows its zig-zag path through the waveguide. Scattering losses may also occur within the bulk of the waveguide as a result of imperfections in the structure. These may be caused by contaminants in the material or crystal defects, and the loss is proportional to the number of imperfections. Scattering is the predominant mechanism for losses in glass and dielectric waveguide structures. Absorption losses can be highly significant when semiconductor materials are used. In this case, photon energy is absorbed by the electron and hole charge carriers in the semiconducting structure. Such losses can be reduced by choosing a propagating wavelength that is not absorbed by the semiconductor material. Radiation losses occur when light is emitted into the substrate and superstrate material surrounding the waveguide and is no longer guided. These losses are particularly important in curved waveguides, where the propagating optical field is distorted.

2.1.2 Waveguide couplers

These are devices for converting the light beam into modes guided within a thin film waveguiding layer. They may also be used to transfer the light energy from one waveguide to another, to act as mode converters or as filters. Couplers also provide the means for making use of the high bandwidth of optical integrated circuits by coupling many optical signals into one waveguide. A range of coupling mechanisms are available to suit different applications. These include transverse or end-fire couplers, prism couplers, grating couplers and fibre-to-waveguide couplers. The operation of any coupler is defined in terms of the coupling efficiency and the coupling loss for individual modes in a selective coupler, or an overall measurement for a multimode coupler.

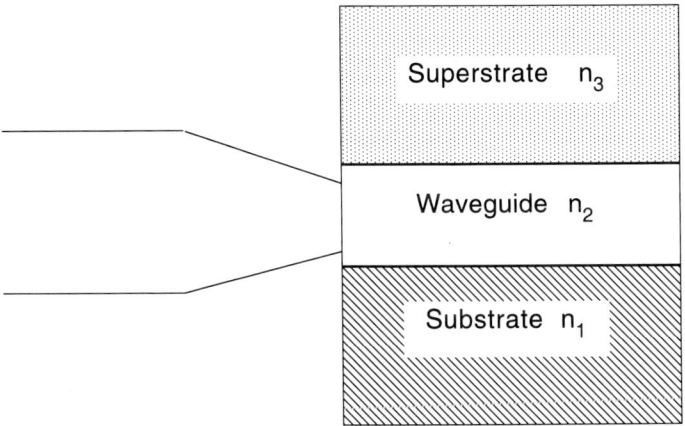

Figure 2.4 Coupling into a waveguide—end fire.

The transverse or end-fire coupling method is illustrated in Figure 2.4. In these couplers, the beam is focused directly into the waveguide and, therefore, an exposed area of the waveguide is required. Efficient coupling occurs when the profile of the incident beam matches that propagating within the waveguide. This is generally achieved using a lens and in principle can be very efficient. However, the alignment of components and their surface quality is critical, and deficiencies in these areas often lead to large losses.

The mechanism of prism coupling is illustrated in Figure 2.5. The thin air gap (which is a function of the state of the surface) between the prism and

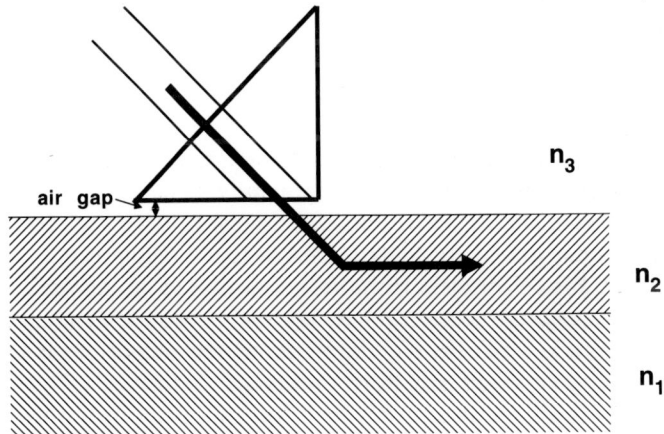

Figure 2.5 Coupling into a waveguide—prism coupling.

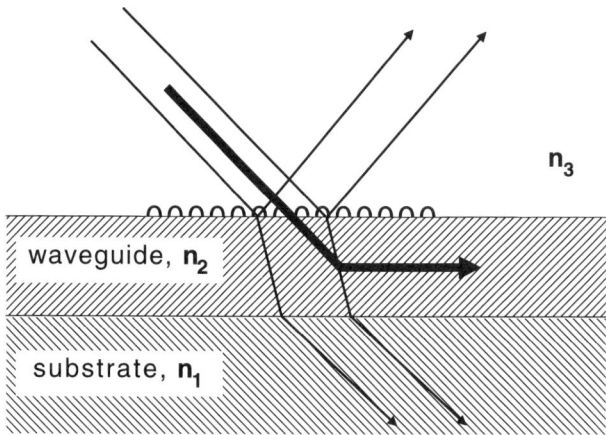

Figure 2.6 Coupling into a waveguide—grating coupling.

the waveguide allows overlap of the prism and waveguide modes and hence coupling of light energy from the prism into the waveguide. For this to be achieved, the refractive index of the prism must be higher than that of the waveguide and the correct coupling angle must be used. Varying the coupling angle allows different waveguide modes to be selected. Prism couplers can be used to couple light both in and out of a waveguide. The efficiency of prism coupling depends on the shape of the beam, which is about 80% for a Gaussian shape. In addition, the geometry of the prism and the waveguide must be optimised for efficient coupling. In practice, this means that prism coupling is generally used in laboratory applications but it is less useful for routine practical applications such as sensors.

Grating couplers offer a more robust mechanism for practical applications (Figure 2.6). Once again, there is matching between the incident optical beam mode and a waveguide mode, but the perturbations caused by the presence of a grating allow direct coupling into the waveguide without the air gap required in prism coupling. A major drawback of grating couplers is that efficiencies are generally rather low because of loss of energy through the waveguide into the substrate. This problem can be improved to some extent through careful design of the gratings. A major advantage of grating couplers is that the grating can be fabricated directly onto the waveguide structure and is, therefore, robust. Grating couplers require relatively sophisticated fabrication techniques, but these are available in modern semiconductor production.

The options for connecting fibres to waveguides include butt coupling of the fibre to the waveguide, coupling to a tapered waveguide and grating couplers (Hunsperger, 1991). For coupling between waveguides, the most common method is to make the guides close together and allow coupling to occur by optical tunnelling in a similar way to that occurring in prism coupling.

2.1.3   Optical modulators and switches

Optical modulators and switches use electrical signals to control the propagation of light in optical waveguides. Many devices can operate as either switches or modulators. In general, modulators are devices that superimpose information onto an optical wave by varying its properties, and switches alter the spatial position of the light, or turn it 'on' or 'off'. For example, a modulator might use an applied electrical signal to increase or decrease the intensity of light, or to change its phase, whereas a switch would only alter the spatial location of the light. A schematic of an optical modulator structure is shown in Figure 2.7. The structure is similar to that of a planar waveguide (Figure 2.2), except that a voltage is applied across the waveguide between the substrate and the superstrate (McMeekin, 1992). The materials must now be chosen for their electro-optic properties in addition to those for waveguiding. Other physical effects may also be used in light modulators and switches, and these include acoustooptic and magnetooptic effects. The suitability of a switch or modulator for a particular application is defined in a similar way to that of an optical waveguide but with some additional parameters of interest. For modulators, the depth of modulation and the power required for modulation to occur are important, whereas for switches, the isolation or cross-talk and the switching time are important parameters. The use of closely spaced strip waveguides for optical coupling has already been mentioned and the addition of

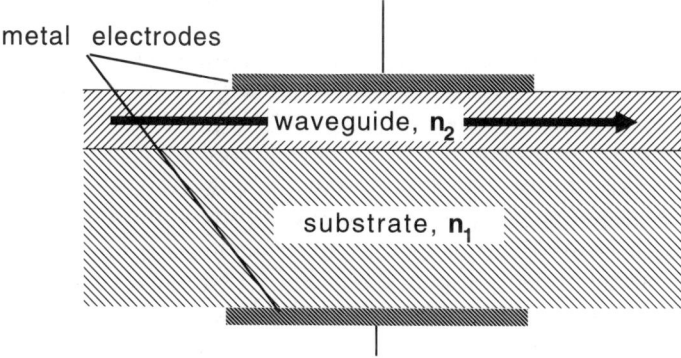

Figure 2.7 An integrated optical modulator structure.

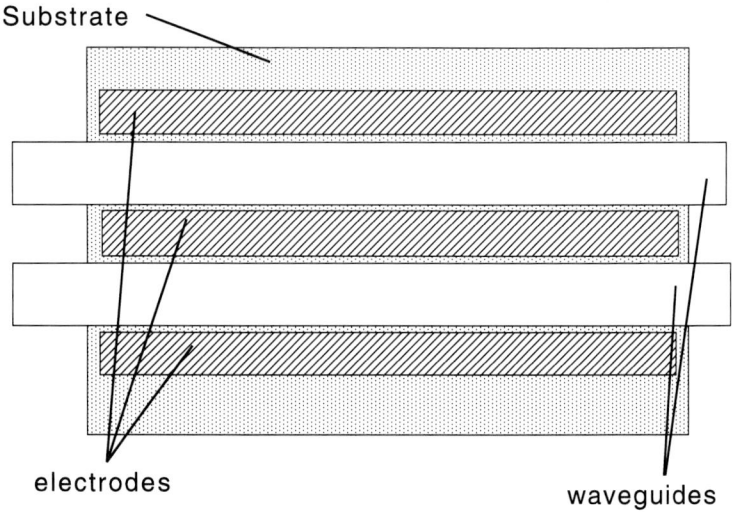

Figure 2.8 A stripe waveguide structure for coupling and modulation.

electrodes to these structures allows them to operate as optical modulators (Figure 2.8).

Switches and modulators are widely used in applications such as telecommunications but, as yet, there are very few examples of their use in the area of sensors.

2.1.4   Integrated optic light sources

For laboratory use, it is convenient to use stand-alone light sources such as gas lasers and lamps. In integrated optics, however, the light sources must be incorporated within a single device structure, so solid-state light emitting diodes (LEDs) and semiconductor lasers are commonly used. These structures make use of the energy levels and associated interactions that occur within semiconductor crystal materials on an electronic scale. Emission of light in semiconductor materials occurs when charge carriers (electrons and holes) combine to produce energy at optical frequencies (photons). This process may either occur spontaneously or be stimulated by the presence of an external light energy source, as illustrated in Figure 2.9. A simple example of this type of light source is a p–n junction diode, as shown schematically in Figure 2.10. This device contains many positive and negative charge carriers that are separated by an energy barrier. When a voltage is applied to the device, the barrier is lowered and the charge carriers can combine to emit photons. Whilst LEDs emit light spontaneously once the energy barrier has been removed, semiconductor

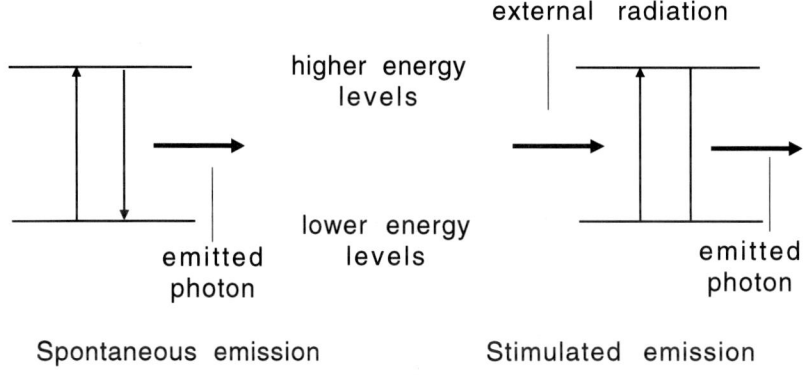

external radiation

higher energy
levels

lower energy
levels

emitted
photon

emitted
photon

Spontaneous emission   Stimulated emission

Figure 2.9 Integrated optical sources. (a) Spontaneous and (b) Stimulated emission of light.

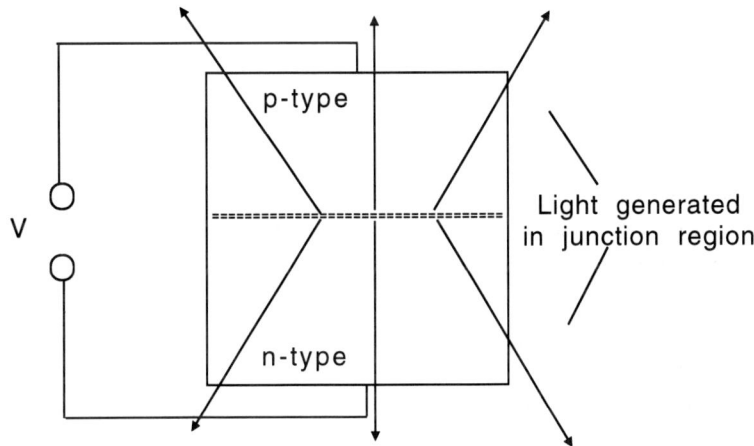

p-type

V

Light generated
in junction region

n-type

Figure 2.10 Schematic structure of an integrated optical source, a p–n junction light-emitting diode.

lasers emit light after stimulation by an energy 'pump' of photons or electrons. A simple semiconductor laser structure is shown in Figure 2.11. Whilst this simple structure could be made from one type of compound semiconductor material, such as gallium arsenide, more complex structures using multiple layers of different compound semiconductors are now routinely used and have improved performance. However, it should be noted that the wavelength spectrum of all semiconductor light emitters is relatively simple, compared with gas lasers for example, and would generally contain one broad emission peak corresponding to the energy gap between the charge carriers, as shown in Figure 2.9.

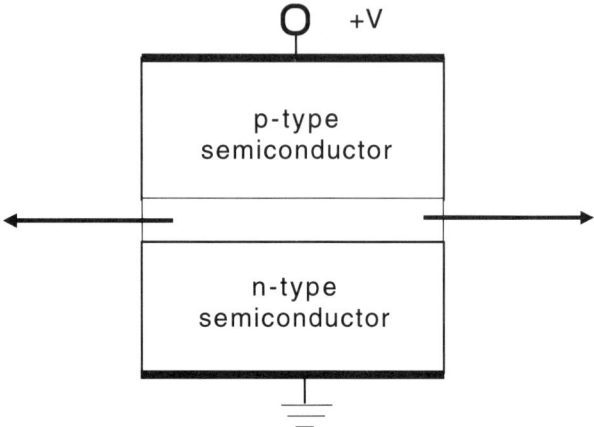

Figure 2.11 Schematic structure of an integrated optical source, a semiconductor laser.

2.1.5   Integrated optical detectors

The most common optical detector used in integrated optics is the semiconductor photodiode. In this device, light, absorbed by the semiconductor crystal structure, generates charge carriers in discrete energy levels. There is a flow of charge carriers to these levels and a corresponding electric current is measured in proportion to the intensity to the incident light. A semiconductor waveguide photodiode is illustrated schematically in Figure 2.12. The wavelengths detected by the device are determined by the crystal structure for the material used. The detector will be sensitive to light with higher energy (or shorter wavelength) than the energy required to create charge carriers within the crystal. This energy can be modified to some degree by the addition of impurity atoms. In

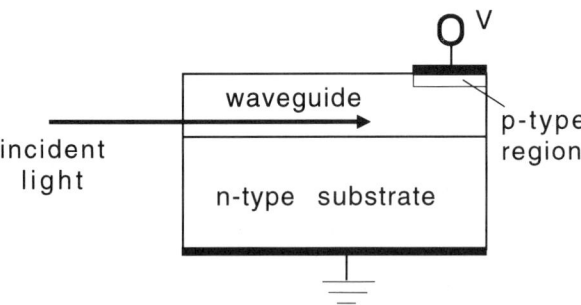

Figure 2.12 Schematic structure of an integrated optical detector, a waveguide photodiode.

addition to the detectable wavelengths, important parameters characterising photodetectors are the speed and the efficiency of response.

2.2   Fabrication of integrated optic devices

Whilst the precise fabrication details for any integrated optic device
will depend on its application, several types of device process will
generally be used. Firstly, thin films are fabricated and these may be
amorphous materials or crystalline structures; secondly, the properties of
these thin films may be modified by addition of other materials or by
modification of crystal structures. Finally, in order to create integrated
optic devices, each of the thin films must be made in a pattern corresponding to the component required (e.g. waveguide, light source, detector, etc.).
These processes are described later in the section but first the choice of
substrate materials for integrated optic devices is discussed. Detailed
descriptions of modern semiconductor fabrication techniques are readily
available (de Cogan, 1990; Fraser, 1990; and Morgan and Board, 1990), as
are details specific to integrated optic devices (Hunsperger, 1991; Tamir,
1975).

2.2.1   Materials for integrated optic devices

The choice of substrate material will depend on the application required,
the cost and the complexity of the fabrication process. In a fully integrated
circuit, a single substrate will be used for all devices (sources, waveguides,
modulators, detectors, etc.) and, therefore, the choice of material will be a
compromise to obtain optimum circuit performance. In addition, the choice
of materials will be limited since those that can be used in active devices
such as light sources are quite limited. These substrates include gallium
arsenide and a range of other compound semiconductors, principally indium
phosphide. Fabrication of circuits using these materials is both difficult and
expensive, and examples of fully integrated optical circuits are generally still
limited to laboratory demonstrations (De La Rue, 1993). More commonly,
devices employ some degree of hybridisation by using two or more substrate
materials bonded together to form a circuit, or even a partially integrated
circuit with some separate components such as light sources and detectors.
This is particularly true in sensor applications where cost is often very
important and the sensor function may dictate the need for hybrid devices.
Additional choices of substrate for hybrid circuits include quartz, lithium
niobate and silicon. In hybrid devices, where a wider range of substrate
materials is available, considerations such as cost and ease of fabrication
become important. Fused quartz or silica, for example, provides a cheap
material that is relatively straightforward for processing but is less suitable

than the more expensive lithium niobate for modulation and switching devices. For sensor applications where the sensing element is disposable, a cheap material such as glass is desirable and a hybrid structure would be used.

2.2.2   *Fabrication and modification of thin films*

Many of the components of integrated optic circuits require the fabrication of thin films. This section reviews the range of fabrication processes available and their suitability in specific application. The thin films used in optical integrated circuits must have a known thickness and refractive index and be of good optical quality, i.e. free from unintentional imperfections and impurities. The boundaries between the different layers should also be as smooth as possible. Thin films may be made either by depositing a layer of material onto a substrate or by causing a chemical or physical change to occur in a layer of the substrate material. Options for adding layers of material include sputtering, vacuum deposition, spin coating, dipping and epitaxial growth, while changes in 'near surface' substrate properties can be effected by diffusion of atoms, ion implantation, ion exchange or proton bombardment. In the sputtering technique, atoms or molecules are removed from the surface of a target made from appropriate material by bombarding it with high energy ions in a vacuum. The atoms or molecules removed from the target are then allowed to deposit onto a substrate within the vacuum chamber. The thickness of the sputtered thin film can be controlled by the duration of the ion bombardment, and resulting films are very uniform and contain few contaminants. Thin films can be produced by deposition from solution; the most common methods are spin coating and dipping. The material to be deposited is held in solution with an appropriate solvent which is removed after film application. Materials for which this technique is used are polymers such as polyurethane with organic solvents such as xylene. The principal disadvantages of this technique are that film purity and levels of contamination are considerably worse than those in sputter deposition. However, the process is very inexpensive and requires no sophisticated equipment. It is, therefore, an attractive option for many sensor applications and, in particular, where a disposable sensing element is required. For fully integrated optical sensors using semiconductor substrate materials, thin films are often produced by epitaxial growth. In the simplest version of this technique, the crystal structure of the substrate and the thin deposited layers are very similar and the chemical composition of the thin films can be precisely tailored with refractive indices and wavelengths to suit the specific application. Epitaxial films may be grown from a number of sources, including liquid, vapour and molecular beams. As discussed earlier, gallium arsenide and other compound semiconductors are used for fully integrated optical sensors. Epitaxially grown layers of aluminium gallium arsenide ($Al_x Ga_{1-x} As$)

allow the formation of active and passive devices on the substrate. In liquid-phase epitaxy, layers are grown from a melt in a tube furnace. Furnaces are usually constructed so that a number of different layers can be grown within one furnace. Growth of layers from the vapour phase produces films with lower levels of contamination. The technique of molecular beam epitaxy allows the greatest control over purity, layer thickness and layer composition, but it is a very expensive technique. The constituent atoms for the layer composition required are accelerated towards the substrate material in a high-vacuum chamber. Shutters allow each of the source material molecular beams to be switched on or off as required. The ability to produce well controlled and very thin layers makes epitaxy the technique of choice where very small 'nanoelectronic' devices are being made (Goldstein, 1992). It is, however, a rather slow technique that is not suited to mass production.

The alternative to depositing or growing layers of material is to change the properties of a thin layer of the substrate material. Two inexpensive techniques in this area are diffusion and exchange of dopant atoms. These techniques must be used where it is necessary to maintain a crystal structure and epitaxial growth is not suitable. Both diffusion and exchange techniques are carried out in furnaces at temperatures of up to about 1000°C and the process may be accelerated by the application, for example, of an electric field. Dopant atoms may be supplied in the form of a gas, liquid or solid source. Diffusion is commonly used for modification of the properties of lithium niobate by diffusion of titanium and has also been used for glass substrates. For glass substrates, however, diffusion is rather slow, and the faster ion exchange process is more common. Ion exchange occurs when the substrate is immersed in a bath of the molten salt that contains the exchange ion. In the case of glass, for example, this may be sodium, silver or potassium. Ion exchange may also be assisted by the application of an electric field. A multistage ion-exchange process has also been used in the formation of buried waveguides where different exchange ions are used in each step (De La Rue et al., 1994). Somewhat greater control over the layer composition can be obtained by using the technique of ion implantation. This is a vacuum process in which dopant ions are accelerated towards the substrate in an electric field. Control of the accelerating electric field allows the depth of implanted dopant atoms to be quite precisely tailored. However, a disadvantage is resulting damage to the crystal lattice, which must often be reduced by subsequent heat treatment leaving a useful refractive index difference.

2.2.3 Patterning processes

The techniques described in the previous section for fabrication of thin films are essentially planar techniques with only limited lateral spatial control

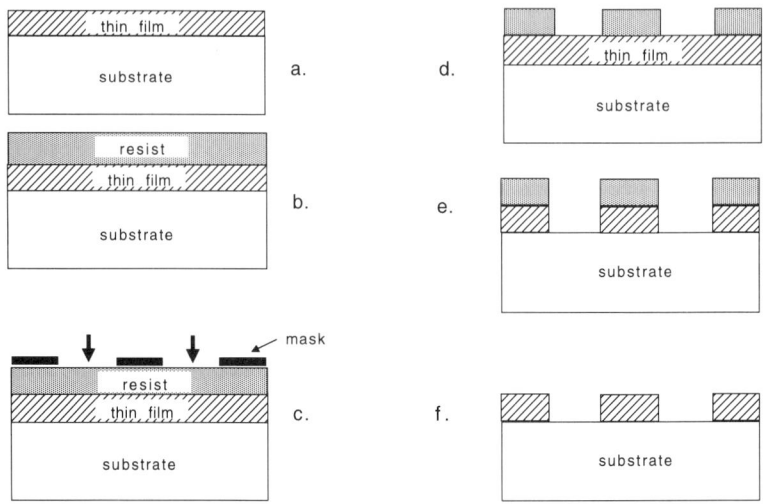

Figure 2.13 The photolithographic process for device patterning.

available through the use of masks and slits. Patterning on the substrate and thin-film layers is required for the production of integrated optic devices and fully integrated devices. The technique used for this patterning is photolithography, which is commonly used in silicon integrated-circuit fabrication. The principal steps in this technique are illustrated in Figure 2.13. Analogous to the photographic process, photolithography uses chemical changes in a light-sensitive material (a 'resist') to transfer a pattern from a mask to the underlying material. Firstly, the resist material is coated onto the substrate in a thin layer, usually by spin coating (b). The resist is then exposed to light of a specific wavelength through the mask containing the required pattern (c). This results in a chemical change in the exposed areas of the resist (usually in the solubility). The chemical composition of photoresists depends largely on the substrate being used; it is often a mixture of organic molecules in which bonds are either formed or broken on exposure to light. Either the exposed or the unexposed resist is then removed leaving the other area to protect the underlying material (d). The exposed pattern of the substrate surface may now be modified, as long as that process does not damage the resist material. Processing options include doping of exposed areas, etching into the substrate or deposition of an additional layer. Once processing is complete, the remaining resist material is removed leaving the substrate (e) with areas modified in the pattern required (f). Critical parameters in the quality of the photolithographic process are mask quality, wavelength and technique used to expose the resist, and the effective of removal of exposed resist. Smaller and more intricate patterns can be made by careful mask design and by using shorter wavelengths for resist exposure. For very-high-resolution work, the resist is exposed by direct writing using a laser or electron beam, removing the need for a mask. This direct writing technique is required, for example, in the production of the finely spaced lines in grating couplers.

2.3 Sensor techniques in integrated optics

The operation of an optical sensor lies in detecting a change in an optical signal as a result of a change occurring at the sensing site. A range of optical parameters can be used such as absorbance, refractive index changes, path-length scattered light or fluorescence. The parameter used must be sensitive to changes in the substance to be studied such as its concentration, orientation or chemical reaction. The parameter chosen and the method of detection depend on the application required and will determine the sensor performance.

2.3.1 Evanescent waves

Many of the techniques described in this section rely on the penetration of the optical evanescent wave into the sample area for detection to occur (see section 1.3.3). A simple treatment of light propagation in waveguides suggests that the optical wave is completely confined within the waveguiding layer and that no signal crosses the boundary either into the substrate or into the superstrate. In fact, the electromagnetic field penetrates into both regions when the refractive index of the guiding layer is greater than that of the adjacent layers. The depth of penetration depends on the refractive indices of the two layers and is of the order of the wavelength of the propagating light. This effect is illustrated in

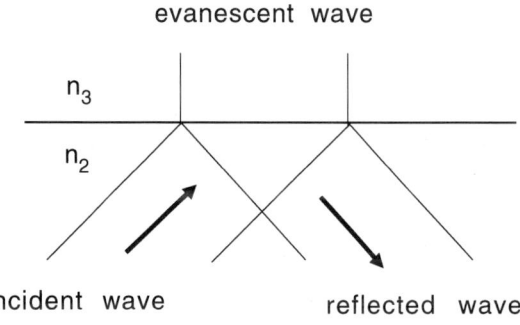

Figure 2.14 The evanescent wave at a waveguide surface.

Figure 2.14 and can be effectively exploited in optical sensor design (Harrick, 1967). Optical materials and transmission wavelengths can be tailored to allow specific sensor regions to be probed or reactions occurring in close proximity to an interface to be studied alone, even in the presence of otherwise interfering effects. The small magnitude of the evanescent wave in the sample can result in rather small signals being detected, and this can be improved by the use of multiple reflections from the interface of a long prism or waveguide (Kronick and Little, 1975).

2.3.2 Spectroscopy

Spectroscopic devices make use of the absorption or emission of specific frequencies by chemical species. These result in a unique 'fingerprint' that can be used both qualitatively and quantitatively for the purposes of identification and measurement of specific molecules either singly or in mixtures. In integrated optic devices, the optical spectrum is generally collected using a reflectance technique. These devices make use of evanescent wave penetration (see also section 2.3.1) to monitor materials that lie within about a wavelength of the device surface. The sample material is placed in contact with the sensor and the reflected wave is attenuated by absorption close to the surface (Sutherland and Dahne, 1987). Alternatively, the incident optical signal may be used to excite a fluorescent material in contact with the surface. This may be either natural fluorescence of the species or fluorescence of a label attached in a specific reaction. As fluorescence occurs at longer wavelengths than the incident light, signals can be separated using appropriate filters. A number of device options have been used in sensor applications and two of these are illustrated in Figure 2.15.

2.3.3 Ellipsometry

When a beam of monochromatic light is reflected at a surface, there are changes in both the amplitude and phase of the two perpendicularly

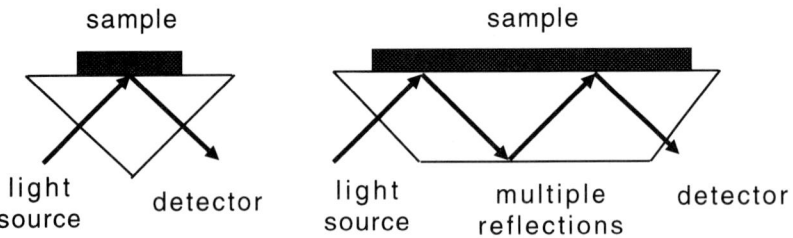

Figure 2.15 Spectroscopic sensor options.

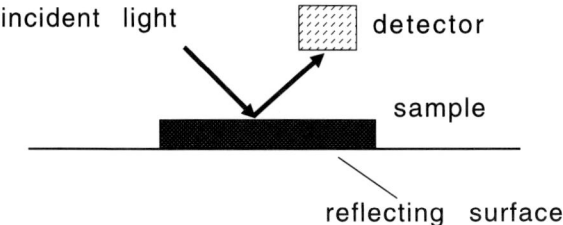

Figure 2.16 Ellipsometric optical sensor.

polarised components. Ellipsometry measures a change in the state of polarisation of light when it is reflected at a surface. The state of polarisation is determined by the relative phases and amplitudes of the two components and is sensitive to the properties of the reflecting interface. Therefore, changes in refractive index or film thickness at the interface will result in changes in the state of polarisation. Ellipsometry is widely used for the measurement of optical properties and thicknesses of very thin films, especially in the semiconductor industry.

Use of ellipsometry in sensor applications relies on the measurement of these properties for layers of a material of interest at a solid sensor surface, as illustrated in Figure 2.16 (Rothen, 1974). However, the majority of applications have involved measurement of changes in film thickness, a parameter that proved to be rather insensitive even to quite large changes. Such measurements cannot be practically applied to many sensor environments where mixtures of many components, often unknown, may be present.

The most significant limitation in the application of ellipsometry to date has been the requirement for large, expensive and complex instruments that are impractical for many uses such as field measurements.

2.3.4 Surface plasmon resonance

Surface plasmons are specific electromagnetic modes that are excited by an optical evanescent wave and propagate along the surface of a metal. When an optical signal is reflected at the metal–substrate interface, the reflected intensity varies with the angle of incidence. A shift in maximum intensity occurs when additional material is present at the interface and is essentially a mass-sensing mechanism. A schematic device is illustrated in Figure 2.17 (Pockrand et al., 1977; Nylander et al., 1982–3; Flanagan and Pantell, 1984).

2.3.5 Light scattering

Light scattering has been used extensively for the study of molecular size and shape. Both static and dynamic measurements can be made and either

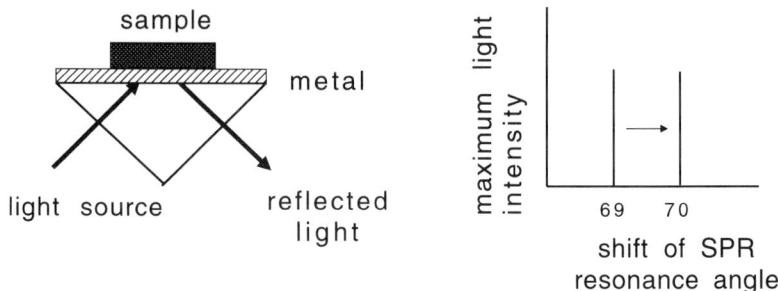

Figure 2.17 A schematic device for sensing surface plasmon resonance (SPR).

may be incorporated into an integrated optical system (Cohen and Benedict, 1975). Light scattering can be measured either by the attenuation of the incident signal as light is scattered away from the beam path or by measuring the intensity of the light scattered by the sample at a particular angle to the incident beam, as illustrated in Figure 2.18. A particular difficulty with light-scattering techniques for sensors is its non-specific nature. All molecules will scatter light to some extent and it is difficult to discriminate between components in a mixture. Therefore, most applications for sensing have used optical markers for this purpose. In addition, light scattering generally requires sensitive, bulky, expensive equipment, skilled operators and extensive sample preparation, which severely limits its application, especially for field use.

2.3.6   Optical biosensors

Optical biosensing does not refer to a specific integrated optical technique but rather to a method of transduction that can be used with integrated

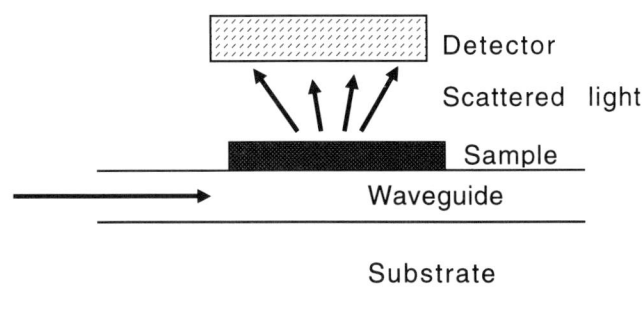

Figure 2.18 A schematic device for a light scattering sensor.

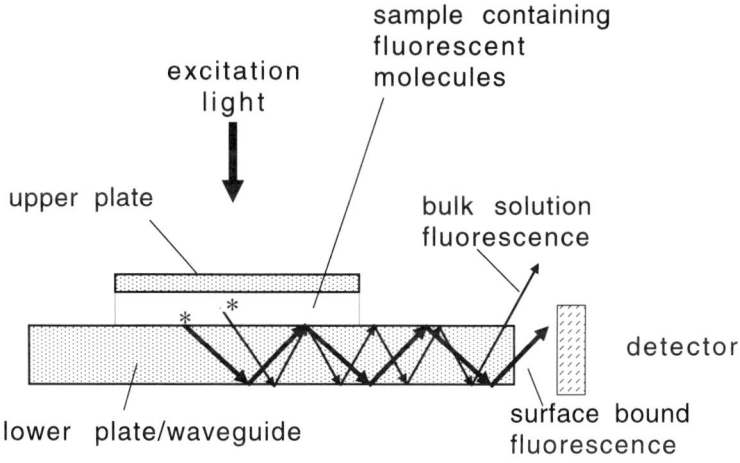

Figure 2.19 A schematic optical biosensor: the fluorescence capillary fill device.

optical systems. The method uses specific binding reactions of biological molecules to immobilise sample material at the sensor surface (Badley *et al.*, 1987; and see, for examples, chapter 7). Figure 2.19 shows an example of a biosensor which uses a waveguide and molecules immobilised on the surface to study a binding reaction (Zhou *et al.*, 1990). It is a popular method of transduction, often used in combination with other techniques, and can reduce interference from other molecules present in a mixture.

2.4   Applications of integrated optic devices for environmental sensing

Whilst there are very many laboratory demonstrations of integrated optic sensors, as yet there are very few commercially available instruments. In this section, the range of species of interest is discussed with examples of sensor devices.

Environmental sensors in integrated optics may be used for gas, liquid or solid materials. Gases to be monitored originate largely from industrial and transport sources, although there is increasing concern about gaseous emissions from intensive farming methods. Amongst the very wide range of gases monitored, some of the most important are carbon dioxide, carbon monoxide, oxides of nitrogen and sulphur, ammonia and a wide range of organic chemicals including hydrocarbons and aromatics. In most cases, very high sensitivities are required at the parts per million (ppm) or parts per billion (ppb) levels (see chapter 3). Usually, complex mixtures of gases are present, which makes monitoring of an individual gas with high sensitivity an even more difficult challenge. For example, carbon monoxide and carbon dioxide concentrations are required, each with a sensitivity of about 10 ppm, but both of these are present in emissions from petrol engines. Similarly, complex mixtures of hydrocarbons of very similar chemical structure are produced by boiler fuel combustion. Most of the techniques described in section 2.3 can be used for monitoring gaseous emissions and, in particular, biosensors, surface plasmon resonance and spectroscopy are effective. However, where complex mixtures are present, spectroscopic techniques are particularly suitable since a unique molecular fingerprint can often be identified.

In the liquid phase most sensing applications are in the area of water contamination (see volume 2, chapter 2) where unwanted chemicals may pollute drinking water supplies and pose a threat to wildlife. The range of species to be monitored is generally wider than for the gas phase, including all of those mentioned above together with phosphates, organometallics and heavy metals plus biological species such as bacteria. Existing and new legislation for monitoring water quality has made the need for these sensors particularly acute. For drinking water, allowed levels of contamination are extremely low and often sensors are required to operate at the parts per billion level. Mercury, for example, has a concentration limit of 1 ppb in drinking water. Most of the techniques described in section 2.3 have been used for liquid samples, while the most popular are, once again, biosensors, surface plasmon resonance and spectroscopy.

For monitoring solid materials, light scattering is a convenient choice for measuring particle size and range, while reflection spectroscopy can identify specific chemical composition. A wide variety of solid materials are monitored, including contaminants in solids such as soil or foods. Particularly important are airborne particles, including carbon from vehicle emissions, dusts and biological materials such as pollen.

The need to monitor one or more species in a complex mixture has led to the increasing use of mediating layers to assist in the selection of an individual chemical and to reduce background noise and interference. These layers may trap a particular species by means of a specific reaction, or they may contain a specific label (e.g. a fluorescent label) that will react only with the species of interest. This technique has generally been used in optical biosensors but is increasingly being used to enhance the performance of, for example, spectroscopic and surface plasmon resonant sensors, which use the evanescent wave effect to monitor films at waveguide interfaces.

2.4.1   An integrated optic biosensor

This is a one-step sensor device requiring only the addition of a sample into the sensor cell and is capable of detecting several different analytes in a single test. Fabrication is based on an ordinary glass microscope slide and it is, therefore, cheap to produce and could be disposable (Laybourn *et al.*, 1993; De La Rue *et al.*, 1994). So far, the device has been used as an immunosensor to monitor antibodies at the parts per million level in a range of medical applications. Immunosensors detect the specific reactions between antibodies and antigens. However, the design of the device is completely general and it could be readily applied to a wide range of optical evanescent measurements. The design is also robust and compact and, therefore, suited to both laboratory and field instruments.

Figure 2.20 shows a schematic of the patterned waveguide sensor. The glass sensor substrate contains a buried waveguide which is located immediately below a series of wells etched into the surface of the glass. A cell is constructed above the wells and when the test sample is introduced at an edge, capillary action draws the liquid in to fill the cell. Waveguides are produced by an ion-exchange process because of its simplicity, low cost and reasonably low propagation loss. A particular feature of this structure that allows for multianalyte sensing and also reduces interference from unwanted species is the buried waveguide structure. This was achieved using a double-exchange process into soda-lime glass, firstly with a potassium nitrate melt and secondly with sodium nitrate. The correct selection of temperature and time for each ion-exchange step allows the production of a single-mode waveguide buried at a depth of about 7 μm. The photolithographic process described in section 2.2 is used to etch wells into the surface of the glass towards the buried waveguide. Different

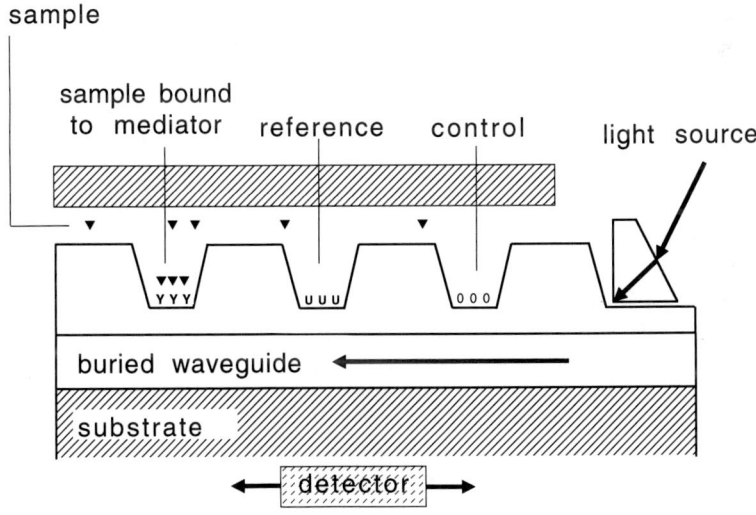

Figure 2.20 An integrated optical immunoassay sensor.

chemicals or mediating layers are deposited at the bottom of each well, allowing a different chemical species to be monitored in each case. The wells can also be used for reference and control signals. When light is coupled into the buried waveguide, the evanescent field of this light penetrates into the wells, thereby exciting the immobilised layer. Any signal emitted by the immobilised layer is then detected by a photodetector placed beneath the well.

A number of devices based on similar principles are under development. A portable device based on surface transverse wave technology has been designed to monitor the herbicide atrazene in the range 0.06 ppb to 10 ppm. The sensor has a rapid response and the surface is regenerated, allowing multiple measurements (Tom-Moy *et al.*, 1994). Another example of an optical immunosensor, this time for monitoring of gas-phase species, uses a total internal reflection sensor and detection of fluorescent emission (Celebre *et al.*, 1992). A further immunosensor device based on the surface plasmon resonant effect has been shown to detect pesticides at concentrations of about $1\ \mu g\ ml^{-1}$ (Brecht *et al.*, 1994).

2.4.2 An integrated optic gas sensor

Figure 2.21 shows a schematic of an integrated optic sensor for ammonia with sensitivity in the parts per million range. The gas is detected by absorption in the sensitive layer on the surface of the waveguide. The waveguide is produced by electric field-assisted ion exchange. The surface layer is a sol–gel preparation of silica containing a dye which is sensitive to ammonia. Light coupled into the waveguide, in this case via a fibre, has an evanescent wave in the sensitive layer. The evanescent field is attenuated by the dye in proportion to the surrounding ammonia concentration and then coupled back into the waveguide. The beam is reflected back along the fibre

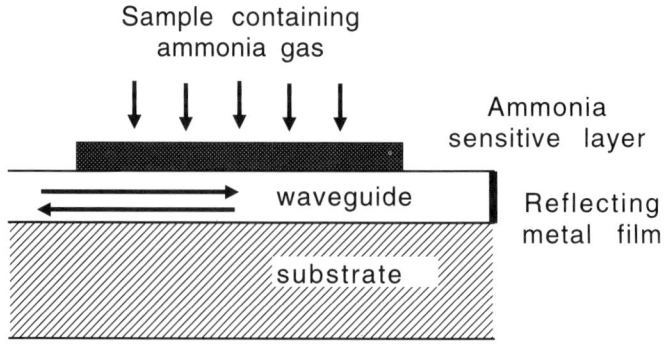

Figure 2.21 An integrated optical gas sensor.

by an aluminium coating on the end of the waveguide. The attenuation of the incident beam gives a measure of the ammonia concentration that is linear in the range of 5–100 ppm (Klein and Voges, 1993). A related sensor using interference effects in substrates such as silicon has been shown to be sensitive to nitrogen dioxide at the parts per million level. (Miller *et al.*, 1991).

2.5 Conclusions

Integrated optics is a relatively new technique in environmental sensing and, as yet, there are few examples of fully integrated sensors. The reasons for this are that, for full optical integration, the range of substrates available is very limited, these materials are expensive and sensor systems using them are difficult to fabricate. However, the potential advantages of integrated optics in terms of robustness and miniaturisation to produce multianalyte and portable devices have made it the subject of much active research. Indeed, in some cases, integrated optics may be the technique of choice for the high-sensitivity routine monitoring likely to be required by future legislation specifying maximum pollutant levels, in water for example.

Research in integrated optics for communications applications is already well advanced and the results of this work may also have applications in sensor devices. Integrated optics for environmental sensors is an emerging technology that is likely to become increasingly important in the future.

References

Badley, R.A., Drake, R.A.L., Shanks, I.A., Smith, A.M. and Stephenson P.R. (1987) Optical biosensors for immunoassay – the fluorescence capillary fill device. *Philosophical Transactions of the Royal Society, London*, **316**, 143–160.

Brecht, A., Kraus., G. and Gauglitz, G. (1994) Theoretical and experimental detectivity of the RIFS-transducer in affinity sensing. *Biosensors '94: The Third World Congress on Biosensors*, June 1–3, New Orleans, USA. Elsevier, London.

Celebre, M., Domenici, C., Francesconi, R., Ahluwalia, A. and Schirone, A. (1992) A comparative study of efficiencies of fiber optic and prism TIRF sensors. *Measurement Science and Technology*, **3**, 1166–1173.

Cohen, R.J. and Benedek, G.B. (1975) Immunoassay by light scattering spectroscopy. *Immunochemistry*, **12**, 349–351.

de Cogan, D. (1990) *Design and Technology of Integrated Circuits*. Wiley, London.

De La Rue, R.M., Magill, J.V., Laybourn, P.J.R., Zhou, Y., Cushley, W. and McSharry, C. (1994) Fabrication of optical waveguide immunosensors. In *Microengineering and Optics: IEE Electronics Division Colloquium*, February 17, Digest No: 1994/043, pp. 12/1–12/4.

De La Rue, R.M. and Marsh, J.H. (1993) Integration Technologies for III–V Semiconductor Optoelectrics Based on Quantum Well Intermixing. *SPIE Critical Reviews of Optical Science and Technology*, CR45, 259–288.

Flanagan, M.T. and Pantell, R.M. (1984) Surface plasmon resonance and immunosensors. *Electronics Letters*, **20**, 968–970.

Fraser, D.A. (1990) *The Physics of Semiconductor Devices*. Oxford University Press, Oxford, UK.

Goldstein, L. (1992) Main Aspects of MOCVD and MBE Growth Technologies. In *Waveguide Optoelectrics* (ed. J.H. Marsh and R.M. De La Rue), pp. 99–122. Kluwer, Netherlands.

Harrick, N.J. (1967) *Internal Reflection Spectroscopy*. Wiley Interscience, New York.

Hunsperger, R.G. (1991) *Integrated Optics: Theory and Technology*. Springer-Verlag, New York.

Klein, R. and Voges, E. (1993) Integrated optic ammonia sensor. *Proceedings of Symposium on Chemical Sensors, 183rd Meeting of the Electrochemical Society*, May 16–21, Hawaii. Electrochemical Society, USA.

Kronick, M.N. and Little, W.A. (1975) A new immunoassay based on fluorescence excitation by internal reflection spectroscopy. *Journal of Immunological Methods*, **8**, 235–242.

Laybourn, P.J.R., Zhou, Y., De La Rue, R.M., Cushley, W., McSharry, C. and Magill, J.V. (1993) An integrated optical immunosensor. In *Uses of Immobilized Biological Compounds* (ed. G.G. Guilbault and M. Manscini), pp. 463–470. Kluwer, Netherlands.

McMeekin, S., De La Rue, R.M. and Johnstone, W. (1992) The transverse electrooptic modulator (TEOM): fabrication, properties and applications in the assessment of waveguide electrooptic characteristics. *IEEE Journal of Lightwave Technology*, **10**, 163–168.

Miller, L.S., Newton, A.L., Sykesud, C.G.D. and Walton, D.J. (1991) Optical gas sensing using LB films. In *Sensors; Technology, Systems and Applications*, (ed. K.T.V Grattan), pp. 139–143. Adam Hilger, London.

Morgan, D.V. and Board, K. (1990) *An Introduction to Semiconductor Microtechnology*. Wiley, London.

Nylander, C., Leidberg, B. and Lind, T. (1982–3) Gas detection by means of surface plasmon resonance. *Sensors and Actuators*, **3**, 79–88.

Pockrand, I., Swalen, J.D., Gordon, I. and Philpott, M.R. (1977) Surface plasmon resonance spectroscopy of organic monolayer assemblies. *Surface Science*, **74**, 237–244.

Rothen, A. (1974) Ellipsometric studies of thin films. In *Progress in Surface Science and Membrane Science* (ed. J.F. Danielli and M.D. Rosenberg), pp. 81–118. Academic Press, New York.

Sutherland, R. and Dahne, C. (1987) IRS devices for optical immunoassays. In *Biosensors* (ed. A.P.F. Turner and I. Karube), pp. 655–678. Oxford Science, New York.

Tamir, T. (ed.) (1979) *Integrated Optics*. Springer-Verlag, New York.

Tom-Moy, M., Baer, R.L., S-Solomon, D. and Doherty, T.P. (1994) Environmental measurements using surface transverse wave (STW) biosensors. In *Biosensors '94: The Third World Congress on Biosensors,* June 1–3, New Orleans, USA. Elsevier, London.

Zhou, Y., Magill, J.V., Laybourn, P.J.R. and De La Rue, R.M. (1990) The use of ion-exchanged waveguides in integrated optical molecular biosensors. In *Biosensors '90: The First World Congress on Biosensors*, May 2-4, Singapore. pp. 164–165. Elsevier, London.

3 Laser-based sensors

K.W.D. LEDINGHAM and M. CAMPBELL

3.1 Introduction

Some of the applications in this chapter overlap with others elsewhere in this book, particularly with Chapter 5 dealing with gas sensors and analysers, with Volume 2, Chapter 3 dealing with air pollution and with Volume 2, Chapter 7 on radiation pollution. However, the emphasis here will be on the different techniques rather than the applications. This book deals with sensors systems. The word sensor often conjures in the mind of the reader a very small device which is hopefully cheap to manufacture. On the contrary, in this chapter, we shall deal with laser-based sensors, which are often bulky and often very expensive. None the less, they are still validly called sensor systems and often have sensitivity and selectivity levels which cannot be equalled by any other techniques.

Under the heading laser-based sensors, the number of possible systems and techniques which can be described is legion. In this chapter, we have been limited to only a few of the possible laser sensor systems because of lack of space. Those we have chosen are ones with which we are most familiar and we apologise to the many experimental groups whose work we cannot review.

3.2 Laser mass spectrometry

Laser mass spectrometry is an ultra-trace analytical technique that can be applied to any sample, whether solid, liquid or gas, stable or radioactive, down to the few atom or molecule sensitivity level. It uses tuned or untuned lasers, pulsed or continuous, in conjunction with mass spectrometers (magnetic sector, quadrupole, time-of-flight) if mass or isotropic information is required, or with conventional nuclear detectors, e.g. proportional, ionisation or ion mobility spectrometers, if a simpler arrangement is all that is necessary. In addition, it can now be coupled with great effect to gas chromatographic techniques.

Although the technique has been used in a great many different analytical situations, only the applications to environmentally sensitive atoms and molecules will be reviewed in this chapter. These are: the detection and identification of explosives at trace level; the detection of some urban pollutants, e.g. NO_x and benzene, and the ultra-trace detection of radio-toxic isotopes.

3.2.1 Resonance ionisation mass spectrometry

Generally, laser mass spectrometry is an ultra-sensitive analytical technique with a great many applications (Lubman, 1990; Vertes *et al.*, 1993). Resonance ionisation mass spectrometry (RIMS) is a particular case of the more general technique that is relatively new: developed since the mid-1980s. The physics on which it is based is, however, much older. Goppert-Mayer (1931) derived the basic expression for the two-photon transition rate (the simplest RIMS process) from second-order perturbation theory as long ago as 1931. A sensitive laser spectroscopic technique was proposed by Ambartzumian and Letokhov (1972), more than twenty years ago, but it is only in the last decade that the pioneering work of Letokhov (1987) and Hurst and Payne (1988) has come to fruition.

In RIMS, there are many arrangements of lasers and mass spectrometers. The lasers are used for two purposes. Firstly, they can gasify or atomise (desorption/ablation) solid samples, and, secondly, they ionise resonantly or non-resonantly the emitted neutral atoms or molecules. In this chapter, many of the samples discussed are in the gaseous phase, which means that they can be admitted to the vacuum vessel effusively or via a beam valve. The mass spectrometers can be of magnetic, quadrupole or, much more commonly, time-of-flight (TOF) types. If the plume consists of molecules rather than atoms, the process is usually called resonance-enhanced multiphoton ionisation (REMPI).

The reasons that led to the development of RIMS or more generally to the post-ionisation of neutrals are, firstly, that the sensitivity is likely to be increased over laser-induced mass analysis (LIMA) and secondary ion mass spectrometry (SIMS) (two similar techniques which analyse the ions produced from solid-phase samples in ablation or sputtering) since the number of neutral atoms or molecules is between two and three orders greater than the number of ions; and secondly, RIMS can eliminate isobaric interferences, which can plague other mass spectrometric techniques.

3.2.2 Factors determining the experimental arrangement

Atoms of each element have a unique set of excited states that may be reached from the ground state by the absorption of one or more photons at the correct frequency, providing certain optical selection rules are satisified. An excited atom would normally decay back to the ground state with a characteristic time of about 10 ns. However, before decaying, the atom can absorb another photon, which may take the atom to a higher

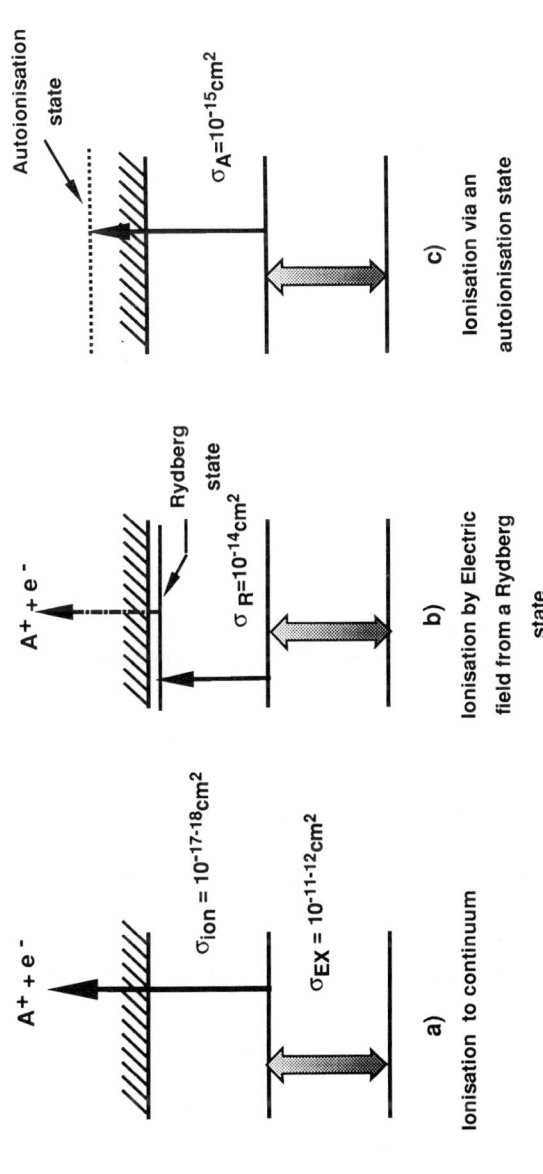

Figure 3.1 (a) Ionisation to continuum. An electron in its ground state absorbs a photon and is raised to an excited state. The ionisation by absorption of a second photon has a small cross-section. (b) The atom is excited to a Rydberg state and is finally ionised by a pulsed electric field with high efficiency. (c) The final step is to an autoionisation state with a large cross-section.

excited state or cause ionisation.

The cross-section for resonant absorption of one photon between bound levels is typically of the order of $10^{-12}\,cm^{-2}$ and a laser fluence of about $0.1\,\mu J\,cm^{-2}$ is sufficient to cause saturation excitation of the atoms in the laser beam. The cross-sections for photoionisation are normally 10^{-6} times smaller, and laser fluences of about $100\,mJ\,cm^{-2}$ are required for saturation. Saturation is defined to mean that every atom or molecule in the laser beam is excited or ionised. This two-photon ionisation process is shown in Figure 3.1(a). Most modern tuneable lasers require moderate focusing to reach these values. Two other ionisation procedures have been employed to alleviate this problem of high fluences (Bekov and Letokhov, 1983). These are shown in Figure 3.1(b) and (c). In Figure 3.1(b) the atom is excited to a Rydberg state and is finally ionised by a pulsed electric field with high efficiency. In Figure 3.1(c), the atom is ionised via autoionisation states. The rate-limiting step in each of these cases has a cross-section some two orders of magnitude larger than that in Figure 3.1(a) and requires much lower fluences to reach saturation.

The laser line width typically used in RIMS measurements is between 0.1 and $1.0\,cm^{-1}$ and hence all the isotopes of an element are ionised simultaneously. Separation of the various isotopes is achieved in the mass spectrometer.

All atomic transitions between the ground state and some excited states, with the exception of He and Ne, may be resonantly excited with commercially available tuneable dye lasers. Hurst and Payne (1988) have proposed five basic ionisation schemes according to the relative energy positions of the intermediate states to the continuum, with each element in the periodic table being ascribed to one of these schemes. With a single dye laser and a frequency doubling facility, 39 elements can be ionised, enabling, in principle, computer-controlled element changes in a matter of seconds (Thonnard et al., 1989). For complete elemental coverage, however, a high-power pump laser such as an excimer or Nd:YAG laser is required and two tuneable dye lasers, one of which is frequency doubled.

RIMS is an acronym primarily associated with the detection of atoms, which usually have a simple set of energy levels. The same techniques can be used with molecules, but molecular structure is considerably more complex since each electronic level has an associated set of vibrational and rotational levels. For molecules, the bound–bound and the bound–continuum transitions are about the same order of magnitude at about $10^{-18}\,cm^{-2}$ (Boesl et al., 1981).

The Glasgow University RIMS instrument (Towrie et al., 1992; Ledingham and Singhal, 1992), which is described here in detail (Figure 3.2), is a typical system, which is now available commercially. The spherical sample chamber is 30 cm in diameter with as many ports as possible facing the centre of the chamber, the point at which the sample stub is held. The

a)

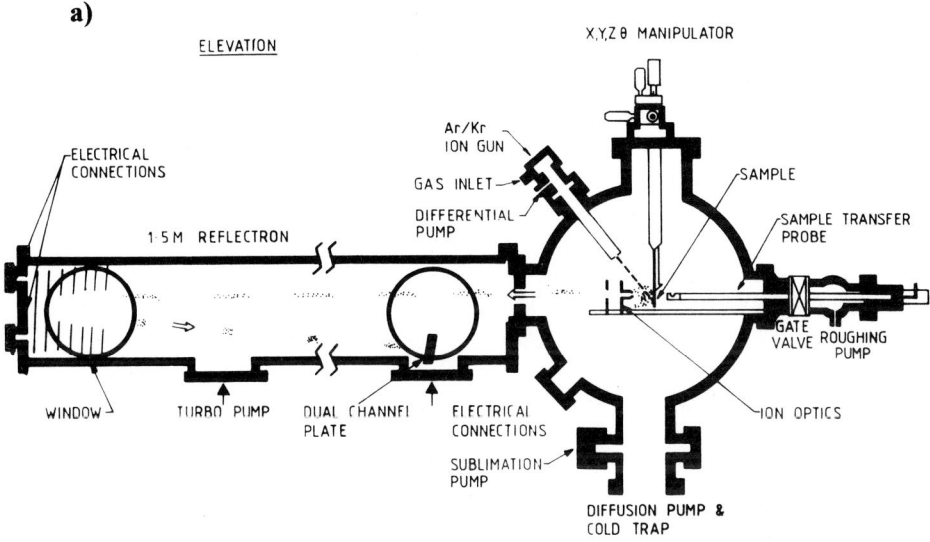

b)

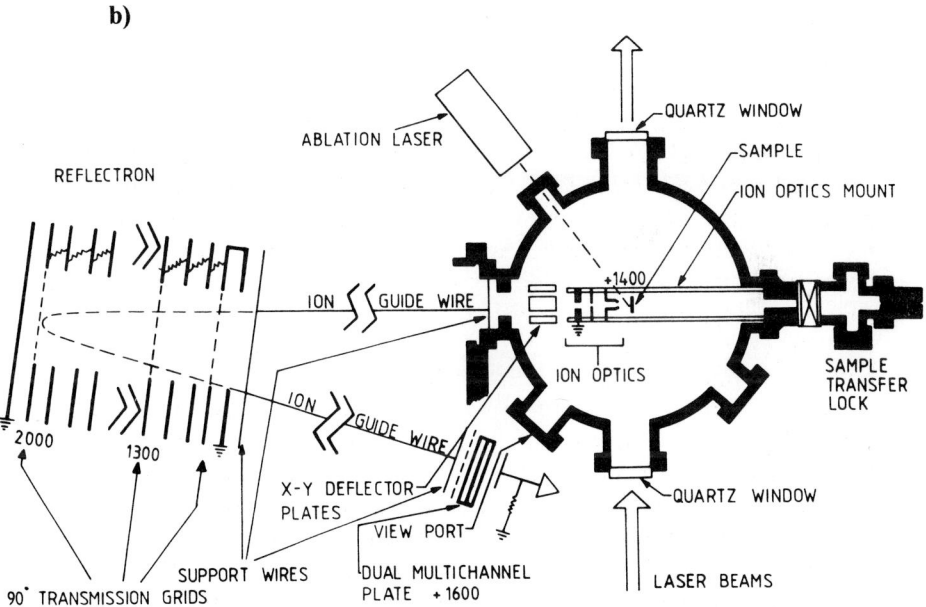

Figure 3.2 (a) Elevation view of the Glasgow post-ablation mass spectrometer. (b) View of the spectrometer showing the electrostatic ion reflector in greater detail. An initial spread of ion energies is compensated by the reflection since the high-energy ions penetrate deeper into the reflecting field and hence spend a longer time than the lower-energy ions. This improves the resolution and the guide wire increases the transmission. A recent modification permits controlled fragmentation by a further laser beam at the turn-round point.

sample is mounted on an $xyz\theta$ manipulator and can be inserted and withdrawn from the sample chamber using a rapid transfer probe. Fast sample exchanges (5 min) can be made without disruption of the main chamber pressure. The ion-extract optics and the electrostatic reflector are shown in Figure 3.2(b). The sample is maintained at a voltage of about 2000 V with the first extraction electrode at 1400 V. The reflectron TOF has an overall drift length of 3 m.

The principal factor that limits the resolution of a conventional TOF is the spread of the initial ion/neutral energies in the ablation process. This spread of ion energies can be compensated using a reflectron TOF mass spectrometer in which the high-energy ions penetrate deeper into an electrostatic ion reflector and hence experience a longer flight time than the ions of lower energy. The full width half maximum (FWHM) value of the present system is about 1000 for ions of about 40 amu. A thin wire, 0.005 cm in diameter, follows the ion path through the flight tube, providing an electrostatic guide for the ions and increasing the transmission of the mass spectrometer. The wire operates at about -10 V.

Ablation is carried out using a Nd:YAG laser operated at one of the harmonic frequencies, although, normally, second harmonic 532 nm is the preferred wavelength. If sputtering is used, the system is fitted with a Kratos Penning ion gun. Both inert gas and oxygen ions can be used with energies up to 15 keV and with beam currents up to 5 μA.

The post-ablation ionisation laser system consists of a Spectron Nd:YAG laser powering two dye lasers, one of which can be frequency doubled. The lasers operate at repetition rates of 10 Hz and with pulse duration of about 10 ns. The tuneable laser pulse energies depend on the dyes used but are normally between 100 μJ and 2 mJ. The time between the ablation and ionisation systems can be varied between 0.1 μs and 10 μs, although it is typically 1 μs. The ionising laser beam is introduced into the sample chamber parallel and as close to the sample stub as possible to maximise the overlap with the ablation plume.

The ions, after passing down the TOF, are detected in a dual channel plate detector and passed to a data acquisition device that measures and stores mass spectra and laser pulse energies on a pulse-to-pulse basis. A LeCroy 2261 transient recorder or a LeCroy 9310 digital oscilloscope coupled to a COMPAQ 386/25 computer forms the basis of the system. Ion signals from the multichannel plate detector are digitised by the transient recorder, which provides 640 time channels (11 bit resolution) each of between 10 ns and 100 ns width. Both wavelength-dependent spectra and mass spectra at specific wavelengths can be taken with the data acquisition system.

The RIMS instrument described above is routinely used for trace analysis and for studies of the characteristics of laser ablation from solid samples. Figure 3.3(a) shows a RIMS signal for gold in copper at 10 ppm level, accumulated over 10^4 shots. Figure 3.3(b) indicates the linearity of the technique over a wide concentration range of impurity concentration for

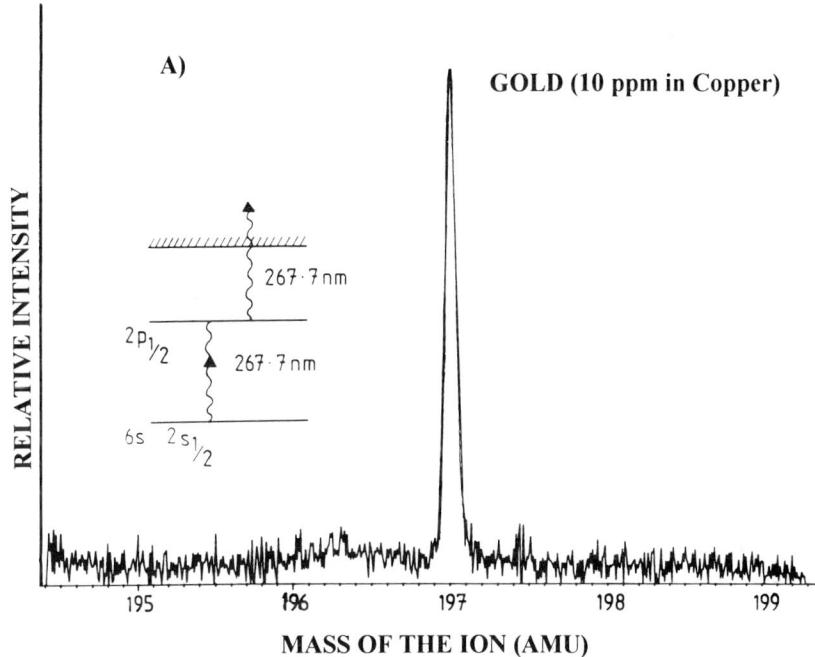

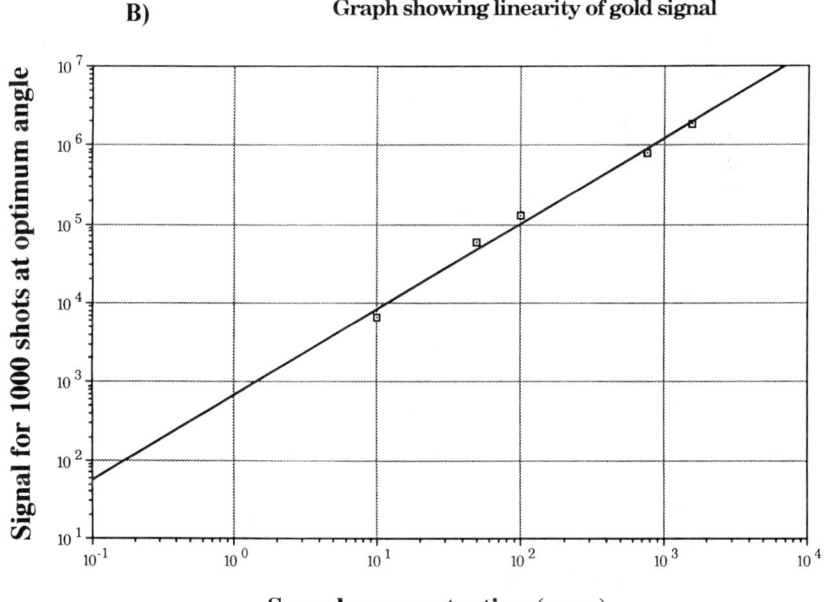

Figure 3.3 (a) RIMS signal for gold in copper at 10 ppm, accumulated over 10^4 shots. A two-photon scheme at 267.7 nm is used. (b) RIMS signal versus gold in copper concentration showing the linearity of the process.

gold in copper (McCombes *et al.*, 1990). RIMS trace detection down to parts per billion are now common in semiconductors, and, indeed, sensitivities down to parts in 10^{12} have been reported for specific elements (Bekov *et al.*, 1985). Atom Sciences Inc. can now make noble gas detection measurements to levels below 100 atoms in 5 min, which makes solar neutrino, small meteorite and ground water dating experiments feasible (Thonnard *et al.*, 1992). Attogram detection limits and isotopic selectivity greater than 10^9 has been demonstrated for rare radionuclides using cw laser RIMS (Bushaw, 1992). These sensitivities are only mentioned to give the reader some idea of the sensitivities for solid samples. One of the possible important future applications of RIMS is in the ultra-trace analysis of urban particulate matter: the *so-called* PM_{10} problem.

3.2.3   Ultra-trace detection of explosive molecules using REMPI

The increasing aspirations of a number of the new nation states have led to the formation of many terrorist groupings whose use of explosives and chemical warfare agents are limiting world-wide security and international travel in particular. The advanced laser-based sensor systems described in this chapter have been mobilised along with many other detection procedures to counter the transfer of these materials across national boundaries.

The most common REMPI scheme for analytical purposes using molecules is resonant two-photon ionisation (R2PI), in which one photon excites the molecule to an electronic state and a second photon causes molecular ionisation. Since molecules normally have ionisation potentials between 7 eV and 13 eV, R2PI requires two UV photons. Tuneable radiation of dye lasers with frequency doubling can provide UV radiation down to wavelengths of about 210 nm. In contrast to atoms, molecules in excited states can undergo different photophysical and photochemical transformations. Photodissociation can compete with photoexcitation and photoionisation processes. For larger laser fluences, the fragmentation of the molecule is more extensive. The fragmentation pattern gives important information regarding molecular structure. In addition, the wavelength dependence of the parent or fragment ions can also be used as a fingerprint for trace detection of molecules.

In a series of recent experiments at Glasgow, and also in America, REMPI has been used to detect and distinguish nitroaromatic and explosive molecules with great sensitivity. All high explosive materials, e.g. DNT, TNT, EGDN, PETN, RDX and SEMTEX (an RDX and PETN mixture), contain a number of NO_2 functional groups, with the more dangerous explosives containing greater numbers of the group. It has been shown that these molecules can be fragmented with great efficiency to yield NO_x ($x = 1, 2$) neutral molecules (Simeonssen *et al.*, 1993; Marshall *et al.*,

1994). On subsequent resonant multiphoton absorption of two or three photons at specific wavelengths, particularly 226 and 215 nm, these NO_x molecules can be ionised and detected with great sensitivity in a TOF. The very low vapour pressures from the explosives were admitted effusively to the acceleration region of the TOF mass spectrometer through a 0.5 mm diameter hole in the centre of the sample stub using a precision leak valve and capillary tube arrangement. The laser–sample interaction occurred at a distance of 1 mm from the sample stub and the ions passed into a 1.2 m field-free drift space where they were detected by a standard Thorn-EMI 18-dynode electron multiplier. The TOF mass spectrometer was operated in a Wiley–McLaren configuration with a mass resolution of 220 at $m/z = 77$.

The spectra for a number of explosives at 226.3 nm is shown in Figure 3.4 with all the spectra showing a prominent peak at $m/z = 30$ (NO), which is characteristic of all nitroexplosive materials; to some degree the different families of explosives can be distinguished by the other hydrocarbon fragments. It is possible that the 215 nm wavelength might provide a greater yield of hydrocarbon fragments and hence a greater degree of selectivity among the different explosive types (Simeonssen et al., 1993). The sensitivity levels for the detection of different types of explosive using laser mass spectrometry is typically between 10 pg and 100 pg.

In a similar fashion, REMPI has been used by Syage (1990) to detect the presence of organophosphonate compounds. These molecules are extensively used in chemical warfare agents, and lasers have been used to fragment and ionise the molecules with great sensitivity using the PO radical in this case as the fingerprint, in an identical fashion to the NO fragment in explosive materials.

3.2.4 Trace detection of urban impurities

The trace detection of urban pollutants, e.g. NO_x $(x = 1, 2)$ and benzene, has become of increasing importance. The recent Earth Summit in Rio emphasises just how serious the environmental problems of the world really are. The combustion of fossil fuels and, in particular, automobile emissions have increased the presence of harmful airborne pollutants, which has serious implications for the health and quality of life in general of urban society. Although the clean air acts of the 1950s and 60s have transformed for the better the situation with respect to emissions of sulphur dioxide and smoke within urban areas, the massive recent increase in road traffic have again reversed the situation. The car with its emissions of the oxides of nitrogen and a number of different hydrocarbons, along with UV light in the atmosphere, have conspired to form a harmful photochemical fog with possible health implications, particularly respiratory diseases and leukaemia. The main reason for the greatly increased monitoring of air

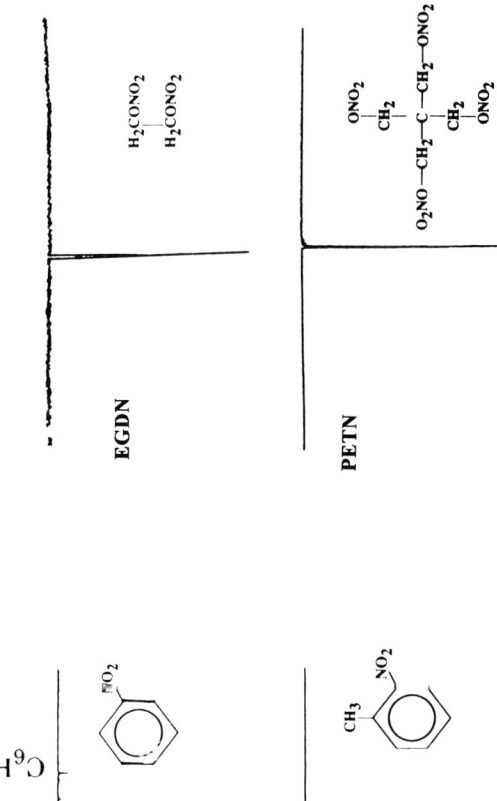

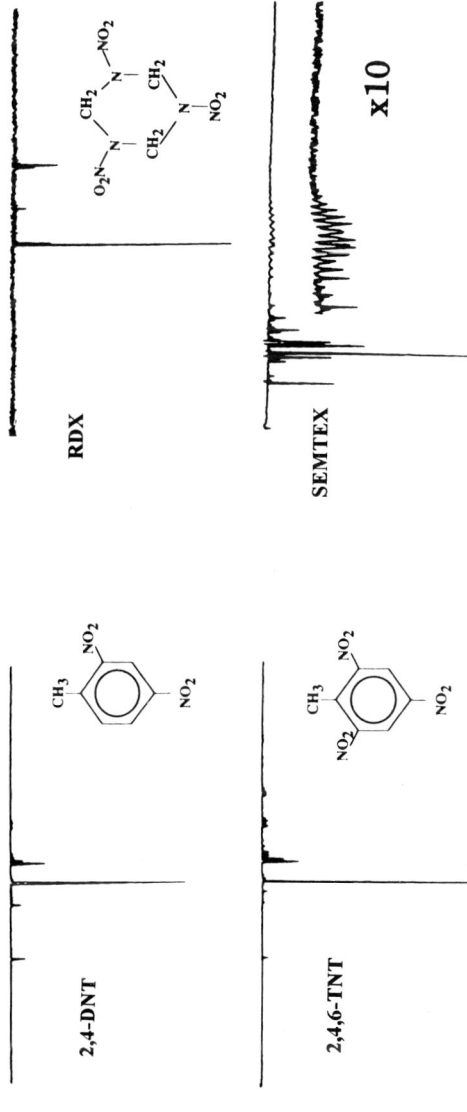

Figure 3.4 The TOF mass spectrum of a number of explosive molecules at 226 nm. All spectra show a prominent peak at $m/z = 30$. To a degree, the different types of explosive can be distinguished by the hydrocarbon fragments. Simeonsson *et al.* (1993) have shown that at 193 nm the fragmentation of the hydrocarbon clusters is much greater.

pollution is to comply with EC and USA directives. At present, a number of different analytical techniques (chapter 3, volume 2) are being used to monitor different pollutants, but a considerable advantage of the laser-sensor REMPI approach is that most, if not all of the culprits, can be monitored with a single laser system by just changing the wavelength. Many papers have been written recently about urban air quality, and the analytical challenges and sensitivity requirements can be found in Finlayson-Pitts and Pitts (1993), Harrison (1994), QUARG (1993; 1994; 1995) and Department of Health (1993).

Automobile emissions have increased the presence of nitrogen oxides in the atmosphere and these have important implications for the urban environment. NO is the most likely molecule produced, but this can easily be oxidised to NO_2 and can combine with water vapour to produce HNO_3 (acid rain). The average concentrations of NO_2 in American and European cities and in Glasgow, in particular, is about 35 ppb (Logan, 1983) with about the same average concentrations of NO. At times, however, these concentrations can exceed 150 ppb, with the recommended European safety level at 135 $\mu g\,m^3$ (65 ppb).

The technology for the routine measurements of nitrogen oxides in the atmosphere is fairly well advanced and a number of approaches are used. The most popular are chemiluminescence, chemical-based diffusion tubes and ionic liquid chromatography. These procedures are fairly complicated and do not lend themselves easily to computer-controlled, real-time analysis, which would be the basis of a comprehensive and rigorous procedure for the monitoring of an entire town or city.

The Glasgow experimental approach is based on the technique developed to detect explosives (Peng et al., 1995). The principle of detection of the NO molecules is via a resonant $(1+1)$ photon ionisation process, $A^2\Sigma(v=0) \leftarrow X^2\Pi(v=0)$, at 226 nm or a $(2+1)$ ionisation process $C^2\pi(v=0) \leftarrow X^2\Pi(v=0)$ at 380 nm. The ions are detected in an ionisation chamber calibrated with ordinary air + 1 ppm NO_2 (BOC special gases) or nitrogen + 1 ppm NO at atmospheric pressure. The laser source consists of a Lumonics excimer pumped dye laser system where coumarin 47 laser dye is used to span the wavelength range 448–460 nm. UV radiation in the range 224–230 nm is obtained by frequency doubling the dye output. For the spectra taken at 380 nm, BBQ dye is used. The laser pulse energies used at 226 nm and 380 nm are about 100 μJ and 3 mJ, respectively, and the beams were focused to about $0.3\,mm^2$. The ionisation chamber consists of two parallel plates of dimensions 2×2.5 cm with a separation of about 1.5 cm placed in a cast metal box. A variable high-voltage supply was connected to the plates via an Ortec 142H preamplifier and the system was operated in the range 100–400 V. For wavelength-dependent spectral measurements, signals from the joulemeter and the ionisation chamber were analysed using 2 SRS 250 gated boxcar integrators coupled to a Macintosh IIfx.

(a)

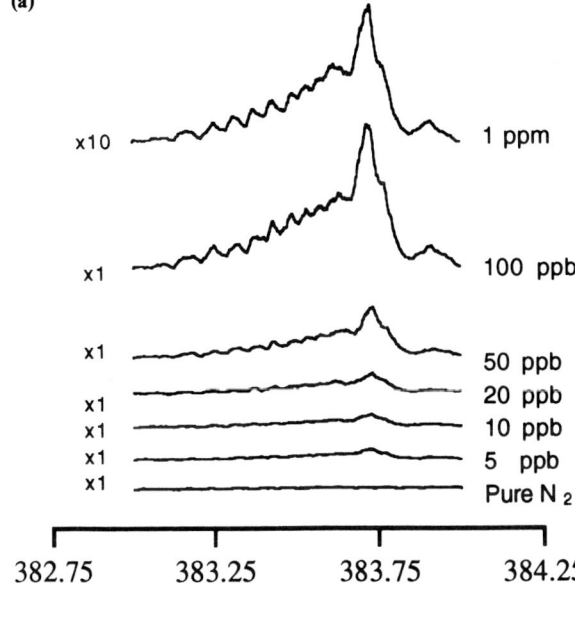

(b)

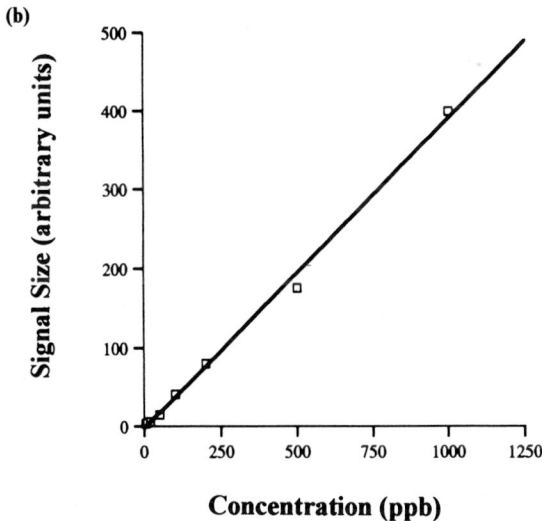

Figure 3.5 (a) $(2+1)$ spectrum at 380 nm for NO in nitrogen at different concentrations in a simple ionisation chamber. (b) Signal size of NO as a function of concentration showing the linearity of the procedure.

The results for NO in nitrogen gas using a $(2 + 1)$ ionisation scheme at 383 nm are shown in Figure 3.5(a). The spectrum at 1 ppm was obtained for a calibrated sample which flowed through the counter at a known rate and the other concentrations were obtained by diluting the sample (as much as 1000/1) with pure nitrogen gas. The spectra are averaged over 10 shots. Figure 3.5(b) is a graph of NO concentration as a function of signal size; it can be seen that the graph is linear over a wide range of concentrations. The lowest level of detection for NO is about 100 ppt, in agreement with the measurements of Guizard et al. (1989). The results obtained at 226 nm were not quite as sensitive, with about 1 ppb being the lowest detectable level. It can be seen that with sensitivities of 1 ppb or better being readily attainable for the detection of NO, the European safety levels are easy to detect.

A series of spectra were taken for NO_2 samples in air at both 226 nm and 383 nm laser wavelengths. For NO_2, photodissociation to $NO + O$ is energetically permitted at wavelengths less than 400 nm. Therefore, it takes three and four photons to photodissociate NO_2 and then to ionise the NO via the same pathways as before. Hence the detection sensitivity for NO_2 is intrinsically not so high as for NO, but the most sensitive wavelength is probably at 383 nm, the maximum of the absorption spectrum and on an NO resonance. At present, the detection limit for NO_2 is about 10 ppb at the laser wavelength of 383 nm, but sensitivities down to a 1 ppb are confidently expected by averaging laser shots and by increasing the laser flux.

It has been shown that NO can be detected with a laser-based procedure using a simple ionisation chamber system at both 226 and 383 nm down to sensitivities below 1 ppb levels. The levels for the detection of NO_2 are not quite so low, 10 ppb being presently attainable. This is expected to be improved to about 1 ppb by averaging over many laser shots and increasing the laser flux. Thus for both oxides of nitrogen, the presently attainable sensitivities are far below the 65 ppb European safety levels. At present, a single analysis takes about 3 min and hence hundreds of samples can be measured per day.

This procedure has been repeated for benzene (Marshall et al., 1995) in air samples at laser wavelengths in the region 246–264 nm. The sensitivity levels are about 5 ppb and are confidently expected to reach levels below 1 ppb with pulse averaging techniques. At present, the recommended level is 5 ppb (QUARG, 1994).

The perceived advantage of the laser-based procedure over present systems is one of ease of use, the possibility of computer control and real-time measurements, as well as extension to many other automobile-produced pollutants that now require to be monitored. In a recent paper, Weickhardt et al. (1994a) have shown how laser mass spectrometry can be applied in time-resolved fashion to multicomponent analysis of automobile exhaust gases. Twenty five typical exhaust gases were analysed in a wavelength range 260–550 nm. This group has also studied extensively many other highly toxic substances, such as the substituted aromatic compounds (Weickhardt *et al.*, 1994b).

Therefore, with a single laser system tuneable in the range 200–400 nm, a number of organic and inorganic pollutants can be detected with great sensitivity and selectivity, without recourse to purchasing a number of different monitors.

3.2.5   Trace detection of radiotoxic isotopes by RIMS

RIMS has become an important technique for ultra-trace environmental analysis and, in particular, it is well suited for the detection of long-lived radiotoxic isotopes. The applications of RIMS to the trace detection of radioactivity has been reviewed recently (Urban *et al.*, 1992; Payne *et al.*, 1994; Wendt *et al.*, 1995); in this chapter only the application to the detection of plutonium will be described, although the Mainz group (Wendt *et al.* 1995) have also developed sophisticated techniques for the detection of technetium and strontium isotopes.

The experimental arrangement for the trace detection of the actinides including plutonium is shown in Figure 3.6(a). The samples are thermally desorbed from a filament in the source region of a TOF mass spectrometer. The neutral atoms are then ionised with three tuneable dye lasers, pumped by two copper vapour lasers at repetition rates of 6.5 kHz. The dye laser beams are coupled into the ionisation region of the TOF either with prisms or by optical fibres. The ions produced by resonant laser excitation and ionisation in the interaction region are accelerated to an energy of 3 keV, pass down a drift length of 2 m and then are detected in a channel-plate detector. It was found that the best excitation scheme was $\lambda_1 = 586.49$ nm, $\lambda_2 = 665.57$ nm and then the final step via an autoionising state at $\lambda_3 = 577.28$ nm. The main losses limiting the overall efficiency are the temporal overlap between the continuous atomisation source and the pulsed lasers and the spatial overlap. A final typical efficiency for the detection of plutonium is 1×10^{-5}. Considering the background, this yields a detection limit of 2×10^6 atoms with a 1 h detection time. For ^{239}Pu, this is an improvement of more than two orders of magnitude over conventional α spectroscopy (4×10^8 atoms in a typical measuring time of over 16 h). Figure 3.6(b) shows a comparison for the detection of 10^9 atoms of ^{239}Pu by α-spectroscopy (top) and by RIMS (bottom).

In addition to rapid and sensitive analysis, the RIMS method permits the determination of isotopic abundances with an accuracy of only a few per cent, which is not possible by direct radiometric methods.

Before concluding the section on RIMS and REMPI, it should be pointed out that although this technology is accepted as a technique with great

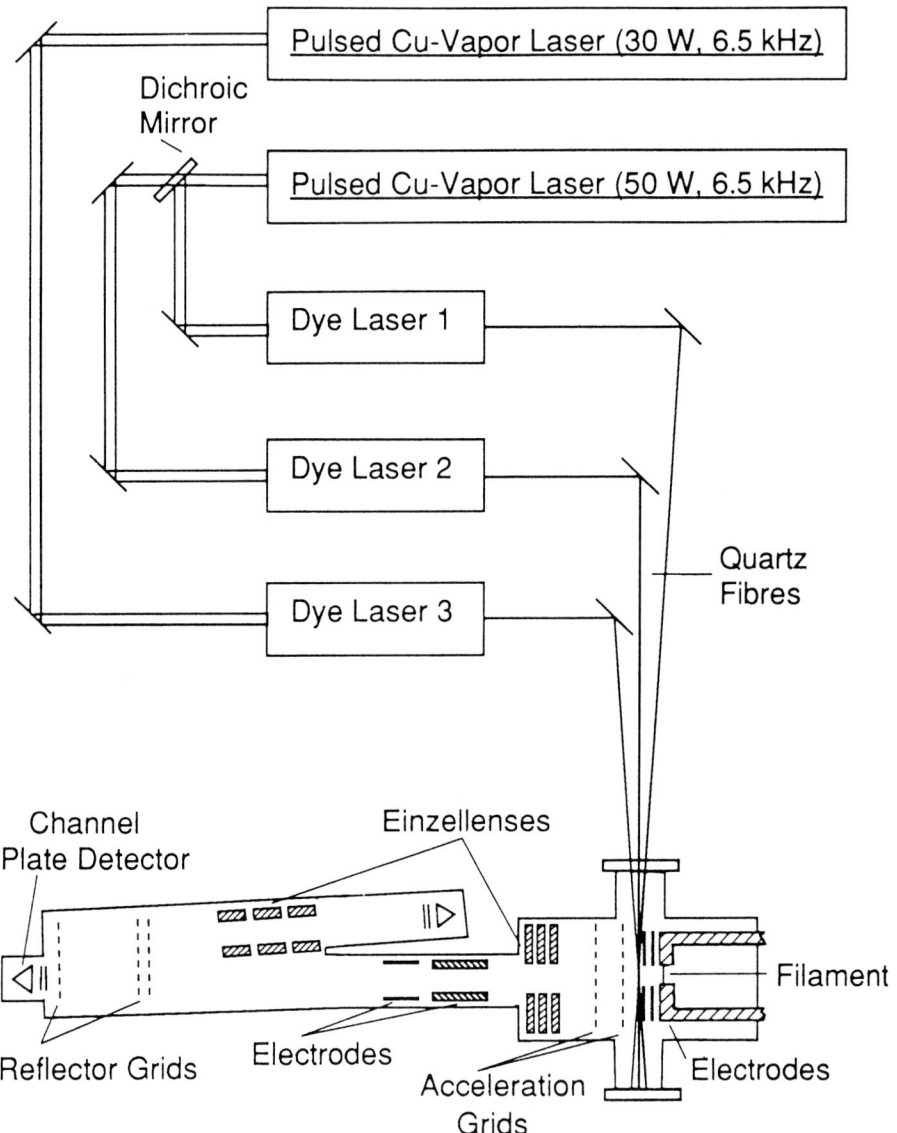

Figure 3.6 (a) Experimental arrangement for the trace analysis of plutonium and other actinides evaporated from a filament. The atoms are resonantly excited and ionised using three dye lasers pumped by two copper vapour lasers. The ions are mass selectively detected with a reflectron TOF.

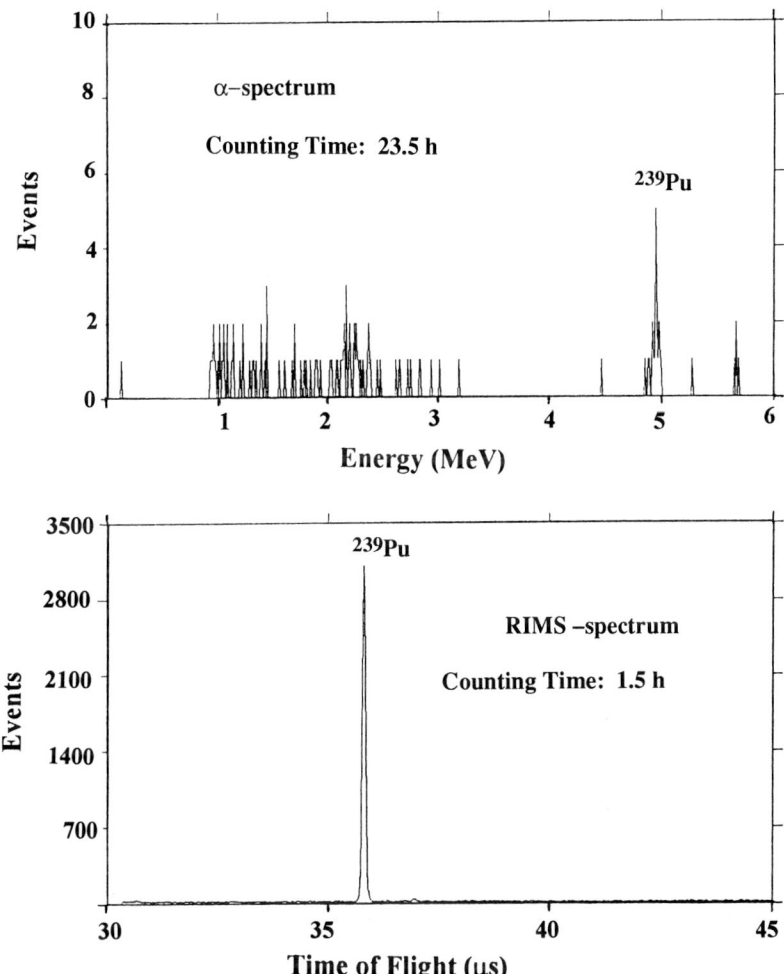

Figure 3.6 (b) Detection of 10^9 atoms of ^{239}Pu. The upper graph is a conventional α-spectrum obtained in 23.5 h and the lower is a RIMS spectrum of the same sample with a counting time of 1.5 h (Wendt *et al.*, 1995; reproduced by permission of the authors).

potential, its general acceptance as a sensor for environmental purposes is being impeded because of the inherent bulkiness and operating difficulties of dye lasers. However, the development of new widely tuneable light sources, e.g. solid-state lasers (Ti–sapphire, alexandrite and optical paramatric oscillators), high-power semiconductor diode lasers and the emergence of cheap, easy to use, solid dye lasers is likely to overcome this difficulty (Ledingham and Singhal, 1991).

3.3   Laser remote sensing

The urban monitoring described in section 3.2.3 dealt with very small discrete sample volumes taken at street level. Advanced laser techniques are also required to monitor over much longer distances to evaluate pollution on a local, regional or even a global scale and at different altitudes above the earth's surface.

3.3.1   LIDAR

LIDAR is a laser-based remote sensing technique for the monitoring of tropospheric and stratospheric pollutants over distances ranging from metres to hundreds of kilometres. LIDAR is an acronym for light detection and ranging and is the optical equivalent of RADAR (radio beam). The idea of using laser beams for remote detection is not new and was described and used in the 1960s just after the invention of the laser (Inaba and Kobayasi, 1969). This section cannot possibly attempt to cover all the LIDAR applications, presently one of the real growth areas in laser sensor technology, with recently over a hundred articles per year being published. Further details are given in Volume 1, Chapter 5. For more extensive sources of detailed information the reader is directed to Demidov (1995), Svanberg (1991), Sigrist (1994), Carts (1994) and Measures (1984).

Conventional continuum light sources, e.g. high pressure Xe lamps, can and are being used extensively for atmospheric monitoring over hundreds of metres. The considerable advantages, however, afforded by modern laser systems (e.g. high powers at specific and variable wavelengths from the UV to the IR, short pulses from micro- to picoseconds, low divergence of the laser beam allowing large amounts of optical energy to be transported over long distances) makes them the preferred optical sources for LIDAR.

A typical LIDAR system is shown in Figure 3.7. A pulsed laser beam emits short pulses of light directed into the atmosphere and the backscattered light is detected in a receiver, usually placed alongside the laser, after a certain delay in radar-like fashion. The optical systems O1 and O2 contain optical components or units that correct or modify the emitted beam or serve to collect or wavelength analyse the collected signal. The divergence of a typical laser system is 0.5 mR, but if one wishes to reduce this, then O1 can be a beam expander that increases the laser diameter and hence reduces the divergence. The optical system O2 may contain a simple lens or a sophisticated wide aperture telescope, depending whether the ranging distance R is a hundred metres or kilometres. The velocity of light is about $300 \, \text{m} \, \mu\text{s}^{-1}$ and hence the pulse is received back from a ranging distance of 150 m, 1 µs $(2R/c)$ after the emission of the pulse. Time-resolved information, obtained from the time delay, is one of the strengths of LIDAR and the range resolution $\Delta R = t_p c/2$ is defined by

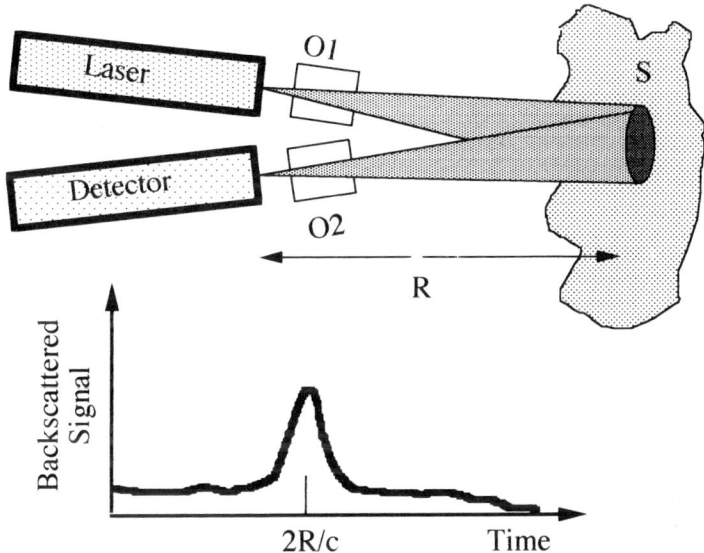

Figure 3.7 Most LIDAR systems consist of the following components. A laser beam is corrected or modified by optics (01), hits a target a distance of R from the laser and the backscattered light is collected by further optics (02) before the detector. The backscattered signal as a function of time is shown with a peak at $2R/c$ (c, velocity of light).

the duration of the pulse t_p. Therefore, a laser pulse duration of 1 ns has a range resolution of minimally 15 cm.

When an object, which may be clouds or emissions from a chimney etc., intercepts the laser beam it generates a backscattered signal. This can be a chemically unspecific signal, e.g. Rayleigh or Mie scattering from aerosols in the atmosphere where the scattering is elastic and the emitted and detected wavelengths are the same. It can also be chemically specific, e.g. Raman scattering or laser-induced fluorescence where the scattering is inelastic and the scattered wavelength is different (usually red shifted) from the emitted signal. The backscattered signal always contains information about the scattering object, thus permitting remote analysis to be made of the physical and chemical properties of the target. According to Demidov (1995) a simplified version of the LIDAR equation is given by

$$P_{\text{det}} = \frac{S}{\pi R^2} P_{\text{las}} \tau \varepsilon \gamma \qquad (2.1)$$

where P_{las} is the power of the emitted laser light, S is the illuminated object area, R is the ranging distance (the distance between laser and object), ε is the object's response to the laser light (which also contains the concentration of the measured substance), τ is the laser propagation properties through O1 and O2 and γ is the geometric form-factor of the laser light and backscattered signal through the ambient air, which will contain a typical Beer–Lambert exponential absorption factor.

One of the problems with LIDAR is how to eliminate the constant flux of solar photons, which can provide a large background in the detector. This is normally eliminated or certainly considerably reduced by gating the detector open for only a very short time (the time duration of the pulse) and increasing the signal-to-noise ratio within this time by only permitting the required wavelength (laser wavelength) to be transmitted to the detector by using filters or monochromators.

Another problem is how to make the LIDAR procedure quantitative. We cannot use normal calibration procedures because of the remote nature of the technique and hence we must use *internal* parameters to conduct quantitative measurements. Therefore, for remote measurements, the ratio or difference of two signals is required (one a calibration signal). The calibration standard can be one of several kinds: the precise timing of the laser pulse or the time delay between two events, the quasi-constant water Raman signal (for aqueous LIDAR measurements) or the nitrogen Raman signal (for atmospheric LIDAR measurements).

3.3.2 DIAL

A variation of the LIDAR technique, DIAL (differential absorption LIDAR), is a perfect example of the principle of relative measurements. The substance is probed by two close-lying laser wavelengths, one corresponding to an absorption band in the substance (λ_1) and the other positioned just off the absorption peak (λ_2). Because the detected signals at both of these wavelengths contain the same dependence on $P_{las}, S, R, \tau, \gamma$ in the LIDAR equation, the ratio of those two detected signals $\{P_{det}(\lambda_1)/P_{det}(\lambda_2)\}$ only contains a dependence on ε which is proportional to the concentration of the substance. Further information is given in section 3.5 and in Volume 1, Chapter 5.

3.4 Applications of LIDAR

3.4.1 Mie scattering LIDAR

Aerosols, for example particulate pollutants, as well as water vapour and ice crystals (in clouds) can be detected by elastic Rayleigh or Mie scattering in which the scattered signal has the same wavelength as the emitted laser light (λ_{las}). This is compared with the detected nitrogen Raman signal at a different nearby wavelength (λ_N). The concentration of the aerosol is proportional to $P_{det}(\lambda_{las})/P_{det}(\lambda_N)$.

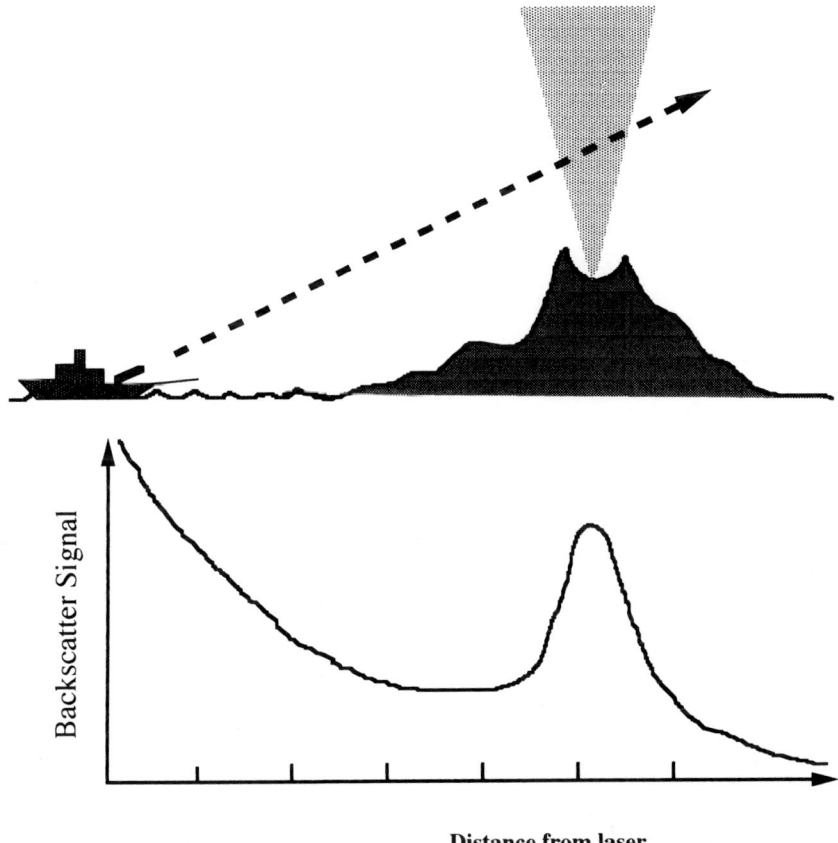

Distance from laser

Figure 3.8 LIDAR system operating via Mie scattering from the plume of a volcano. The smoothly falling $1/R^2$ signal is from the uniform background particulate matter.

In Figure 3.8, a LIDAR particle monitoring measurement is illustrated in which a laser beam is directed through the atmosphere at and through volcanic emissions. The smoothly falling $1/R^2$-dependent signal is from the uniform background particulate distribution in the atmosphere. On passage through the plume, the scattered intensity increases dramatically. This signal only indicates the presence of particulate matter without revealing its chemical nature.

Pure Mie scattering measurements has been carried out at many LIDAR stations world-wide to monitor volcanic eruptions, which can pump tremendous quantities of aerosols into the troposphere and stratosphere. The volcano Pinatubo (erupted 1991) has been monitored extensively by Osborne *et al.* (1992) to show how the stratospheric dust layer developed.

3.4.2 Raman LIDAR

In contrast to Rayleigh–Mie elastic scattering of light, Raman inelastic scattering provides detailed information about the chemical nature of the scatterer (Svanberg, 1994). In the vibrational Raman effect, the incident photon is absorbed into a virtual state of the molecule which de-excites by emitting a characteristically red-shifted photon (Stokes' shifted) to an excited vibrational level of the ground state. Thus, the backscattered photon is shifted by an amount characteristic of the absorbed gas. The required concentration of the gas is compared with the Stokes' component of the nitrogen molecules for internal calibration. These Raman peaks are normally well resolved. The main difficulty with the Raman effect is that it is inherently very weak, being about three orders of magnitude weaker than Rayleigh scattering for the same molecule. Hence, Raman LIDAR is only useful where intense lasers are available or where fairly major constituents in the atmosphere are being monitored, e.g. oxygen, water, carbon dioxide, sulphur dioxide or ozone.

3.4.3 Fluorescence LIDAR

In fluorescence LIDAR, the laser is tuned to some absorption transition in the atmospheric, aqueous or terrestrial species. The fluorescence light is detected in the receiver and again compared with, usually, the Raman nitrogen signal, from which quantitative results can be obtained. Many molecules and, in particular, a number of aromatic molecules fluoresce efficiently using UV light; hence, this technique is frequently applied for the detection of petrochemical pollutants on sea water surfaces (Pantani and Cecchi, 1991). If a nitrogen laser (337 nm) is used in, for example, an airborne LIDAR system, the fluorescence signal from the different oils is detected at between 400 nm and 500 nm, with the comparison water Raman peak at about 380 nm. To some degree, the different types of oil can be distinguished by the fluorescence wavelength fingerprint and hence the oil's provenance can be identified.

Phytoplankton is one of the most important species in natural waters (see Volume 2, Chapter 2). It is a form of algae, it is photosynthetic and serves as a food for zooplankton, which, in turn, supports fish colonies. Therefore, phytoplankton is one of the basic factors of life in the oceans that determines the biological productivity of the sea environment. These algae contain various photosynthetic pigments, the most important of which is chlorophyll a. These pigments absorb the laser light and emit fluorescence characteristic of chlorophyll a. Figure 3.9 shows the backscattered spectrum from ocean water where the exciting laser was the second harmonic of the Nd:YAG laser (532 nm). The peaks of the chlorophyll fluorescence and the Raman water scattering are clearly visible, from which

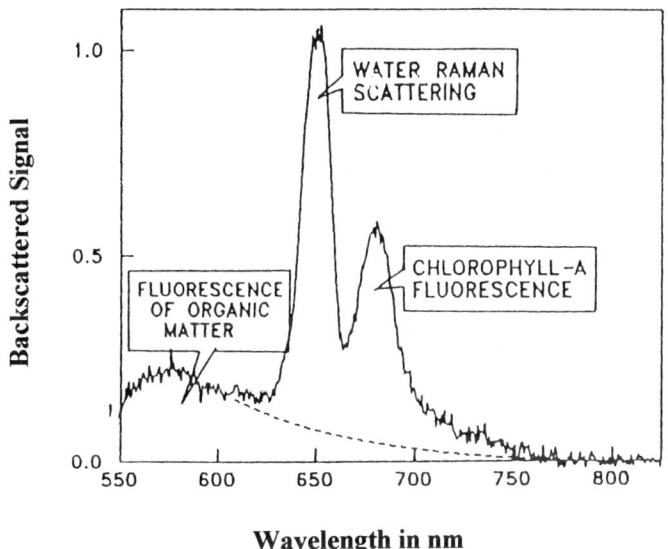

Wavelength in nm

Figure 3.9 A typical LIDAR response from ocean water excited by the second harmonic of a Nd:Yag laser (532 nm). A strong water Raman peak is visible with a chlorophyll fluorescence peak. (Chekalyuk *et al.*, 1992).

the concentration of the chlorophyll content can be calculated and hence the biological productivity of the water (Chekalyuk *et al.*, 1992).

Similar fluorescence spectra can be obtained from nitrogen laser excitation of different plants and trees from terrestrial surfaces. To some degree, the different types of tree can be distinguished as well as distinguishing different types of vegetation. The chlorophyll *a* spectrum in plants show two peaks, at 685 and 730 nm. The ratio of these two peaks can be a measure of the environmental stress on the plants and the photosynthesis efficiency (Pentani and Cecchi, 1991; Svanberg, 1995).

3.5   DIAL

Although we have given DIAL a separate section in this review, there is no real fundamental difference between DIAL and any other forms of LIDAR. The only reason that it is considered to be different is that two different laser wavelengths are used in the measurement and, because of this, tuneable rather than fixed frequency lasers are normally used. The tuneable lasers in the optical part of the spectrum are normally YAG or excimer pumped dye lasers or some of the new solid-state lasers like alexandrite or

Ti–sapphire. In the long optical wavelength to IR region, carbon dioxide or semiconductor diode lasers are popular.

If a gaseous effluent is being monitored, laser light is alternatively transmitted at a wavelength (λ_{on}) where the pollutant strongly absorbs and at a neighbouring wavelength (λ_{off}) which is off resonance. The on-resonance signal is attenuated on passing through the cloud, whereas the off-resonance signal is not. The off-resonance signal will normally exhibit the normal $1/R^2$ dependence, with the on-resonance signal exhibiting additional losses when the pollutant is encountered. These two spectra are divided and if they are identical, a value of unity is obtained at all distances. However, the normal DIAL ratio is not zero but exhibits variations of ratio as a function of the range, from which gas concentrations can be calculated. From such measurements have been obtained sophisticated ozone profiles at different altitudes between 10 m and 10 km above sea level over the Antarctic, which demonstrated the stratospheric ozone hole formation (Browell, 1991).

3.6   Laser absorption spectroscopy

Absorption spectroscopy is based on the selectivity of the wavelengths of light absorbed by different chemical compounds in a sample cell and involves monitoring the variation in the intensity of absorption as a function of its wavelength.

There are two basic categories of absorption technique that are commonly in use in spectroscopy. These involve the measurement of either (i) transmitted power through a sample or (ii) absorbed power, directly or indirectly, by the sample. A block diagram of the general experimental arrangement is shown in Figure 3.10.

In normal practice (Andrews et al., 1990), the laser source would be tuneable over a range of wavelengths that are selected using a monochromator of very narrow bandpass. The light from the laser source can be pulsed or CW and the latter may be modulated or unmodulated. In the transmission method, which is based on conventional spectrophotometry principles, the sample cell can be either single pass or multipass and it may or may not form an intracavity system. Simplified absorption spectra may be obtained using a molecular beam to reduce molecular rotation (M.N.R. Ashford, personal communication). The sensor would typically be a photodetector and the data acquisition system would be a y–t recorder or digital oscilloscope.

In the absorbed power technique, several calorimetric methods are readily available, which utilise thermocouple, thermal lens and photoacoustic (or acousto-optic as it is sometimes called) technologies. These techniques will now be discussed in more detail.

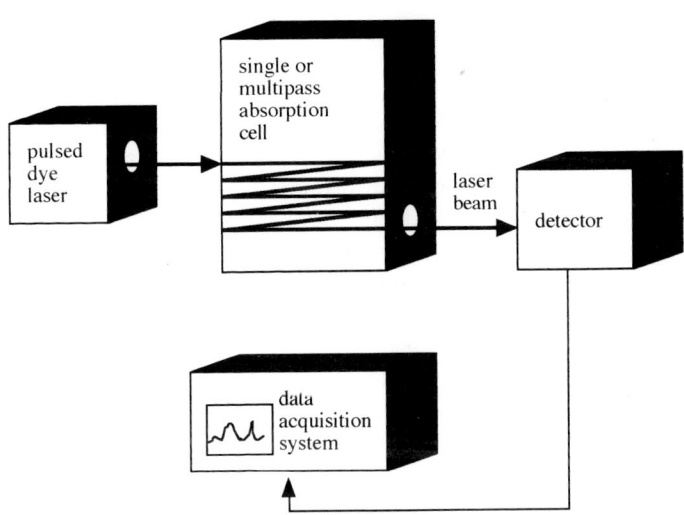

Figure 3.10 Schematic experimental arrangement for laser absorption spectroscopy.

3.6.1 Transmission methods

These are basically difference techniques where the measurement of small absorbances requires the detection of a very small difference between two large signals (Campbell *et al.*, 1994). Reductions in the signal-to-noise ratio in these systems, caused by shot noise associated with the light source, can be a limiting factor but may be reduced by modulating the light source. The pathlength within the sample may be greatly increased using multipass cell geometry, which has already been very successfully exploited in, for example, gas-phase IR analysis with very high resolution (Harris and Lylte, 1983). The reader is also referred to the DOAS section in Volume 1, Chapter 5.

Single-pass, long-path cells can be used for gases in laser-based systems because of the spatial coherence of the incident light. The measurement of liquid absorbance using hollow glass and quartz fibres has been well established (Stone, 1978) with pathlengths in the range 4.5 m to 130 m. The small diameter of the fibres ($\ll 2$ mm) means that very small volumes of liquid give very long pathlengths (e.g. 100 ml of sample can have lengths of ~ 30 m). There are, however, problems associated with hollow fibres, such as cleaning and refilling, but these can be overcome if shorter lengths of disposable fibre are employed.

A major drawback with absorbance measurements is reflections from the container–sample interface, which tend to exist even when the refractive indices are very similar.

3.6.2 Direct laser absorption methods

3.6.2.1 Diode laser spectroscopy. This method is ideally suited to the analysis of gas-phase species by high-resolution IR spectroscopy. For example, the photolysis of carbon disulphide using 193 nm light from an ArF laser, results in CS radicals that readily produce a diode laser absorption spectrum (Kanamori and Hirota, 1987).

3.6.2.2 Difference frequency laser spectroscopy. IR transitions in the wavelength range 2.2–4.2 mm can be studied using light from a difference frequency laser source. This involves mixing the outputs of a CW dye laser and a single mode Ar ion laser using a lithium niobate crystal. Jet cooling is often employed to reduce rotational complications associated with C–H, O–H and N–H type molecules (McIlroy and Nesbitt, 1990a,b).

3.6.2.3 Intracavity absorption techniques. These techniques are particularly useful for applications involving very weak absorptions in the visible and IR regions of the spectrum. Two of these methods, which are characterised by an increase in the effective absorption pathlength, will now be briefly outlined.

The first system, employing a simple absorption cell that is an integral part of a tuneable laser cavity, is now well established (Demtroder, 1981). The basic intracavity arrangement is illustrated in Figure 3.11. Sensitivity to very weak absorption levels can be dramatically increased by operating close to the lasing threshold, since any additional reduction in intracavity intensity greatly reduces the system gain, which, in turn, causes a severe reduction in output beam intensity.

The second system forms the basis of cavity ring down spectroscopy and is well suited to investigations such as hydrogen cyanide absorption. Unlike the intracavity method, the high sensitivity is achieved using an external resonator or ring down cavity (RDC), which contains the material of interest. The system is similar to that shown in Figure 3.11 but with the

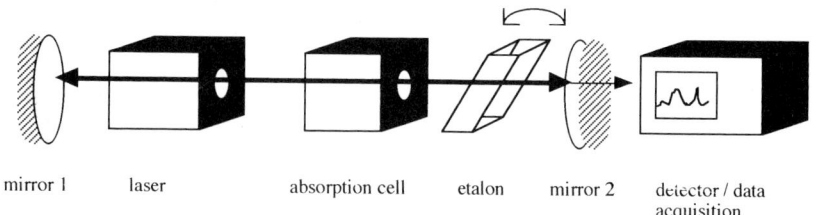

mirror 1   laser   absorption cell   etalon   mirror 2   detector / data
                                                                        acquisition

Figure 3.11 Basic layout of an intracavity spectrometer.

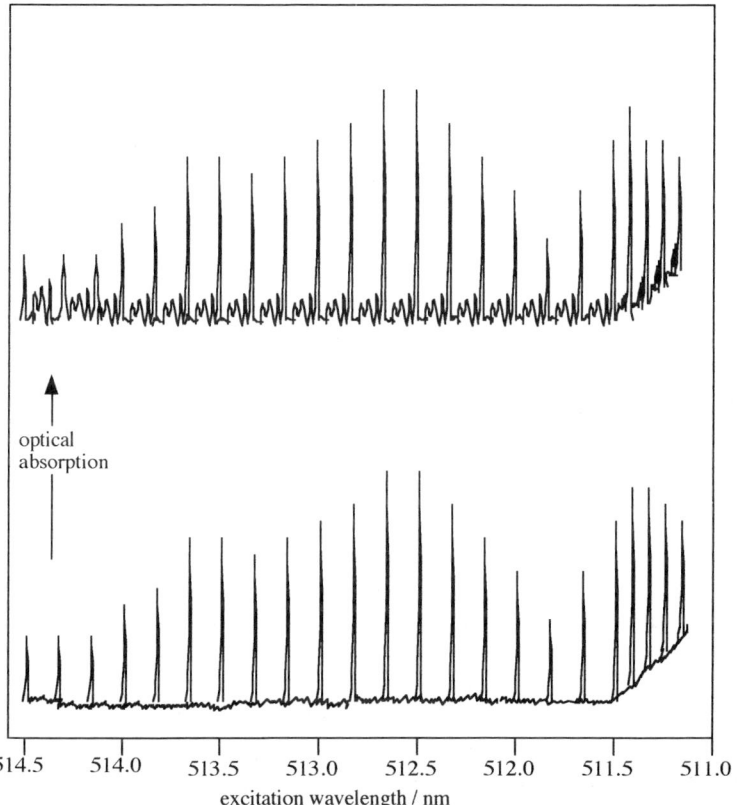

optical
absorption

514.5   514.0   513.5   513.0   512.5   512.0   511.5   511.0
excitation wavelength / nm

Figure 3.12 Typical ring down cavity spectra before and after baseline correction. (The 205 HCN overtone band data are taken from Romanini and Lehmann (1993).)

sample cell external to the cavity. A spectrum typical of the type obtained from a ring down cavity system is shown in Figure 3.12.

An excimer laser such as XeCl (308 nm, 10 ns, 10 mJ and 20–40 Hz) is suitable for very weak absorption spectra (Romanini and Lehmann, 1993). A small fraction of the laser beam is used to obtain calibration lines emitted by a hollow cathode neon lamp and relative frequency markers from a $1 \, cm^{-1}$ free spectral range etalon. The photomultiplier tube and associated amplifier ensure that peak intensities are of the order of several millivolts. A boxcar averager is used to process the data in order to simplify the RDC scheme without loss of system sensitivity.

3.6.3   Fixed frequency laser IR spectroscopy

There are two established methods that enable fixed frequency IR lasers to tune to resonances by application of an external electric or magnetic field. These methods have continued to be used despite the introduction of high-power tuneable lasers for the following reasons. They are extremely sensitive to low concentrations of reactive species, they provide very high resolution in saturation spectroscopy with high-power lasers and the calibration problems normally associated with tuneable lasers are avoided since shifts are measured relative to the positions of accurately measured laser frequencies.

3.6.3.1 Laser Stark spectroscopy. Laser Stark spectroscopy (LSS) is sometimes called laser electric resonance spectroscopy and it is an extremely important investigative tool for the spectra of polar and non-polar atoms and molecules (Duxbury, 1985). The underlying principle is the Stark effect (Stark, 1913), which causes transitions to shift between ro-vibronic levels in molecules under the influence of an external electric field so that laser light, whose frequency is already close to the zero field resonant frequency, becomes resonant for that molecule. The amount of shift is dependent on the value of the permanent electric dipole and the applied electric field. A typical intracavity system is shown in block diagram form in Figure 3.13, although it is also common practice to place the cell outside the cavity. LSS has been further enhanced by the development of double resonance methods based on (i) low-pressure molecular beams, (ii) microwave-IR double resonance, (iii) tuneable lasers, and (iv) modulated carbon dioxide lasers.

3.6.3.2 Laser magnetic resonance spectroscopy. Laser magnetic reson-ance spectroscopy (LMRS) utilises the Zeeman effect for tuning the fixed frequency laser light to the modified ro-vibronic transition levels. As the method relies on the presence of a permanent magnetic dipole moment

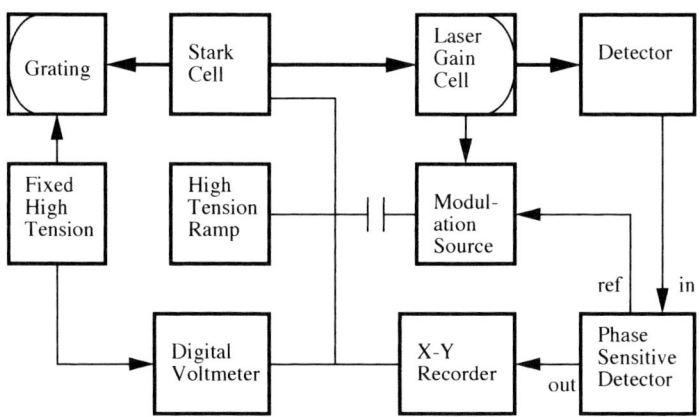

Figure 3.13 Block diagram arrangement for laser Stark spectroscopy.

within the sample, it is extensively used for probing atoms and radicals which possess one or more unpaired electrons in the wavelength range 40–1000 μm (Evenson, 1981). This wavelength is relatively easy to achieve if a carbon dioxide laser is used to pump a far-IR laser. Increased signal-to-noise ratios, and hence sensitivities, may be achieved using modulated IR and lock-in amplifiers.

3.6.4 Photoacoustic spectroscopy

The basis for photoacoustic spectroscopy (PAS) was established more than a century ago by Alexander Graham Bell, founder of the Bell Telephone Co. (Bell, 1880). As the name suggests, it involves the generation of acoustic waves through the generation and absorption of modulated electromagnetic radiation. Recently, because of the advent of intense light sources such as lasers and arc lamps and the development of sensitive detectors such as piezoelectric sensors, the technique has been resurrected for the spectroscopy of gases, opaque specimens and powdered materials (Tam, 1983).

PAS is a specialised absorption spectroscopy that can be used, for example, to measure the temporal variations in ammonia concentrations over a period of several days, as shown in Figure 3.14. It relies on the fact that the localised heating that occurs when a medium absorbs energy is

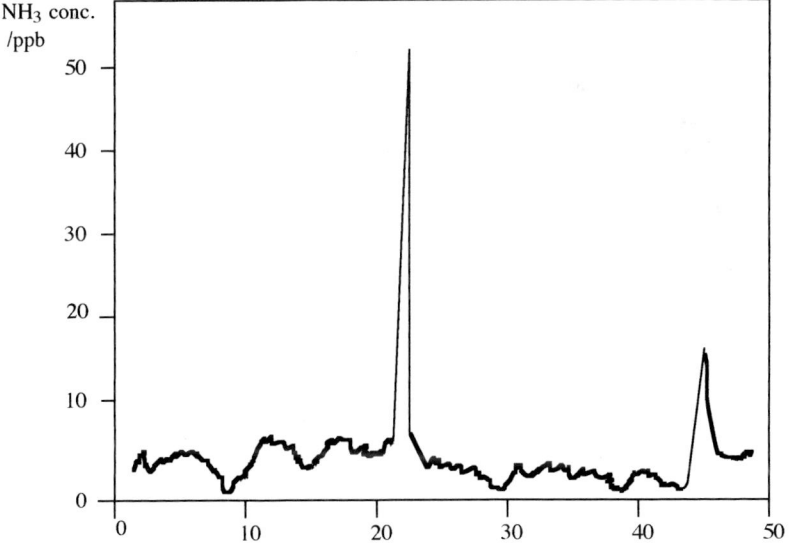

NH₃ data taken in 6 minute intervals over a 50 hour period

Figure 3.14 Temporal variation of ammonia concentration, based on data taken from Rooth *et al.* (1990). Data were taken at 6 min intervals over a period of 50 h.

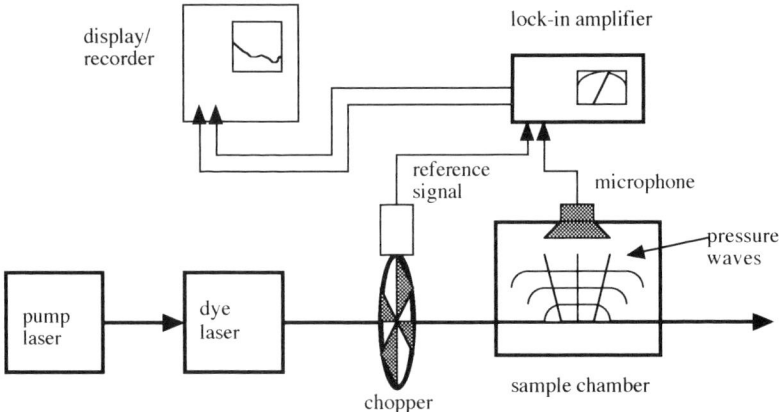

Figure 3.15 Schematic arrangement for modulated CW laser photoacoustic spectroscopy.

usually accompanied by a corresponding increase in pressure. The resultant pressure waves can be used to monitor the effect using an acoustic detector situated on a wall of a sample cell. Two laser-based systems that are in common use will now be briefly described.

3.6.4.1 Modulated CW system. The modulated CW system utilises a CW laser and mechanical chopper, as shown in Figure 3.15. This technique, and the pulsed system described below, is suited to gases and liquids although the heat capacity and heat transfer characteristics of the media have to be taken into account when designing the system. For example, high heat-capacity solvents such as water tend to dampen the acoustic signal and so organic solvents are usually preferred.

The tuneable CW laser is mechanically chopped to produce a modulated beam at a reference frequency f. A lock-in amplifier, referenced at this frequency, ensures that the acoustic signal has a high signal-to-noise ratio through a reduction in $1/f$ noise. The signal is obtained as a function of laser wavelength to enable the photoacoustic spectrum of a gas to be obtained (Andrews, 1990).

3.6.4.2 Pulsed system. The arrangement for pulsed PAS is shown in Figure 3.16. A tuneable pulsed laser beam is transmitted through a sample cuvette containing the analyte of interest in an appropriate solvent (Schurig *et al.*, 1993). An ultrasonic sensor such as a piezoelectric crystal is attached to one of the sides of the cuvette lying parallel to the beam. The piezoelectric sensor and cuvette are often referred to as the photoacoustic cell. Uniform heating along the path of the laser beam is ensured by keeping the thickness of the cuvette walls and the width of solution as small as

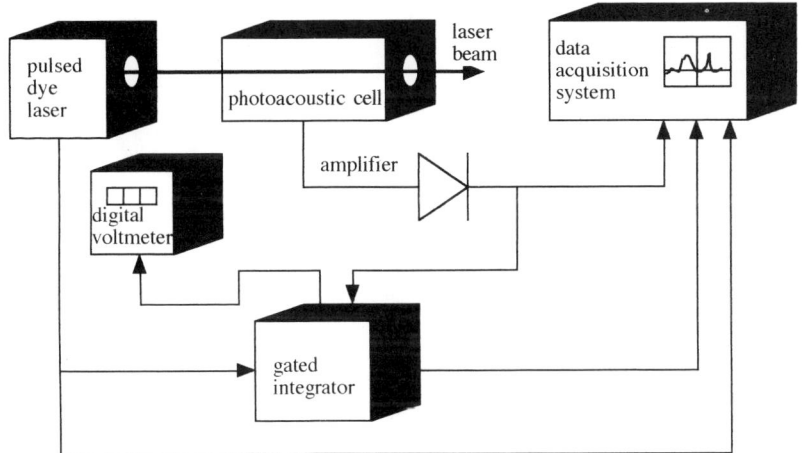

Figure 3.16 Block diagram for pulsed laser photoacoustic spectroscopy.

practicable. During each laser pulse, the analyte absorbs light energy and produces thermal energy as part of its relaxation process. The resulting thermal expansion produces a transient stressed region in the solvent, which, in turn, produces a transient signal at the piezoelectric sensor. The gated detection system enables any peak within the transient to be selected so that any variation in peak intensity, as a result of solution absorption, may be taken as a measure of the analyte concentration (Beitz *et al.*, 1990; Russo *et al.*, 1990; Torres *et al.*, 1990; Bennet *et al.*, 1992). One application of the system is the monitoring of ammonia, which is used in the reduction of nitrogen oxide gases (Hammerlich, 1990). Instead of a tuneable dye laser, a carbon dioxide laser provides more than 80 lines that are scanned in windows of 230 MHz using a blazed grating. A repetition rate of 700–900 Hz enables a sensitivity of better than 500 ppb to be achieved.

3.6.5   Thermal lens spectroscopy (TLS)

Thermal lens spectroscopy (TLS) is an ultrasensitive absorption technique that is particularly applicable to rare earth and actinide species in aqueous solution. The variation in the intensity profile of a laser beam across its diameter gives rise to inhomogeneous heating effects in the sample solution. Thermal gradients, caused by the passage of the beam, produce de-focusing effects resulting from changes in the refractive index of the sample. By scanning a dye laser and monitoring the variation in intensity of the transmitted light through a pinhole, it is possible to obtain an absorption spectrum as a function of laser wavelength.

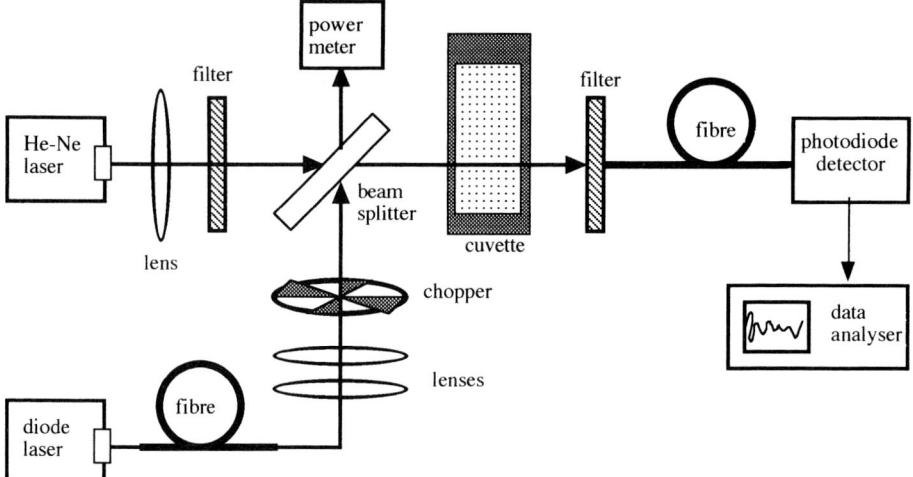

Figure 3.17 Schematic arrangement for thermal lens spectroscopy.

The use of high-power tuneable dye lasers is not, however, practicable in, for example, remote environmental monitoring. A suitable system has been developed (Rojas *et al.*, 1991; 1992) using a GaAlAs laser for excitation (794.6 nm) and a He–Ne at 632.8 nm for probe purposes. For compactness, a 1 m length of 100 μm core diameter step index fibre of numerical aperture 0.3 is used to couple the semiconductor diode to a chopper and a 10 m length of 200 μm core diameter graded index fibre was used to take the light from the cuvette to the photodiode. The system has been used for aqueous solutions of Pu^{6+} in a process stream environment and the basic layout is shown in block form in Figure 3.17.

References

Ambartzumian, R.V. and Letokhov, V.S. (1972) Selective two-step (STP) photoionization of atoms and photodissociation of molecules by laser radiation. *Appl. Optics*, **11**, 354–358.

Andrews, D.L. (1990) *Lasers in Chemistry*. Springer-Verlag, Berlin.

Bekov, G. I. and Letokhov, V. S. (1983) Laser atomic photoionization spectral analysis of element traces. *Appl. Phys.*, **B30**, 161–167.

Bekov, G. I., Letokhov, V. S. and Radaev, V. N. (1985) Laser photoionization spectroscopy of ruthenium traces at the level of 1 part in 10^{12}. *J. Opt. Soc. Am.*, **B2**, 1554–1560.

Bell, A.G. (1880) *Am. J. Sci.*, **20**, 305.

Boesl, U., Neusser, H.J. and Schlag, E.W. (1981) Multi-photon ionization in the mass spectrometry of polyatomic molecules: cross sections. *Chem. Phys.*, **55**, 193–204.

Bushaw, B.A. (1992) Attogram measurement of rare isotopes by CW resonance ionization mass spectrometry. *Resonance ionization spectroscopy 1992*, Institute of Physics Conf. Ser. **128**, pp. 31–36. IOP, Bristol.

Browell, E.V. (1991) Differential absorption lidar detection of ozone in the troposphere and lower stratosphere. In *Optoelectronics for Environmental Science* (ed. S. Martelucci and A.N. Chester), pp. 77–88. Plenum Press, New York.

Campbell, M., Lynch, F. and Brown, G. (1994) An inexpensive portable ratiometer for absorbance measurements. *Int. J. Electronics*, **76** (1), 153–161.

Carts, Y.A. (1994) Optical sensors gather data to map atmospheric gases. *Laser Focus World*, **April**, 74–80.

Chekalyuk, A.M., Demidov, A.A., Fadeev, V.V. and Gorbunov, M.Yu. (1992) In *XVIIth Congress of the International Society for Photogrammetry and Remote Sensing*, **29** (B7), 878.

Demidov, A.A. (1995) Laser remote sensing, in *An Introduction to Laser Spectroscopy* (ed. Andrews, D.L. and Demidov, A.A.) Plenum, N.Y., pp. 133–147.

Demtroder, W. (1981) *Laser Spectroscopy*. Springer-Verlag, Berlin.

Department of Health (1993) *The Oxides of Nitrogen: Advisory Group on the Medical Aspects of Air Pollution Episodes*. HMSO, London.

Duxbury, G. (1985) Laser Stark spectroscopy. *Int. Rev. Phys. Chem.*, **4**, 237–278.

Evanson, K.M., Saykally, R.J., Jennings, D.A., Curl, R.F. and Brown, J.M. (1980) *Chemical and Biochemical Applications of Lasers* (ed. Moore, C.B.), Vol. 5, New York Academic Press.

Finlayson-Pitts, B.J. and Pitts, J.N., Jr (1993) Atmospheric chemistry of tropospheric ozone formation: scientific and regulatory implications, *Air and Waste*, **43**, 1091–1100.

Goppert-Mayer, M. (1931) *Ann. Phys. LPZ.* **9**, 273

Guizard, S., Chapoulard, D., Horani, M. and Gauyacq, D. (1989) Detection of NO traces using resonantly enhanced multiphoton ionization: a method for monitoring atmospheric pollutants. *Appl. Phys.*, **48**, 471–477.

Hammerlich, A., Olafson, A. and Henningen, J. (1991) in *Optoelectronics for Environmental Science* (ed. Martelluci, S. and Chester, A.N.), Plenum Press, New York and London, pp. 123–127.

Harris, T.D. and Lytle, F.E. (1983) in *Ultrasonic Laser Spectroscopy* (ed. Kliger, D.S.), Chapter 7, pp. 369–433. Academic Press (London).

Harrison, R. (1994) A fresh look at air. *Chem. Britain*, **30**, 987–992.

Hurst, G.S. and Payne, M.G. (1988) *Principles and Applications of Resonance Ionization Spectroscopy*. Adam Hilger, Bristol, UK.

Inaba, H. and Kobayasi, T. (1969) *Nature*, **224**, 170–172.

Kanamori, H. and Hirota, E. (1987) Infrared laser kinetic spectroscopy of a photofragment CS generated by the photo-dissociation of CS2 at 193 nm: nascent vibrational-rotational-translational distribution of CS, *J. Chem. Phys.*, **86** (7), 3901–3905.

Ledingham, K.W.D. and Singhal, R.P. (1991) Assessment of titanium-sapphire lasers and optical parametric oscillators as sources of variable wavelength for resonant ionization mass spectroscopy. *J. Anal. Atom. Spectrom.*, **6**, 73–77.

Ledingham, K.W.D. and Singhal, R.P. (1992) Laser mass spectrometry. In *Applied Laser Spectroscopy – Techniques, Instruments and Applications* (ed. D.L. Andrews), pp. 365–400. VCH Publishers, New York.

Letokhov, V.S. (1987) *Laser Photoionization Spectroscopy*. Academic Press, London.

Logan, J.A. (1983) Nitrogen oxides in the troposphere: global and regional budgets. *J. Geophys. Res.*, **88**, 10785–10807.

Lubman, D.M. (ed.) (1990) *Lasers and Mass Spectrometry*. Oxford University Press, New York.

McCombes, P.T., Borthwick, I.S., Jennings, R., Ledingham, K.W.D. and Singhal, R.P. (1991) Resonance ionization mass spectrometry applied to trace analysis of gold. In *Optogalvanic Spectroscopy* (ed. R.S. Stewart and J.E. Lawler), *Inst. Phys. Conf. Ser.*, **113**, pp. 163–168. IOP, Bristol, UK.

McIlroy, A. and Nesbitt, D.J. (1990a) Vibrational mode mixing in terminal acetylenes: high resolution infrared laser study of isolated J states. *J. Chem. Phys.*, **92** (4), 2229–2243.

McIlroy, A. and Nesbitt, D.J. (1990b) High resolution slit jet infrared spectroscopy of hydrocarbons: quantum state specific mode mixing in CH stretch-excited propyne. *J. Chem. Phys.*, **91** (1), 104–113.

Marshall, A., Clark, A., Ledingham, K.W.D., Sander, J., Singhal, R.P., Kosmidis, C. and Deas, R.M. (1994) Detection and identification of explosives using laser ionization time-of-flight techniques. *Rapid Commun. Mass Spectron*, **8**, 521–526.

Marshall, A., Ledingham, K.W.D. and Singhal, R.P. (1995) Trace detection of benzene vapour in a simple ion chamber employing laser induced REMPI in the 246-264 nm wavelength region. *Analyst*, **120**, 2069–2073.

Measures, R.M. (1984) *Laser Remote Sensing: Fundamentals and Applications*, Wiley, New York.

Osborne, M.T., Winker, D.M. Woods, D.C. and De Coursey, R.J. (1992) Lidar observations of the Pinatubo volcanic cloud over Hampton, Virginia, *Nasa Conf. Publ.*, **3158**, 91–94.

Pantani, L. and Cecchi, G. (1991) Fluorescence lidars in environmental remote sensing. In *Optoelectronics for Environmental Science* (ed. S. Martellucci and A.N. Chester), pp. 131–148. Plenum Press, New York.

Payne, M.G., Lu Deng, Lu and Thonnard, N. (1994) Applications of resonance ionization mass spectrometry. *Rev. Sci. Instrum.*, **65**, 2433–2458.

Peng, W.X., Ledingham, K.W.D., Marshall, A. and Singhal, R.P. (1995) Urban air pollution monitoring: Laser-based procedure for the detection of NO_x gases. *Analyst*, **120**, 2537–2542.

QUARG (1993) *Diesel Vehicle Emissions and Urban Air Quality*. Quality of Urban Air Group, HMSO, London.

QUARG (1994) *Improving Air Quality: a Discussion Paper on Air Qualities Standards and Management*. Department of the Environment, HMSO, London.

QUARG (1995) *Air Quality: Meeting the Challenge, The Government's Strategic Policies for Air Quality Management*. Department of the Environment, HMSO, London.

Rojas, D., Silva, R.J., Spear, J.D. and Russo, R.E. (1991) Dual beam optical fiber thermal lens spectroscopy. *Anal. Chem.*, **63** (18), 1927–1932.

Rojas, D., Silva, R.J. and Russo, R.E. (1992) Thermal lens spectroscopy using a diode laser and optical fibers. *Rev. Sci. Instrum.*, **63** (5), 2989–2993.

Romanini, D. and Lehmann, K.K. (1993) Ring down cavity absorption spectroscopy of the very weak HCN overtone bands with six, seven and eight stretching quanta. *J. Chem. Phys.*, **99** (9), 6287–6301.

Rooth, R.A., Verhage, A.J.L. and Wouters, L.W. (1990) Photoacoustic measurement of ammonia in the atmosphere: Influence of water vapour and carbon dioxide. *Appl. Opt.*, **29**, 2663–2665.

Sigrist, M.W. (ed.) (1994) *Air Monitoring by Spectroscopic Techniques.* Wiley, New York.

Simeonssen, J.B., Lemire, G.W. and Sausa, R.C. (1993) Trace detection of nitrocompounds by ArF laser photofragmentation/ionization spectrometry. *Appl. Spec.*, **47**, 1907–1912.

Stark, J. (1913) Sber. Preuss. Akad. Wiss., **47**, 932.

Syage, J.A. (1990) Real-time detection of chemical agents using molecular laser mass spectrometry. *Anal. Chem.*, **62**, 505A–509A.

Svanberg, S. (1991) *Atomic and Molecular Spectroscopy*. Springer-Verlag, Berlin.

Svanberg, S. (1994) Differential absorption lidar (DIAL). In *Air Monitoring by Spectroscopic Techniques* (ed. M.W. Sigrist), pp. 85–161. Wiley, New York.

Svanberg, S. (1995) Fluorescence lidar monitoring of vegetation status. *Phys. Scripta*, **T58**, 79–85.

Tam, A.C. (1983) *Ultrasensitive Laser Spectroscopy*, Ch-1, pp. 1–108, Academic Press, London.

Thonnard, N., Parks, J.E., Willis, R.D., Moore, L.J. and Arlinghaus, H.F. (1989) Resonance ionization of neutral atoms with applications to surface science, noble gas detection and biomedical analysis. *Surf. Int. Anal.*, **14**, 751–759.

Thonnard, N., Wright, M.C., Davis, W.A. and Willis, R.D., (1992) The second-generation RIS–TOF noble gas detector: detection limits below 100 atoms in less than 5 minutes. In *Resonance Ionization Spectroscopy 92, Inst. Phys. Conf. Ser.*, 128, pp. 27–30. IOP, Bristol.

Towrie, M., Drysdale, S.L.T., Jennings, R., Land, A.P., Ledingham, K.W.D., McCombes, P.T., Singhal, R.P., Smyth, M.H.C. and McLean, C.J. (1990) Trace analysis using a commercial resonant ionisation mass spectrometer. *Int. J. Mass Spec. Ion Proc.*, **26**, 309–320.

Urban, F.J., Deissenberger, R., Hermann, G., Kohler, S., Riegel, J., Trautmann, N., Wendeler, H., Albus, F., Ames, F., Kluge, H.-J., Krass, S. and Scheerer, F. (1992) Resonance ionization mass spectroscopy of platinum with a reflectron time-of-flight mass spectrometer. In *Resonance Ionization Spectroscopy, Inst. Phys. Conf. Ser.*, **128**, pp. 233–236. IOP, Bristol.

Vertes, A., Gijbels, R. and Adams, F. (ed.) (1993) *Laser Ionization Mass Analysis: Chem. Analysis Ser.*, Vol. 124. Wiley, New York.

Weickhardt, C., Boesl, U. and Schlag, E.W. (1994a) Laser mass spectrometry for time-resolved multicomponent analysis of exhaust gas. *Anal. Chem.*, **66**, 1062–1069.

Weickhardt, C., Zimmermann, R., Schramm, K.W., Boesl, U. and Schlag, E.W. (1994b) Laser mass spectrometry of the di-, tri-, and tetrachlorobenzenes: isomer-selective ionization and detection. *Rapid Communications Mass Spectrom.*, **8**, 381–384.

Wendt, K., Passler, G. and Trautmann, N. (1995) Trace detection of radiotoxic isotopes by resonance ionization mass spectrometry. *Phys. Scripta*, Vol. T58, 104–108.

4 Electrochemical sensors

R.O. ANSELL and A. McNAUGHTAN

4.1 Introduction

At the start of the 19th century much of the development of electro-chemistry revolved around the study of chemical transformations by electrolysis, which then led to the quantitative analysis by extensive electrolysis. The early work of Faraday, Fick, Nernst and others laid the foundations for the development of polarography and voltammetry in the early 20th century, by which time potentiometric and amperometric techniques, with their improved specificity over conductance techniques, were being more widely developed. Around this time the analytical chemist was also becoming more interested in the determination of trace substances in samples and in 1929 Heyrovsky and Berezicky demonstrated the first amperometric titration using a dropping mercury electrode. Since then, steady progress has led to a wide assortment of electrochemical techniques and instruments being developed for use in an ever-widening range of applications.

Many of today's sophisticated electrochemical measuring instruments offer an impressive specification which however, does not always meet the specific needs of the analytical chemist involved in environmental measurements. In particular, the elaborate support facilities that are often required restricts their use in remote locations and where on-site measurements are required. However, some of the more recent developments in electrochemical sensing offer a number of advantages over many of the alternative sensor technologies. It is the aim of this chapter to present some of the more recent developments in electrochemical sensing methods that offer specific advantages in the field of environmental measurement.

4.2 Voltammetric and potentiometric techniques

4.2.1 Background

Electrochemical methods can be conveniently grouped into three main categories: potentiometric, voltammetric and conductimetric. Each group has its own advantages and disadvantages although they can all provide qualitative and quantitative data.

4.2.1.1 Potentiometric techniques. Potentiometric methods require the measurement of the potential of the working electrode with respect to a reference electrode when no current flows. The activity and hence concentration of an ion may be determined from the electrode potential. This method has the advantage that there is no net consumption of the analyte and, apart from stirring to ensure solution homogeneity, the resultant potential difference is independent of mass transport of the analyte to the electrode. One of the most common types of potentiometric sensor is the pH sensor. Ion-selective electrodes fall into this category and offer the ability to selectively monitor certain ions in solution. By incorporating suitable membranes, gas sensing and biosensing devices can also be produced.

4.2.1.2 Voltammetric techniques. Voltammetric techniques are based on the measurement of current when a potential is applied across the electrochemical cell. Voltammetric techniques represent a wide range of methods including AC and DC polarography, linear scan and potential step voltammetry, anodic and cathodic stripping analysis, and amperometry. Developments in semiconductor technology and the electronics industry have enabled voltammetric methods to be developed to such an extent that it is now one of the most sensitive analytical techniques available. In addition, the simultaneous detection of several analytes may also be possible in a single scan within a few seconds. One of the most common voltammetric sensors is the dropping mercury electrode used in polarography.

4.2.1.3 Conductimetric techniques. Conductimetric techniques are based on the measurement of the conductance of a solution by applying an AC potential between two electrodes in the solution. The presence of an ionic species will result in an increase in the conductance of the solution. This method is widely used in the determination of water purity, in titrations and in chromatographic applications.

4.2.1.4 Reference electrodes. Inherent in all electrochemical measurement systems is the basic process of oxidation or reduction of a species at a working electrode. The reaction may be either an anodic process in which the species is oxidised at the electrode with a subsequent loss of electrons or the reaction may be a cathodic process in which a species is reduced at the electrode with a subsequent gain of electrons. In analytical applications, the reaction at only one electrode is normally of interest. A second fixed potential electrode, normally referred to as a reference electrode, is required to complete the electrochemical cell. The internationally accepted primary reference is the standard hydrogen electrode. This is a rather impractical electrode for everyday use and, therefore, measurements are often made with respect to a class II electrode such as the saturated calomel electrode or silver/silver chloride electrode. The saturated calomel electrode (SCE) is the most frequently used reference electrode in electrochemical measurements. However, this electrode contains a mercurous paste and is, therefore, potentially toxic. This may preclude its use in certain applications. Both the silver/silver chloride electrode and the saturated calomel electrode have the disadvantage of the possible leakage of chloride ions from the internal electrolyte into the solution, which may present a problem in some applications.

4.2.2 Applications

The main application of conductimetric measurement systems is in monitoring the purity of water. They can also be used to monitor ions that are difficult to detect by potentiometric and voltammetric methods. One of the main differences in this method compared with potentiometric and voltammetric methods is that conductimetric measurements are concerned with the bulk solution, whereas potentiometric and voltammetric methods are concerned with reactions at the electrode surface. Conductimetric methods are well established and documented and are, therefore, only mentioned for completeness.

Miniature conductivity cells have also been developed using micro-fabricated interdigitated electrode arrays with dimensions of a few square millimetres and digit widths of 5 μm or less. Models which provide useful design rules have been developed for fabricating planar conductivity cells by Sheppard et al. (1993). These sensors offer the possibility of selectivity by coating the electrode with a suitable film or enzyme that responds to the substance of interest. Kolesar has reported a detection limit in the parts per billion range for NO_2 using an interdigitated gate electrode field effect transistor that was coupled to an electron-beam evaporated copper phthalocyanine thin film (Kolesar and Wiseman, 1989). The application of these cells in gas sensing is discussed further in Volume 1, Chapter 5. The main limitation of these thin film sensors is their lack of immunity to interferents.

Fuel cell type electrochemical gas sensors have recently been developed using a sol–gel derived composite carbon–ceramic material capable of detecting SO_2, CO_2, and O_2 (Ovadia and Tsionsky, 1995). Since the potentials of the reduction of O_2 and CO_2 and the oxidation of SO_2 are sufficiently far apart, the sensor is capable of detecting all three gases simultaneously using ordinary polarographic methods.

Potentiometric methods of analysis using ion-selective electrodes are widely used for both anions and cations in solution. Ion-selective electrodes allow the measurement of a wide range of substances including Li^+, Na^+, K^+, Rb^+, Cs^+, NH_4^+, NR_4^+, Ag^+, Ti^+, Ca^{2+}, Ba^{2+}, Cd^{2+}, Cu^{2+}, Pb^{2+}, S^{2-}, F^-, Cl^-, Br^-, I^-, SCN^- and CN^- and also amino acids and complex organic molecules. A comprehensive list of applications and procedures has been collated previously (Cammann, 1979). The most common ion selective electrode is the pH electrode. By combining the pH electrode with a suitable gas permeable membrane, SO_2, CO_2, NH_3 and NO_2 measuring sensors can be constructed. HF, H_2S and HCN can also be measured using the appropriate ion-selective electrode. Modified ion-selective electrodes have also been used to measure pesticides (Kumaran and Tran-Minh, 1992). More recently, an extensive review of potentiometric methods of water analysis was undertaken (Midgley and Torrance, 1991) while applications in soil and water analysis have been considered by Yu and Ji (Yu, 1993).

While the use of ion-selective electrodes is the major application of potentiometric methods, potentiometric stripping analysis can also be used for the determination of heavy metal concentrations in sea water, sludge and sediments (Jagner and Kerstin, 1979). Potentiometric stripping analysis uses a reduction pre-concentration step followed by re-oxidation of the metal using an oxidising agent in the solution. During the oxidation stage, the potential of the working electrode is measured as a function of time. Potentiometric stripping analysis has the advantage that it is less affected by adsorption of organic matter on the electrode; however, long deposition times are required for concentrations below the parts per billion range. Selective deposition of the analyte can provide immunity to inorganic and organic interferents and has been investigated in the determination of antimony in environmental samples using a glassy carbon mercury film electrode (Adeloju and Young, 1995). A gold-coated screen-printed electrode has been developed for the measurement of lead that avoids environmental contamination associated with the disposal of mercury electrodes (Wang and Baomin, 1993). Typical oxidising agents are O_2, Cr^{6+} or Hg^{2+}. It should be noted that when the concentration of the oxidant is high, the stripping process can be so fast that the response time of conventional chart recorders is insufficient and alternative high-speed data acquisition systems are required. Controlled temperature and also convection of the oxidant to the electrode is required to obtain reproducible results. The technique has the advantage of simple and easily automated instrumentation. Voltammetric methods are capable of detecting a wide range of analytes in environmental applications. One of the voltammetric methods that is of significant importance in environmental applications is stripping analysis for the determination of heavy metals, which are often considered to be one of the most harmful pollutants since they are not biodegradable (Wang, 1985). They may be present in air, soil and, increasingly more importantly, the water system, where stripping analysis has been used to measure metals at trace levels in oceans, lochs and rivers. Stripping analysis can provide extremely sensitive determination of heavy metals in the environment and is one of the few techniques which can determine heavy metals directly in natural water. If microelectrode techniques are used, they can offer the possibility of *in situ* analysis without the need for added supporting electrolyte and, therefore, this avoids contamination problems and changes in the physiochemical nature of the species (Ansell *et al.*, 1991, 1993). Stripping analysis has been widely used to measure a range of metals including zinc, lead, cadmium, copper, mercury, nickel, antimony, bismuth, thallium, uranium, and indium. However, in many applications, consideration has to be given to organic interferents and sample pre-treatment procedures such as acid digestion or irradiation with UV light are generally required to destroy the organic matter. The application of stripping analysis in environmental measurements including water analysis, rain and snow analysis, airborne particulate analysis, geological samples, fly ash, rocks, minerals, sediment analysis and also food analysis have been well documented (Wang and Baomin, 1992).

As well as potentiometric stripping methods, anodic and cathodic stripping methods have been developed. Anodic stripping voltammetry consists of a pre-concentration step in which metal ions are reduced at the electrode followed by a stripping step in which the metal is oxidised and removed from the electrode. The peak current produced at the oxidation potential of the metal is proportional to the concentration of the metal in solution. As well as measuring heavy metals in solution, anodic stripping voltammetry has also been used to simultaneously determine heavy metals in airborne particulates. The airborne particulates were collected on membrane filters and, because of the high sensitivity of the method, only relatively small samples were required. Other applications of anodic stripping voltammetry are the measurement of toxic and carcinogenic organic substances such as phenols, amines, benzidines and chlorophenols. Antifouling agents containing organotin fungicides can also be detected using an anodic stripping method in which an organotin free radical is formed and adsorbed onto a mercury electrode.

In cathodic stripping voltammetry, the pre-concentration step involves the oxidation of the analyte on the surface of the electrode as an insoluble film followed by the reduction of the film during the negative going stripping step. This method is particularly suited to the measurement of organic analytes or ions capable of forming insoluble salts. In particular, toxic organics from pesticides, insecticides and herbicides are found in runoff water, which may contaminate soil, plants and water supplies and hence enter the food chain. Certain pesticides containing sulphur or chlorine can be detected by this method by pre-concentrating onto a mercury film electrode. The adsorbed layer is then stripped off during a cathodic voltage scan.

One of the main problems in detecting many of these substances by electrochemical means is the elimination of other electroactive interferents. The use of high performance liquid chromatography with electrochemical detection has been very successful in overcoming this problem. The detection of carbamate pesticides in river water has been reported using a Kelgraf microarray electrode in a liquid chromatographic application (Anderson *et al.*, 1985).

The most common amperometric gas sensor is the Clark oxygen sensor. The use of selective membrane filters has extended the selectivity and sensitivity of amperometric gas sensors. The development of CHEMFET sensors, in which a catalytic coating on the gate alters the drain current as a result of adsorption of a gas, has enabled the detection of NH_3 or H_2S in air with detection limits in the parts per million range.

Electrochemical measurement systems offer a number of advantages over alternative measurement techniques, such as spectroscopic methods, in the environmental applications and in particular in *in situ* measurements or on-line monitoring. The range of applications is vast, and recent developments, in most notably biosensors, microelectronics fabrication techniques and miniaturisation, is likely to extend their use even further (Fleet and Gunasingham, 1992). One of the main difficulties associated with electrochemical sensors is the long term stability and reproducibility of the sensor. A variety of techniques to maintain stability and reproducibility including mechanical cleaning, potential cycling and protective semi-permeable membranes have been described elsewhere (Tenygl, 1984). As a consequence of these difficulties, disposable screen printed electrodes have been developed. This has enabled novel amperometric sensors to become a commercial viability and has had a significant impact on the availability of portable amperometric measurement systems.

4.3 Microelectrode voltammetric sensors

4.3.1 The microelectrode sensor

Microelectrode sensors are small electrodes which are generally considered to have at least one dimension with a size in the 0.1–50 μm range. Microelectrode sensors exhibit a number of advantages over conventional sized electrochemical sensors that makes them suitable for environmental applications. Their small physical size, low cost, low power requirements and simplified measurement systems make them portable and eminently suitable for remote applications. In addition, ohmic drop (iR) losses are reduced and steady-state diffusion conditions are rapidly established, which enables them to be used in novel applications that are not amenable to larger conventional sized electrodes. In particular, they offer the possibility of measurements of environmental significance in, for example, natural water without added supporting electrolyte and in possibly hostile and remote locations. Microelectrodes have also been demonstrated as sensitive detectors in the gas phase (Pons et al., 1986; 1987). The unique properties of microelectrode sensors have been well documented in the literature (Wightman, 1981; Fleischmann and Pons, 1987; Fleischmann et al., 1987; Rolison, 1987; Bond et al., 1989; Wightman and Wipf, 1989).

4.3.2   Principle of operation

Under mass transport controlled conditions, the current associated with the oxidation or reduction of a species in solution is proportional to the concentration of the species. There are three principal modes of mass transport to the electrode.

• Migration is the result of the action of an electrostatic force on the charged particles in solution when they are subjected to a potential gradient. This effect can be minimised by ensuring that there is sufficient supporting electrolyte in the solution to carry the charge.
• Convection is the result of the thermal agitation of the solution, which results in movement of the species in solution. Microelectrodes are very much less susceptible to the effects of convection, compared with conventional sized electrodes, since they rapidly attain a steady-state condition and have a thin stagnant layer of solution adjacent to the electrode surface.
• Diffusion occurs when there is a concentration gradient between the electrode surface and the bulk solution. At larger conventional sized electrodes, a non-steady-state planar diffusion field is established. In analytical measurement systems, diffusion is normally controlled by rotating the electrode in solution. At a disc microelectrode of a few micrometres diameter, a steady-state diffusion field is attained at a stationary electrode in a very short time interval.

The mass transport to a microelectrode is essentially controlled by the nature of its small physical size. The geometric shape and size of the electrode, therefore, plays a significant role in the nature of the response of the electrode.

The equilibrium potential for a reversible electrochemical process is given by the Nernst equation:

$$E = E^{\circ} + \frac{RT}{F} \ln\left[\frac{C_O}{C_R}\right] \tag{4.1}$$

where E is the equilibrium potential; E° the standard electrode potential; F, the Faraday constant; C_O the concentration of the oxidised species, C_R the concentration of the reduced species; R, the molar gas constant and T, the absolute temperature.

If the electrode potential is altered, the subsequent reaction at the electrode will produce a change in concentration at the electrode surface which will re-establish the equilibrium condition. Hence there will be a concentration gradient at the electrode and also an associated electron transfer, which will be proportional to the concentration of the species in solution. The concentration gradient at the electrode surface is obtained by solving Fick's second law, which for the case of a spherical electrode is described in spherical co-ordinates by:

$$\frac{\partial C(r, t)}{\partial t} = D\left[\frac{\partial^2 C(r, t)}{\partial r^2} + \frac{2}{r}\frac{\partial C(r, t)}{\partial r}\right] \tag{4.2}$$

where C is the bulk concentration and r, the distance from centre of electrode, D is the diffusion coefficient, and t is the time.

Applying the appropriate boundary conditions for the application of a potential step that will cause the surface concentration of the analyte to go to zero gives the current i as a function of time:

$$i(t) = \frac{nFADC^\infty}{r_0} + \frac{nFAD^{1/2}C^\infty}{\pi^{1/2}t^{1/2}} \tag{4.3}$$

where r_0 is the radius of the electrode (cm), C^∞ is the bulk concentration mol cm^{-3} and A is the area of the electrode (cm^2).

This indicates the current response following the application of a potential step consists of a time-independent term and a time-dependent term. At short timescales, the diffusion profile is dominated by linear diffusion to the electrode and the current is given by the Cottrell equation:

$$i(t) = \frac{nFAD^{1/2}C^\infty}{\pi^{1/2}t^{1/2}} \tag{4.4}$$

where the terms have the same meaning as before.

At sufficiently long timescales, spherical diffusion to the electrode is dominant and a steady-state current is given by:

$$i(t) = \frac{nFADC^\infty}{r_0} \tag{4.5}$$

where the terms have the same meaning as before.

Depending on the size of the electrode, either the steady-state term or the time-dependent term will be dominant. The time-dependent term will be dominant for a large electrode, which will behave as a planar electrode, while the steady-state term will be dominant for a very small electrode.

Virtually any shape of inlaid electrode will result in a diffusion-limited current provided it is bound around its perimeter. However, the time taken for such an electrode to reach a steady-state condition will depend on the size of the electrode (Bond et al., 1989).

The above description refers to a spherical electrode at which the current density is uniform. However, with the exception of the hanging mercury electrode, spherical electrodes are not easily constructed. An alternative and commonly used electrode geometry is the inlaid disc electrode, which has the advantage of simplicity and ease of construction using a range of electrode materials. At an inlaid disc electrode, the current density is not uniform since the perimeter has easier accessibility for the electroactive species, which results in a higher current density compared with the central region of the disc. Equation 4.3 must, therefore, be modified to account for different electrode geometries (Wightman *et al.*, 1980). It has been shown that for a Nernstian process at an inlaid disc microelectrode, a quasi-hemispherical steady-state diffusion current is readily established. The steady-state diffusion-limited current at a disc microelectrode is given by:

$$i = 4nFr_0DC^\infty \tag{4.6}$$

where i is the diffusion limited current and the other terms have the same meaning as before.

Initially at an inlaid disc microelectrode of a few micrometres diameter, a planar diffusion field is formed, followed by a hemispherical diffusion field a short time after. A spherical or quasi-spherical diffusion field is, therefore, readily established at a disc microelectrode and, provided the electrode radius is small, the time-invariant term will be dominant. The concentration of a species in solution can, therefore, be readily determined by measuring the diffusion-limited current. Historically, microelectrode sensors were used to measure the diffusion coefficients of analytes in solution.

Conventional size electrode measurement systems normally use a third electrode to compensate for iR drop in the solution. The basic three terminal electrochemical cell consists of a working electrode, a reference electrode and an auxiliary electrode often referred to as a counter electrode. The principle of operation of the three electrode electrochemical cell can be explained by referring to Figure 4.1. The potential to be applied to the cell is applied to the non-inverting input of the operational amplifier. The operational amplifier operates in such a way that the output is driven to ensure the potential of the inverting input will be equal to the potential of the non-inverting input. A current is, therefore, driven through the auxiliary electrode, which ensures no current flows through the reference electrode and the potential difference between the working electrode and the reference electrode will be equal to the applied potential. The potential of the working electrode with respect to the reference electrode is, therefore, well defined and will not be influenced by the iR drop in the cell.

Since the current associated with the microelectrode is very small, the impedance associated with the electrode process is significantly larger than the ohmic drop in solution and the counter electrode impedance.

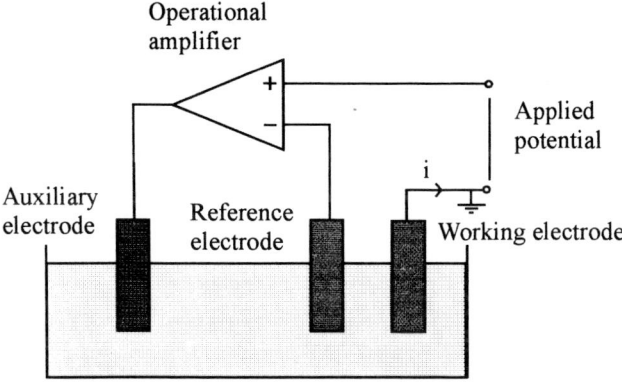

Figure 4.1 Three electrode electrochemical cell.

Consequently a simplified two electrode measurement system consisting of a working electrode and a reference electrode may be used successfully. The two electrode measurement system also has the advantage that it does not require a potentiostat, which can introduce a significant level of noise to the measurement system.

4.3.3   Fabrication techniques

In general, the smaller the electrode the greater the enhancement of the unique microelectrode properties, although at the expense of having to measure smaller currents. The number of publications since the mid-1980s reflects the considerable interest and considerable effort spent in developing methods of fabricating microelectrode sensors (Fleischmann and Pons, 1987; Fleischmann *et al.*, 1987). They have been fabricated with a variety of geometries including disc, micro-ring, thin ring, hemisphere, band, cylindrical, disc arrays and interdigitated arrays, as illustrated in Figure 4.2. They have been constructed using a wide range of materials including gold (Wang and Baomin, 1993; Wu, 1993), carbon paste (Wightman *et al.*, 1980), carbon fibre (Wightman *et al.*, 1980), platinum (Abruna and Pendley, 1990), mercury film (Ewing and Wong, 1990), iridium (Glass *et al.*, 1990), graphite Kel–f (Anderson *et al.*, 1978) and carbon aerogel composites (Wang *et al.*, 1993). Enzyme-modified electrodes have also been widely reported (Hall, 1990) Various fabrication techniques such as sealing wire in glass, photolithographic processes (Glass *et al.*, 1990; Wang and Baomin, 1992), sandwich construction methods (Lay *et al.*, 1990), chemical growth (Ewing and Strein, 1992), encapsulation (Josowicz *et al.*, 1989) and etching techniques (Samuelson *et al.*, 1991) have been widely reported in the

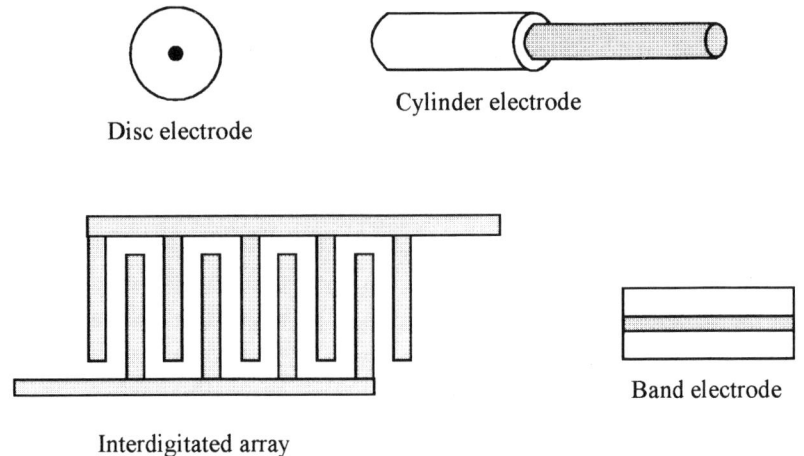

Disc electrode

Cylinder electrode

Interdigitated array

Band electrode

Figure 4.2 Microelectrode geometries.

literature; anyone wishing to construct their own electrodes is encouraged to refer to the many research publications which are available.

One of the most common and easiest methods of microelectrode fabrication, which is reliable, cheap and has been used extensively by the authors, is that of the platinum inlaid disc microelectrode, as shown in Figure 4.3.

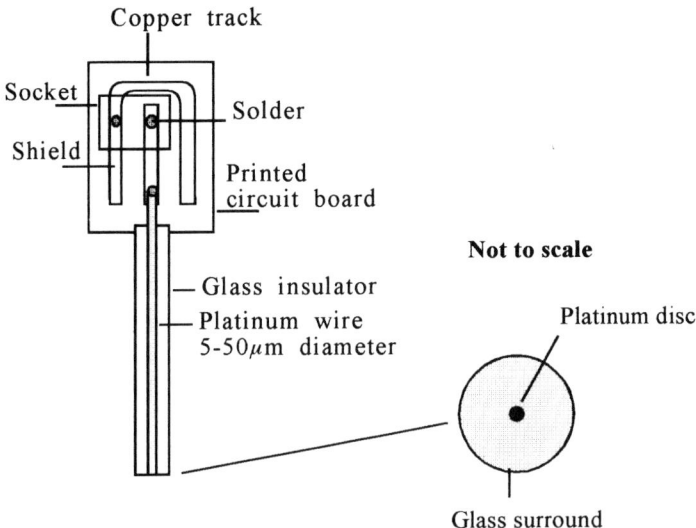

Copper track

Socket

Shield

Solder

Printed circuit board

Not to scale

Glass insulator

Platinum wire 5-50μm diameter

Platinum disc

Glass surround

Figure 4.3 Platinum inlaid disc microelectrode sensor.

High-purity platinum wire (Goodfellows Metals, Cambridge) is readily available down to 10 µm diameter. The wire is placed in a fine glass capillary tube and sealed at one end using a microflame burner. The end is then polished on silicon carbide paper to expose the platinum wire cross-section embedded in a glass insulating surround. A piece of copper clad circuit board at the other end facilitates the electrical connection to the platinum wire.

Smaller diameter electrodes were fabricated by a similar method using Wollaston wire. Wollaston wire consists of a fine core of platinum wire surrounded by silver, which results in an overall diameter of around 25 µm and makes the wire much easier to handle. Microelectrodes of 5 µm diameter were constructed as follows: one end of a 4 cm piece of wire was soldered to a piece of copper-clad printed circuit board. This enables the wire to be handled more easily and also provides a means of making electrical contact to the wire. Approximately 1 cm of the free end was then immersed in 0.1 M nitric acid for 4 min to remove the silver coating. The wire was thoroughly rinsed in nanopure water and then inserted into a soft glass capillary tube of approximately 1.5 mm outside diameter. The glass tube was glued to the printed circuit board at the other end to provide mechanical strength. A few millimetres of the platinum wire was then sealed in the glass tip, as described previously, and then polished on silicon carbide paper.

Microelectrode sensors are also available commercially. EG&G Instruments (Wokingham, Berks) and BAS Technicol (Stockport, Cheshire) can supply gold, platinum or glassy carbon electrodes with a 10 µm diameter wire or fibre sealed in glass. EG&G also market interdigitated microsensor arrays that consist of a microelectrode array which has been micro-fabricated from gold spluttered onto a glass substrate. Interdigitated arrays offer the same diffusion characteristic as a single microelectrode but with an enhanced signal response because of the larger surface area. They are available in monolithic, combined differential and full differential configurations and can also be supplied with either a polypyrrole or polyaniline coating to modify their response. These electroactive polymer films can be further modified by immobilising enzymes or other materials onto the surface to modify the electrode response to specific analytes.

4.3.4 Microelectrode measurement system considerations.

The range of possible applications of microelectrode sensors in electro-analytical measurement systems is vast. However, each application has its own individual complexities that have to be considered when designing a measurement system. It is, therefore, essential to consider each application carefully in order to apply a suitable measurement strategy that will provide the level of sensitivity and resolution required.

Microelectrode analytical measurement systems require the application of a suitable excitation potential and the subsequent measurement of the current associated with the oxidation or reduction of a species in solution. In many cases, there are significant advantages in using microelectrodes. However, there is a trade-off since reducing the size of the electrode also reduces the size of the current to be measured. The ultimate sensitivity will, therefore, be limited by the ability to resolve the signal from background noise. Noise may be introduced into the measurement system by internal noise sources within the instruments, electromagnetically coupled or capacitively coupled noise, and noise resulting from electrochemical effects, such as interferents reacting at the electrode or reacting with products of the reaction being investigated. All these noise sources can be minimised by careful measurement system design and the use of suitable noise reduction techniques (Bixler *et al.*, 1986; Wightman *et al.*, 1991). The effect they have on the measurement is to a large extent dependent on their relative magnitudes; therefore, some of these noise sources only become significant when measuring very low concentrations of analyte.

Microelectrode sensors have the advantage that the very small currents associated with the microelectrode means that a simplified two electrode measurement system may be used rather than the more complex three electrode configuration associated with larger sized electrodes. Also, since there is no need to use a potentiostat, the possibility of the potentiostat introducing noise into the measurement system is eliminated.

When an electrode is immersed in an electrolyte, the charge on the metal electrode is compensated for by oppositely charged ions in solution. This forms a double layer capacitance at the electrode solution interface. When the potential applied to the electrode is changed, then a double layer charging current will flow, in addition to any faradaic currents associated with redox processes at the electrode. However, since the charging current is proportional to the electrode area and the faradaic current is proportional to the electrode radius, the microelectrode exhibits an improved signal-to-noise ratio compared with conventional sized electrodes. Also, since the microelectrode can respond rapidly to changes in the applied potential, they are more suitable for the investigation of fast reactions (Okazaki *et al.*, 1991).

A particular analytical application may preclude the use of certain techniques since the applicability of the measurement systems depends on the analyte, the electrode material and interferents in the system. In order to develop a measurement system for a specific application, considerable work has to be done to investigate each individual situation since different electrode materials, electroactive substances and interferents react in different ways.

Many of the well-established electrochemical measurement strategies such as linear ramp cyclic voltammetry, potential step methods, anodic and cathodic stripping voltammetry, normal pulsed and differential pulsed voltammetry, and constant potential amperometric methods may be used with microelectrode sensors with the advantage of simplified measurement systems. In addition, the smaller double layer capacitance associated with the microelectrode compared with conventional sized electrodes and the rapid establishment of a steady-state diffusion current allows microelectrode sensors to be used with fast scan voltammetric measurement systems (Wightman and Howell, 1984).

Microelectrode sensors lend themselves to *in situ*, miniature and portable measurement systems for environmental applications. The limits of detection of microelectrode sensors in the laboratory are generally in the range 10^{-6} to 10^{-8} mol dm^{-3}. The main limitations of microelectrode sensors in environmental applications is electrode surface fouling and the difficulty in cleaning the electrode surface while still maintaining a high quality seal between the electrode surface and the insulating surround (Bond, 1994). Although elaborate polishing techniques have been reported in the literature, they tend to degrade the seal. As a consequence, disposable low-cost screen-printed microelectrode sensors have been developed (Wang and Baomin, 1993).

4.3.5 Linear ramp cyclic voltammetric measurement systems

Although a wide range of applied potential waveforms may be employed, linear ramp cyclic voltammetry is often the first measurement strategy adopted since it can provide a rapid insight to the redox processes at the electrode and is invaluable in optimising alternative strategies such as pulsed amperometric techniques. It is described here in order to illustrate some of the features of microelectrode sensors. Cyclic voltammetric microelectrode measurement systems have been widely reported in the literature and basically consist of an electrochemical cell, a triangular ramp voltage function generator and a current measuring device. The current flowing through the cell is measured and plotted as a function of the applied potential. By cycling the applied potential between $+1$ V and -1 V, the platinum microelectrode is maintained in a readily reproducible form. A block diagram of a typical microelectrode cyclic voltammetric measurement system is shown in Figure 4.4.

A typical voltammetric response is shown in Figure 4.5. The voltammogram shows a number of different processes that take place at the electrode surface. The hysteresis results from the double layer capacitance, which results in a capacitive current that is proportional to (i) the area of the electrode and (ii) the scan rate of the applied linear ramp. Note the diffusion-limited plateau associated with dissolved oxygen in solution. Encouraging results have been obtained by the present authors in preliminary investigations into the suitability of platinum microelectrodes

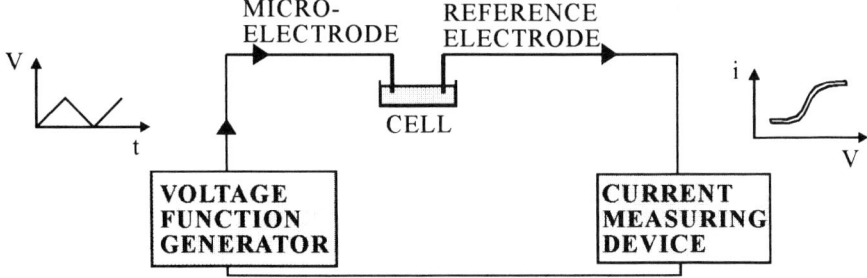

Figure 4.4 Cyclic voltammetric measurement system.

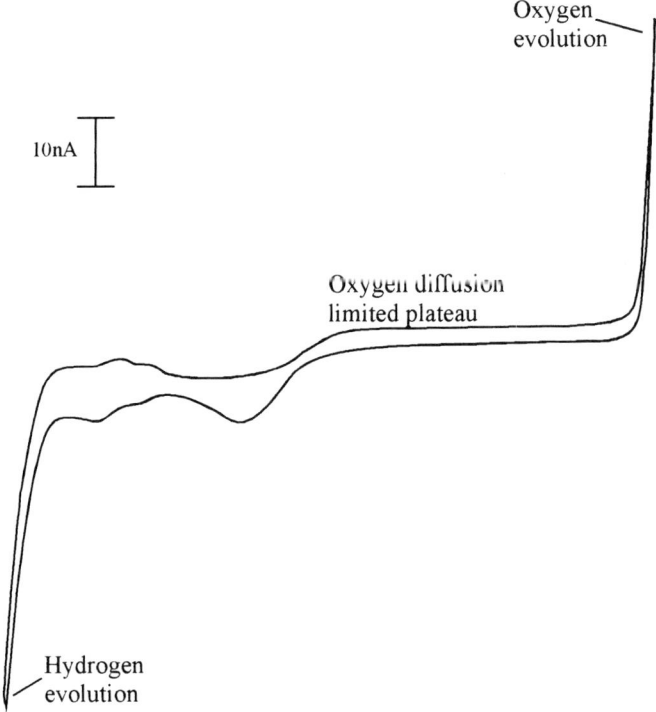

Figure 4.5 Typical cyclic voltammogram. Solution: 1 M KCl; atmosphere: air; working electrode: 25 µm diameter platinum; reference electrode: Ag/AgCl; scan rate: 200 mV s^{-1}; scan limits: +1 V to −1 V.

for the measurement of dissolved oxygen in samples of water and sludge from sea and river beds. Although further development work is required, the potential for microelectrode sensors in these applications is significant.

4.3.6 Anodic stripping analysis

The main application of stripping analysis in environmental applications is the determination of heavy metals. Stripping analysis offers the possibility of the simultaneous determination of several heavy metals down to trace levels, using relatively inexpensive equipment compared with alternative methods such as atomic absorption spectrometry. However, the number of metals detectable is limited to about 30, which is fewer than those detected by spectroscopic methods. Non-metal ions, such as halides, can also be detected using cathodic stripping analysis.

Stripping analysis involves the pre-concentration of the analyte onto the working electrode followed by an electrochemical measurement of the concentrated analyte. The metallic ion is reduced at a negative potential and deposited on the electrode. The pre-concentration step results in a significant increase in concentration of the metal ion at the electrode and hence an improvement in sensitivity over conventional polarographic methods. Reliable and versatile instruments are widely available and are suitable for automatic on-line monitoring. These instruments generally use a mercury electrode, which forms an amalgam with the metal in solution.

In comparison with conventional sized electrodes, the microelectrode sensor allows a stable deposition current without stirring or rotation of the electrode and is particularly suitable for trace metal ion determination in natural lakes and rivers. Since the concentration of the metal ions is lower than that of other ions in solution, the other ions act as supporting electrolyte, allowing the electrodeposition of the ion of interest. The ohmic drop effects in solution are minimised as a result of the small size of the electrode.

Anodic stripping voltammetry requires the electrode to be held at a negative potential relative to the reference electrode such that metallic ions in solution are deposited on the working electrode. This pre-concentration process enables very low concentrations of metallic ions to be determined by extending the deposition time. The second stage of the process requires the potential of the working electrode to be increased in a positive direction and the current that flows through the cell measured. When the working electrode reaches the oxidation potential of the metal deposited on the electrode, the current increases rapidly as the metal is removed from the electrode. The height of the current peak is proportional to the concentration of the metal ion in solution and the oxidation potential provides a means of identifying the metal species. A typical anodic

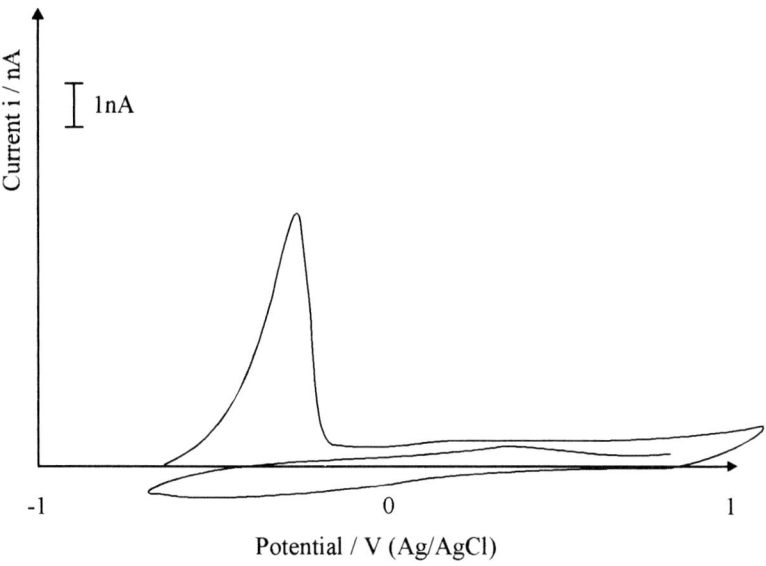

Figure 4.6 Anodic stripping voltammogram for the platinum microelectrode. Deposition time 10 min; lead concentration 0.6×10^{-6} mol dm^{-3}; electrode 25 μm diameter platinum, reference electrode: Ag/AgCl.

stripping voltammogram for a platinum microelectrode in a solution containing lead is shown in Figure 4.6.

The usefulness of the microelectrode technique has been demonstrated by the determination of lead in Glasgow rain water, without the addition of supporting electrolyte (Ansell *et al.*, 1991). Impurities in the water were sufficient to provide supporting electrolyte for the small currents associated with the microelectrode. A diagram of the measurement system is shown in Figure 4.7.

Investigations were made by direct deposition on a 0.005 mm diameter platinum disc microelectrode. Platinum microelectrodes were chosen since mercury electrodes were considered unsuitable for *in situ* analysis. A two-terminal cell was used with a perspex lid, which held the microelectrode and the saturated calomel reference electrode. The cell contained 10 ml of solution and was de-aerated with white-spot nitrogen. Voltammetric measurements were made using a voltage scan unit (Oxford Electrodes). The current flowing through the cell was measured as the potential drop across a 1 MΩ series measuring resistor. The measured potential was amplified by a PAR 113 voltage pre-amplifier (Princeton Applied Research) and the current was plotted as a function of applied potential on an x–y chart recorder. The potential of the microelectrode was held at -0.7 V

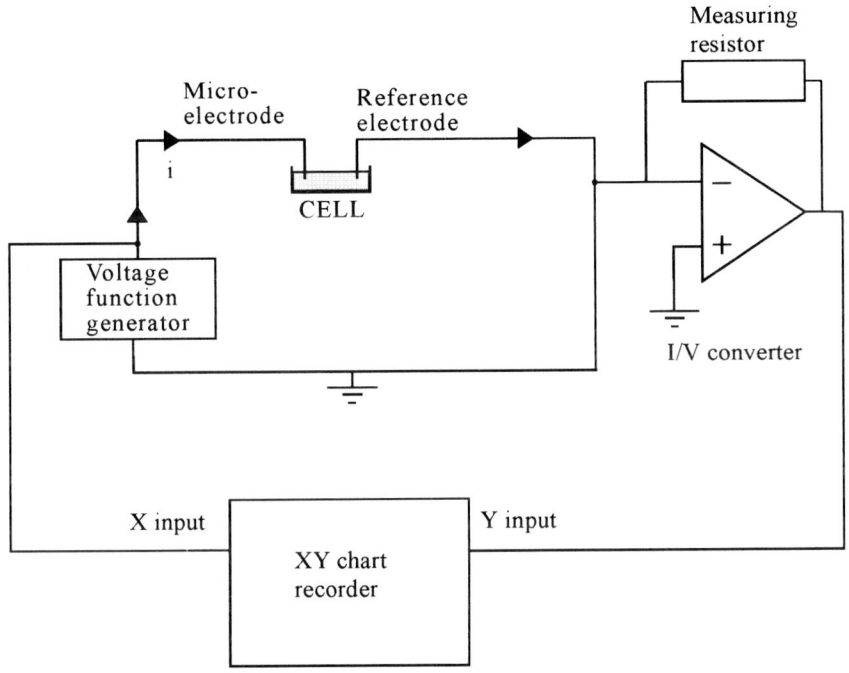

Figure 4.7 Anodic stripping analysis measurement system.

relative to the saturated calomel electrode for 40 min and, following this pre-concentration stage, an anodic scan was initiated using a linear ramp potential with a slope of $200\,mV\,s^{-1}$. Voltammetric peak currents of 400 pA were obtained for lead in a sample of Glasgow rain water. However, the use of platinum microelectrodes is not straightforward in this application. The electrode has to be conditioned prior to deposition by holding the electrode at a potential of $-0.7\,V$ versus the saturated calomel reference electrode for 30 min while outgasing the solution. Since platinum has the ability to chemisorb oxygen, dissolved oxygen influences the response, and extended outgasing is required to measure down to trace levels (Ansell *et al.*, 1993).

Ewing and Wong (1990) have reported a technique for the *in situ* deposition of a lead film on a carbon ring microelectrode for the determination of lead and cadmium without deliberately added supporting electrolyte.

In summary, microelectrode sensors offer the possibility of *in situ* remote sensing, and consistent lead data can be obtained in the laboratory under carefully controlled conditions using equipment that could readily be transferred to the field. The main limitation of this system with platinum microelectrodes arises from the presence of the oxygen interference, which at present, prevents its use for *in situ* applications. However, screen-printed mercury film electrodes (Wang and Baomin, 1992) and mercury-free gold-coated electrodes (Wang and Baomin, 1993) have been reported and used successfully.

4.3.7   Diagnostic techniques

Impedance spectroscopy has been widely recognised as a powerful diagnostic tool for the investigation of electrochemical systems. By measuring the frequency dispersion of the cell impedance, the various rate-determining components may be resolved. Knowledge of these different rate-determining processes is of considerable value when interpreting the response of the microelectrode and is invaluable in the development of an equivalent circuit model of the electrochemical process. Evidence of sufficient supporting electrolyte and mass transport processes to the microelectrode can be readily obtained from impedance spectra. The technique provides a useful method of characterising electrodes and electrolytes. It is generally not used as an analytical measurement system.

Commercial instruments are available and can be programmed to make these measurements over a wide range of frequencies. As a consequence of the very high impedances associated with the microelectrode sensor, very few, if any, commercial instruments are capable of making proper measurements with microelectrodes. However, by adding a suitable buffer amplifier to these instruments, a low-noise impedance measurement system can be constructed for use with microelectrode sensors. A low-noise measurement system including a buffer amplifier and low-pass filtering is shown in Figure 4.8.

In order to make impedance measurements on a microelectrode system, it is necessary to apply a small amplitude signal, of the order of 10 mV or less, to ensure the response can be assumed to be linear, since the response of the microelectrode is essentially non-linear. The microelectrode must also be in a steady-state condition at the potential at which the impedance measurements are to be made. This can be confirmed by applying a step potential and monitoring the current for a period of time. The micro-electrode has the advantage that a steady-state current is rapidly attained.

In order to achieve low-noise performance, care must be exercised in construction. Special consideration has to be given to connections and circuit lay-out since the design of high-impedance measurement systems introduces problems associated with leakage current and noise pick-up. Short rigid cables should be used to connect the electrode to the buffer amplifier to minimise leakage paths, noise pick-up and capacitive loading. Driven shields should be used to minimise noise and stray capacitance at the input of the buffer amplifier and complete shielding of the circuit and

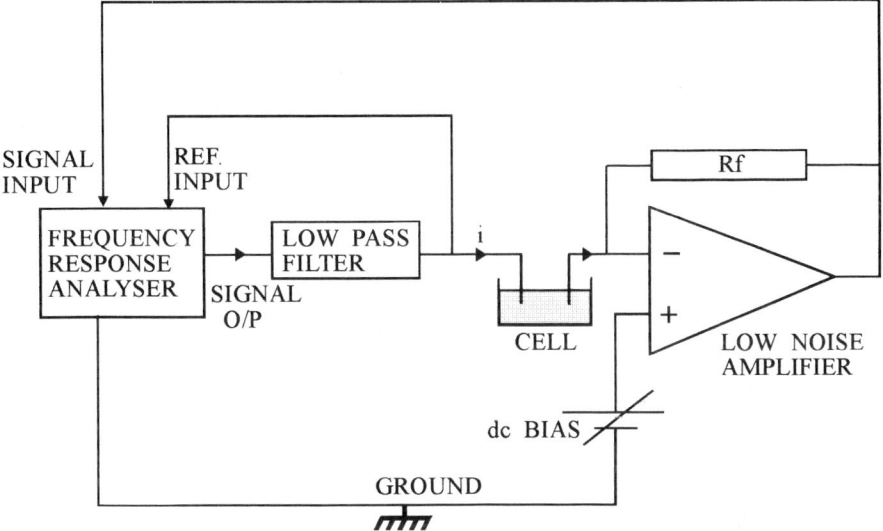

Figure 4.8 Impedance spectroscopy measurement system.

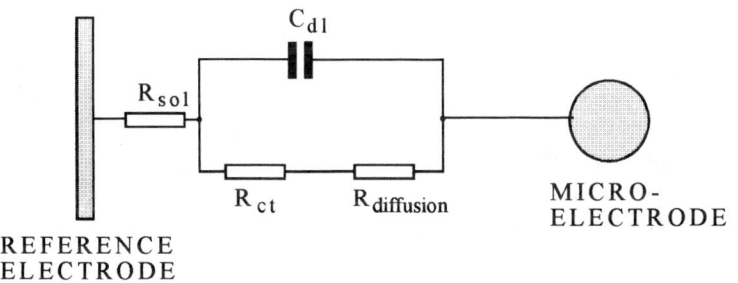

Figure 4.9 Equivalent circuit model.

electrochemical cell can be obtained by placing it in a metal box connected to earth. The measurement system should be verified by using a dummy cell comprising a parallel combination of a resistor and capacitor to represent the double layer capacitance and charge-transfer impedance of a typical electrochemical cell.

An equivalent circuit model of an electrochemical cell, based on the Randles equivalent circuit, is shown in Figure 4.9. R_{sol} represents the solution resistance which can normally be considered as much lower than the electrode impedance if sufficient supporting electrolyte is added. C_{dl} represents the double layer capacitance of the electrode which is in

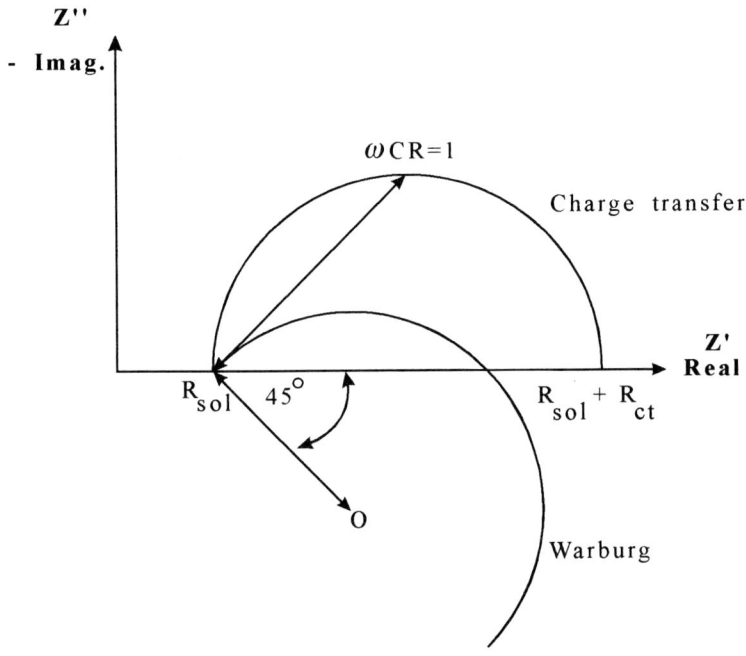

Figure 4.10 Complex plane impedance response.

parallel with a charge-transfer impedance and a microelectrode diffusion impedance. In general, the charge-transfer impedance, the diffusion impedance or a combination will be observed depending on their relative magnitudes. A diffusion-limited current at a conventional sized electrode is represented by a Warburg impedance, which is a frequency-dependent resistance.

The current flowing through a microelectrode electrochemical cell is dependent on a number of parameters, such as the size of the microelectrode, concentration of the analyte or the supporting electrolyte concentration. The expected complex plane impedance response for a charge transfer-limited reaction is a semicircle, which is shifted along the real axis as shown in Figure 4.10. The displacement along the real axis represents the solution resistance and provides a useful means of checking for iR drop due to insufficient supporting electrolyte.

For the case of a spherical diffusion-limited reaction at conventional sized electrodes, the impedance may be represented by a Warburg impedance, which results in a shift of $45°$ below the real axis. Since the microelectrode also exhibits a spherical diffusion field at long timescales, it also results in a $45°$ shift below the real axis. The appropriate equations for

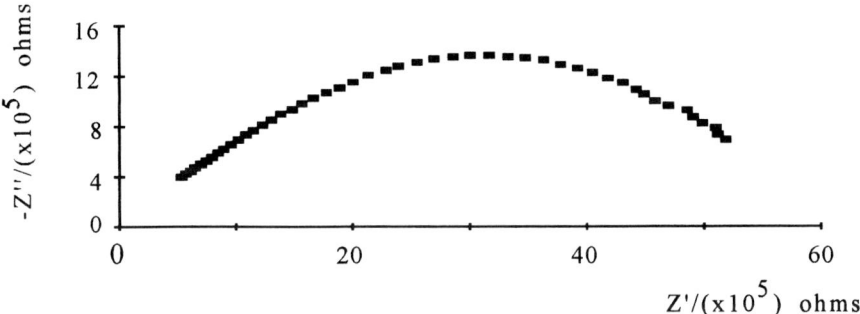

Figure 4.11 Impedance spectrum for 25 μm diameter platinum microelectrode in $1 \, mol \, dm^{-3}$ potassium chloride supporting electrolyte with $5 \times 10^{-3} \, mol \, dm^{-3}$ potassium ferricyanide and $5 \times 10^{-3} \, mol \, dm^{-3}$ potassium ferrocyanide as the electroactive species.

the AC impedance at a disc microelectrode have been developed by Fleischmann and Pons (1988). A typical impedance plot of the real and imaginary components over the frequency range 0.01–50 Hz is shown in Figure 4.11. This shows the characteristic flattened semicircle associated with a diffusion-limited reaction, and the high-frequency intercept indicates that there is no appreciable iR drop as a result of the resistance of the bulk solution.

In summary, AC impedance measurements can be successfully used to investigate and characterise the response of the microelectrode sensor since the iR drop, charge-transfer or diffusion-limited currents, leakage currents or blocking of the electrode may be readily determined.

4.4   Thick film sensors

4.4.1   Chemical environmental sensor arrays

Chemical environmental sensor arrays (CENSAR[tm]) are currently being developed by a research consortium comprising Siemens Plessey Controls Ltd, Unilever UK Central Resources Ltd and the University of Southampton. To date, a multiparameter array has been produced and tested, although it is not yet commercially available. The array contains a temperature sensor, a four electrode conductivity sensor and a pH sensor. A dissolved oxygen sensor is still in development. The array has been developed using thick film technology, which allows repeatable large volume production. The arrays are constructed on an alumina substrate, with the individual sensing elements constructed by printing various layers of inks. The inks are either insulating or contain metal elements. The sensor is aimed at portable instruments for both the field and laboratory water analysis, and for fixed continuous on-line water quality measurements. The array was developed to be multiparameter but either individual elements or multiparameter sensors can be accommodated.

4.4.2 Palintest disposable sensors

A commercially available anodic stripping voltammetric measurement system has been developed by Palintest Ltd (Gateshead, Tyne and Wear) for the analysis of lead and copper in water. The battery operated measurement system uses a disposable pre-calibrated electrochemical sensor and is suitable for remote on-site measurements. On-site measurements are of particular advantage when measuring lead, which is prone to adsorption on the walls of the container being used to transport the sample. The measurement system is automatically controlled by a scanning analyser, which applies the voltage scan to the cell, collects the data, interprets the data and displays the result in $\mu g \, l^{-1}$. The instrument can store up to 300 results in memory and can download to a computer via an RS232 interface.

The single-use disposable sensor consists of a plastic strip that has the three electrodes forming the electrochemical cell printed onto the surface. The electrodes are calibrated during manufacture and a calibration code must be entered into the analyser prior to the measurement being made. The sensor is dipped into the sample, which requires the addition of supporting electrolyte and an appropriate buffer to ensure consistent results. The supporting electrolyte and buffer are supplied in the form of a tablet, which should be dissolved in the sample. The scanning analyser guides the user, controls the measurement and produces a result in 3 min. The measurement system is reported to have a range of 2–$100 \, \mu g \, l^{-1}$ for lead and 70–$2000 \, \mu g \, l^{-1}$ for copper, with a resolution of $1 \, \mu g \, l^{-1}$ for both lead and copper.

4.5 Ion-selective electrodes

Potentiometric measurements using ion-selective electrodes are used to measure selective ion activities at an electrode. In general, potentiometric ion-selective electrodes are limited in sensitivity and versatility compared with amperometric sensors. However, they have the advantage of a simpler measurement system, which may be important for automated or remote applications. The construction principles and characteristics are well documented (Cammann, 1979; Midgley and Torrance, 1991; Yu and Ji, 1993); therefore, only the more recent developments are reported here.

There is no net oxidation or reduction process involved in these measurements and, therefore, there is no depletion of the ion in solution.

The ion activity or concentration can be directly obtained by measuring the potential of the ion-selective electrode relative to a suitable reference electrode. Direct measurement of the e.m.f. requires the use of a high impedance buffer amplifier (10^9 Ω or greater) to ensure no current is drawn from the electrode, since this would alter the concentration of the ion at the electrode surface.

One of the most common types of ion-selective electrode is the glass electrode for pH measurements. Modifications to the glass can enable other ions such as Na^+, K^+ and NH_4^+ to be measured. By changing the glass to a solid-state ionically conducting compound, the electrode can exhibit a Nernstian response to a range of different anions and cations. The fluoride electrode is an example of a solid-state membrane sensor and is well documented in the literature. Solid-state membrane electrodes have the fastest response times of all ion-selective electrodes and are very stable. They can operate successfully for two or three years, although their lifetime can be reduced drastically if they are used in solutions that contain an ion which reacts with the membrane. Interfering ions can form a precipitate on the electrode surface that will increase the response time of the electrode and may alter the stability of the electrode.

A gas-sensing electrode can be formed by using a gas-permeable membrane to isolate the gas from possible interferents. A buffer is then used to trap the gas and convert it to an ionic species that can be detected with an ion-selective electrode. As an example, this technique can be used for the measurement of ammonia gas.

Of interest in environmental applications is the development of ion-selective electrodes modified by a suitable enzyme to form a biosensor. Kumaran and Tran-Minh (1992) have developed an enzyme electrode capable of detecting carbofuran, carbaryl and paraoxon insecticides using a pH electrode modified with butyrylcholinesterase. The presence of the insecticide inhibits the activity of the enzyme, which leads to a change in the electrode potential.

4.5.1   Ion-sensitive field effect transistors

Ion-sensitive field effect transistors (ISFET) or chemically sensitive field effect transistors (CHEMFET) consist of a metal oxide field effect transistor in which the gate oxide of the device has been left exposed and can be placed in direct contact with a solution. The main advantages of ISFETs are the possibility of low-cost mass production and the possibility of a combined sensor and amplifier in a single package.

ISFETs have been developed mainly for pH measurements, although they are not so widely used as conventional glass electrodes. The main limitation of these devices is drift of the output voltage, which limits the precision of the device. At present they are only suitable for non-demanding applications, although they are more robust than glass electrodes. ISFETs may be used with conventional reference electrodes or the reference electrode may be a second ISFET on the same substrate.

ISFETs have also been modified to detect other ions such as Na^+ and Ca^{2+} by applying an ion-selective membrane to the ISFET. A potential is developed across the membrane that is used to control the current flowing through the field effect transistor.

The response of ISFET sensors is poor compared with conventional electrodes. They have a short lifetime and a non-linear response. Their main advantage is their small physical size, which enables then to be used with very small volumes of liquid. The ISFET sensor has not yet found widespread use.

4.6   Summary

Analytical instruments using electrochemical sensors have been developed for the measurement of a wide range of analytes in the laboratory. The theory of electrochemical sensors is well established, and the performance of these devices is suitable for many analytes of interest in environmental applications. The two main areas that require further development are (i) the lowering of detection limits and (ii) the development of miniature, portable systems for *in situ* measurements. Microelectrode sensors offer the possibility of battery operated portable measurement systems for use in remote and hostile locations, such as the sea bed. However, problems with electrode fouling and polishing of the electrode are, at present, restricting their use in 'real' environmental applications. In an attempt to overcome these difficulties, disposable screen-printed electrodes have been developed. They are essentially designed for single measurements and have no advantage in terms of fouling during continuous monitoring. Wang is reported to be developing a novel disposable test strip with gold or carbon working electrodes and a silver reference electrode, which will be connected to the end of a 30 m cable to perform *in situ* monitoring of heavy metals in groundwater (Wang, 1994).

At present, ISFETs have not achieved the performance required for routine use in environmental applications. However, they offer a number of advantages over conventional electrochemical sensors and may be invaluable in environmental applications in the future.

The development of computer-controlled measurement systems and the power of microprocessor-based data analysis systems has enabled electrochemical measurement systems to be made automated; the versatility of these measurement systems is likely to result in electrochemical measurements being more widely used as a powerful and alternative electroanalytical tool.

The present research in the field of electrochemical sensors is likely to have a substantial impact on environmental measurements in the future. These efforts are slowly being translated to practical applications and the successes to date suggest that electrochemical sensors will enable measurements to be made in environmental applications that have so far proved intractable.

References

Abruna, H.D. and Pendley, B.D. (1990) Construction of submicrometer voltammetric electrodes. *Analytical Chemistry*, **62**, 782–784.

Adeloju, S.B. and Young, T.M. (1995) Anodic stripping potentiometric determination of antimony in environmental materials. *Analytica Chimica Acta*, **302**, 255–232.

Anderson, J.L., Anderson, J.E., Tallman, D.E. and Chesney, D.J. (1978) Fabrication and characterisation of a Kel-f–graphite composition electrode for general voltammetric applications. *Analytical Chemistry*, **50**, 1051–1056.

Anderson, J.L., Whiten, K.W., Brewster, J.D., Ou, T.Y. and Nonidez, W.K. (1985) Microarray electrode flow detectors at high applied potentials and liquid chromatography with electrochemical detection of carbamate pesticides in river water. *Analytical Chemistry*, **57**, 1366–1373.

Ansell, R.O., McAleer, H., McNaughtan, A. and Pugh, J.R. (1991) Voltammetric investigations of natural waters using microelectrodes. *Analytical Proceedings*, **28**, 63–64.

Ansell, R.O., McAleer, H., McNaughtan, A. and Pugh, J.R. (1993) The development of techniques for in situ voltammetric analysis of natural waters. *Science of the Total Environment*, **135**, 95–102.

Bixler, J.W., Bond, A.M., Lay, P.A., Thormann, W., van den Bosch, P., Fleischmann, M. and Pons, B.S. (1986) Instrumental configurations for the determination of sub-micromolar concentrations of electroactive species with carbon, gold, and platinum microdisk electrodes in static and flow-through cells. *Analytica Chimica Acta*, **187**, 67–77.

Bond, A.M. (1994) Past, present and future contributions of microelectrodes to analytical studies employing voltammetric detection. *Analyst*, **119**, R1–R21.

Bond, A.M., Oldham, K.B. and Zoski, C.G. (1989) Steady-state voltammetry. *Analytica Chimica Acta*, **216**, 177–230.

Cammann, K. (1979) *Working with Ion-selective Electrodes*. Springer-Verlag, Berlin.

Ewing, A.G. and Wong, D.K.Y. (1990) Anodic stripping voltammetry at mercury films deposited on ultrasmall carbon-ring electrodes. *Analytical Chemistry*, **62**, 2697–2702.

Ewing, A.G. and Strein, T.G. (1992) Characterisation of submicron-sized carbon electrodes insulated with a phenol–allylphenol copolymer. *Analytical Chemistry*, **64**, 1368–1373.

Fleet, B. and Gunasingham, H. (1992) Electrochemical sensors for monitoring environmental pollutants. *Talanta*, **39**, 1449–1457.

Fleischmann, M. and Pons, S. (1987a) The behaviour of microelectrodes. *Analytical Chemistry*, **59**, 1391A–1399A.

Fleischmann, M., Pons, S., Rolinson, D. and Schmidt, P.P. (ed.) (1987b) *Ultramicroelectrodes*, Datatech Systems, Morganton, N.C.

Fleischmann, M. and Pons, S. (1988) The behaviour of microdisk and microring electrodes. Mass transport to the disk in the unsteady state. *Journal of Electroanalytical Chemistry*, **250**, 277–283.

Glass, R.S., Perone, S.P. and Ciarlo, D.R. (1990) Application of information theory to electroanalytical measurements using a multielement microelectrode array. *Analytical Chemistry*, **62**, 1914–1918.

Hall, E.A.H. (1990) Cell-based biosensors. In *Biosensors*, pp. 203–207. Open University Press, Milton Keynes.

Jagner, D. and Kerstin, A. (1979) Potentiometric stripping analysis for zinc, cadmium, lead and copper in sea water. *Analytica Chimica Acta*, **107**, 29–35.

Josowicz, M., Potje-kamloth, K., Janata, P. and Janata, J. (1989) Electrochemical encapsulation for sensors, *Sensors and Actuators*, **18**, 415–425.

Kolesar, E.S. and Wiseman, J.M. (1989) Interdigitated gate electrode field effect transistor for the selective detection of nitrogen dioxide and diisopropyl methylphosphonate. *Analytical Chemistry*, **61**, 2355–2361.

Kumaran, S. and Tran-Minh, C. (1992) Insecticide determination with enzyme electrodes using different enzyme immobilization techniques. *Electroanalysis*, **4**, 949–954.

Lay, P.A., McAlpine, N.S. and Harding, G.L. (1990) Flexible ultra-micro-line electrodes. In *Microelectrodes*, (ed. J Wang), VCH, New York, pp. 235–239.

Midgley, D. and Torrance, K. (1991) *Potentiometric Water Analysis* (2nd edn). Wiley, New York.

Okazaki, S., Nomura, S. and Nozaki, K. (1991) Fabrication and evaluation of a shielded ultramicroelectrode for submicrosecond electroanalytical chemistry. *Analytical Chemistry*, **63**, 2665–2668.

Ovadia, L. and Tsionsky, M. (1995) Electrochemical composite carbon–ceramic gas sensors: introduction and oxygen sensing. *Analytical Chemistry*, **67**, 2409–2414.

Pons, S., Ghoroghchian, J., Sarfarazi, F., Dibble, T., Cassidy, J., Smith, J.J., Russell, A., Dunmore, G. and Fleischmann, M. (1986) Electrochemistry in the gas phase. Use of ultramicroelectrodes for the analysis of electroactive species in gas mixtures. *Analytical Chemistry*, **58**, 2278–2282.

Pons, S., Brina, R. and Fleischmann, M. (1988) Ultramicroelectrode sensors and detectors. Considerations of the stability, sensitivity, reproducibility, and mechanism of ion transport in gas phase chromatography and in high performance liquid chromatography. *Journal of Electroanalytical Chemistry*, **244**, 81–90.

Samuelson, M., Armgarth, M. and Nylander, C. (1991) Microstep electrodes: band ultramicroelectrodes fabricated by photolithography and reactive ion etching. *Analytical Chemistry*, **63**, 931–936.

Sheppard, N.F., Tucker, R.C. and Wu, C. (1993) Electrical conductivity measurements using microfabricated interdigitated electrodes. *Analytical Chemistry*, **65**, 1199–1202.

Tenygl, J. (1984), Electrochemical Detectors. In *Proceedings of Fifth Anglo-Czech Symposium* (ed. J. W. Ryan), Plenum, New York, p. 89.

Wang, J.(1985) *Stripping Analysis. Principles, Instrumentation and Applications.* VCH Publishers Inc., New York.

Wang, J. (1994) Decentralized electrochemical monitoring of trace metals: from disposable strips to remote electrodes. *Analyst*, **119**, 763–766.

Wang, J. and Baomin, T. (1992) Screen-printed stripping voltammetric/potentiometric electrodes for decentralized testing of trace lead. *Analytical Chemistry*, **64**, 1706–1709.

Wang, J. Angnes, and L. Tobias, H. (1993) Carbon aerogel composite electrodes. *Analytical Chemistry*, **65**, 2300–2303.

Wang, J. and Baomin, T. (1993) Mercury-free disposable lead sensors based on potentiometric stripping analysis at gold-coated screen-printed electrodes. *Analytical Chemistry*, **65**, 1429–1532.

Wightman, R.M. (1981) Microvoltammetric electrodes. *Analytical Chemistry*, **53**, 1125A–1134A.

Wightman, R.M. and Howell, J.O. (1984) Ultrafast voltammetry and voltammetry in highly resistive solutions with microvoltammetric electrodes. *Analytical Chemistry*, **56**, 524–529.

Wightman, R.M. and Wipf, D.O. (1989) Voltammetry at ultramicroelectrodes. In *Electroanalytical Chemistry*, Vol. 15 (ed. A. J. Bard), pp. 267–353. Marcel Dekker, New York.

Wightman, R.M., Dayton, M.A., Brown, J.C. and Stutts, K.J. (1980) Faradaic electrochemistry at microvoltammetric electrodes. *Analytical Chemistry*, **52**, 949–950.

Wightman, R.M., Wiedemann, D.J., Kawagoe, K.T., Kennedy, R.T. and Ciolkowski, E.L. (1991) Strategies for low detection limit measurements with cyclic voltammetry. *Analytical Chemistry*, **63**, 2965–2970.

Wu, H.P. (1993) Fabrication and characterisation of a new class of microelectrode arrays exhibiting steady-state current behaviour. *Analytical Chemistry*, **65**, 1643–1646.

Yu, T.R. and Ji, G.L. (1993) *Electrochemcial Methods in Soil and Water Research*. Pergamon Press, Oxford.

5 Gas sensors and analysers

J.R. BATES and M. CAMPBELL

5.1 Gas sensors

Gases encountered in the environment may be harmless, toxic or flammable. Leakage of gases, therefore, may have potentially serious implications for all of us and, in particular, for certain high-risk employees. Where there is a likelihood of hazardous atmospheres, occupational safety legislation requires that employers either supply appropriate breathing apparatus or ensure that the air is safe to breathe. It is for this reason, if no other, that it is necessary to measure the concentrations of potentially hazardous mixtures in a continuous or periodic manner. From the point of view of personnel safety, it is often more important to know that a toxic or flammable gas has leaked into the environment than to be capable of making a precise measurement of very small concentrations of gas.

Nevertheless, many toxic gases have known short-term exposure limits (STELs) and time-weighted averages (TWAs), as indicated in Table 5.1, and it may be necessary to measure precise concentrations as a function of time. It is also necessary to be able to relate the physiological effects of toxic gases to the STELs and TWAs. For example, the physiological effects of two such gases, CO and CO_2, are shown in Tables 5.2 and 5.3, respectively. Flammable gases have known lower explosive limits (LELs) and the values for several common gases are shown in Table 5.4.

In the past, the presence of flammable gases in mines was indicated using canaries and flame lamps. The flame lamp has a small flickering yellow flame that changes to a blue triangular shape in the presence of firedamp (> 90% methane). Because estimation of gas concentration in such circumstances is very subjective, it became necessary to develop portable and fixed installation instrumentation that could produce quantitatively repeatable values on demand. The rigorous nature of inspection and monitoring tasks in the modern oil, gas and chemical-based industries has been a major driving force behind the development of a range of instruments to perform specific measurement tasks within very strict safety regimes.

In general, each gas measuring sensor system will have block diagram arrangement of the type shown in Figure 5.1.

Table 5.1 STELs and TWAs of common gases[a]

Gas	UK[b] STEL (10 min)	UK[b] TWA (8 h)	Germany[c] STEL[(e)]	Germany[c] TWA (8h)	USA[d] STEL (15 min)	USA[d] TWA (8 h)
Carbon monoxide	300	50	60 (30)	30	200	35
Hydrogen sulphide	15	10	20 (10)	10	15	10
Sulphur dioxide	5	2	4 (5)	2	5	2
Nitric oxide	35	25	–	–	–	25
Nitrogen dioxide	5	3	10 (5)	5	1	3
Chlorine	1	0.5	1 (5)	0.5	1	0.5
Chlorine dioxide	0.3	0.1	0.2 (5)	0.1	0.3	0.1
Hydrogen[f]	–	–	–	–	–	–
Hydrogen cyanide	10	–	20 (30)	10	4.7	–
Hydrogen chloride	5	–	10 (5)	5	5	–
Ammonia	35	25	100 (5)	50	35	25
Ozone	0.3	0.1	0.2 (5)	0.1	0.1	–
Ethylene oxide	–	5	–	–	5	1
Phosphine	0.3	–	0.2 (5)	0.1	1	0.3

[a] Courtesy of City Technology Ltd
[b] Health and Safety Executive (HSE)—EH40/95
[c] Deutsche Forschungsgemeinschaft (DFG)-1993
[d] Occupational Safety and Health Administration–Code of Federal Regulations 29 CFR 1910.1000–1910.1200, July 1993
[e] Figures in brackets represent maximum duration of exposure at this level, in minutes
[f] Hydrogen is a flammable asphyxiant; the LEL value is 4%

Table 5.2 Physiological effects of varying levels of CO in air[a]

CO level (ppm)	Resulting conditions/effect on humans
35	US permissible exposure level, 8 h (OSHA)
200	Possible mild frontal headache in 2 to 3 h
400	Frontal headache and nausea after 1 to 2 h
800	Headache, dizziness, and nausea in 45 min; collapse and possibly death in 2 h
3200	Headache and dizziness in 5 to 10 min; unconsciousness and danger of death in 30 min
6400	Headache and dizziness in 1 to 2 min; unconsciousness and danger of death in 10 to 15 min
12 800	Immediate effect unconsciousness; danger of death in 1 to 3 min

[a] Courtesy of City Technology Ltd

Table 5.3 Physiological effects of varying levels of CO_2 in air

CO_2 (% v/v)	Physiological effects
~ 5	Headaches and breathing difficulties
7–10	Increase in depth and rate of respiration in addition to more severe headaches
> 10	Short-term exposure: depression of the respiratory system and the onset of tinnitus, visual disturbances and tremors
	Long-term exposure: death occurs, even in the presence of adequate O_2, because of respiratory system failure
20–30	Almost immediate asphyxiation

Table 5.4 LELs of common combustible gases

Gas	LEL (% v/v in air)
C_9H_{20}	0.8
C_6H_{14}	1.1
C_7H_{16}	1.2
C_6H_{12}	1.3
C_5H_{12}	1.5
C_4H_{10}	1.8
C_3H_8	2.2
C_2H_2	2.5
C_2H_6	3.0
C_2H_4	3.1
H_2	4.0
C_2H_5OH	4.3
CH_4	5.3
CH_3OH	7.3
CO	12.5

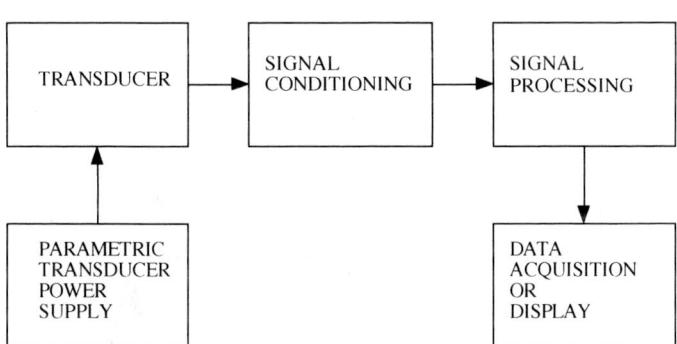

Figure 5.1 Block diagram of the basic sensor system.

5.1.1   Portable gas detectors

These instruments are small and self-contained and are able to produce spot readings in the field using, for example, a hand-aspirated pump and lightweight sampling arm. The reading scales are clear and unambiguous for ease of operator use.

5.1.2   Portable continuous gas monitors

These devices are lightweight but capable of continuous operation when carried around a site. They are designed to warn of any potentially explosive environment into which the operator may have moved as well as to warn of leaks that may occur during on-going work. They are often fitted with audible alarms as well as readout scales.

5.1.3   Fixed continuous gas measurement installations

These are usually located within a site at points where there is a higher probability of gas leakage. As well as operating as gas alarms, these instruments are also able to perform remedial tasks such as gas supply shutdown and extinction.

5.2   Principles of gas detection

Gas detection may be performed using a variety of techniques including catalytic oxidation, thermal conductivity, semiconductor electrical conductivity, modified gate field effect transistors (gasFETs), piezoelectricity, laser-based photoionisation, electrochemistry, biomechanisms, photo-absorption and interferometry. Clearly, several of these techniques are dealt with in detail within this book and, as such, they will only be given a mention here.

An arbitrary definition of a gas sensor is taken to mean a relatively simple and inexpensive device that is intended to be incorporated within an instrument for the purposes of detecting gas concentrations with a lower limit of, say, 100 ppm. A gas analyser, however, is taken to mean a more sophisticated instrument that can measure and record one or more gases in concentrations of lower limit, say, 1 ppm or better.

5.3   Catalytic oxidation gas sensors

This is a very well established technology in gas detection (Firth, 1966; Baker and Firth, 1969; Firth et al., 1973) that relies on the fact that miniature coils of a precious metal such as Pt which has a very high temperature coefficient of resistance, can be made to act as microcalori-

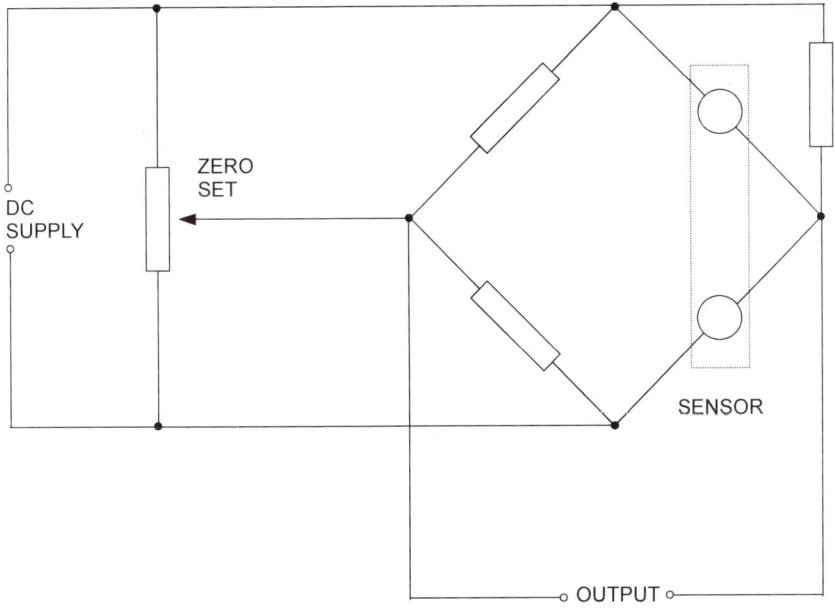

Figure 5.2 Basic circuit diagram for pellistor pairs.

meters. They can be customised to perform sensing via transduction processes including catalytic oxidation and thermal conductivity. These particular sensors are classified as passive (or parametric) transducers and, therefore, require external power supplies. In order to eliminate ambient effects, it is standard practice to incorporate both sensing and compensating coils within a Wheatstone bridge configuration with zeroing capability, as shown in Figure 5.2.

Generally speaking, the fixed resistors in the other two arms of the bridge circuit tend to have values of $\sim 27\,\Omega$ and $\sim 1\,k\Omega$ for loaded and unloaded configurations, respectively, with the sensor coils having cold and hot resistances of $\sim 1\,\Omega$ and $\sim 4\,\Omega$, respectively. The principle of signal transduction is the same for all versions of filament detectors: a change in temperature of the active sensor relative to the compensating one causes the output of the bridge to become non-zero. The graph depicted in Figure 5.3 indicates that the bridge imbalance becomes non-linear for changes in resistance in excess of 10%.

In practice, the catalytic oxidation systems have semi-logarithmic analogue meters or smart digital displays because of the large change in active sensor resistance while those systems based on thermal conductivity, where the change in resistance is much smaller, have near-linear scales, as shown in the inset in Figure 5.3.

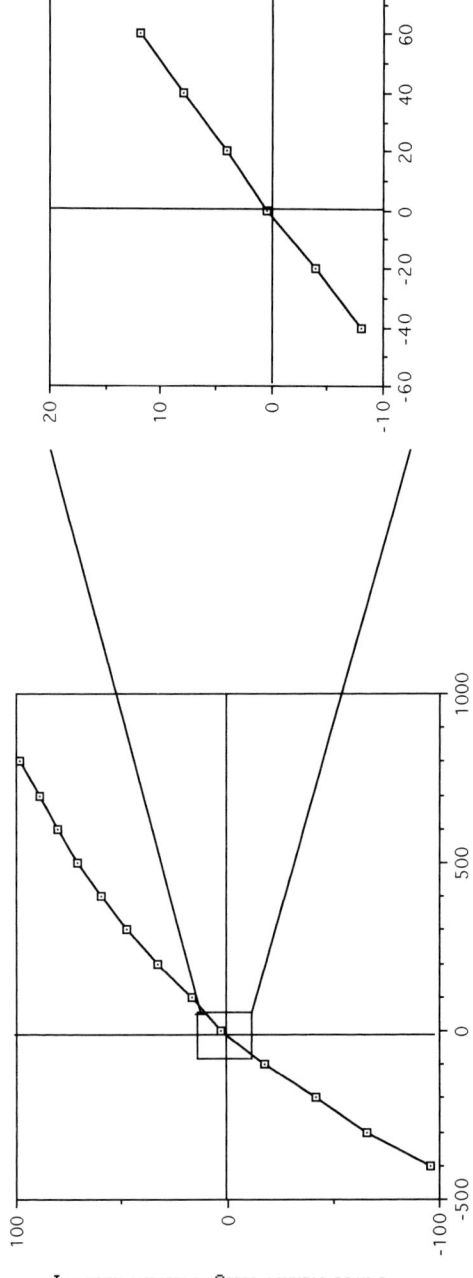

Figure 5.3 Non-linearity of bridge output signal for large changes in resistance. (Inset: linear range for relatively small changes in resistance.)

5.3.1   Unembedded resistance filament sensors

Wire coil transducers can be wound using 25 μm or 50 μm diameter Pt wire and coated using Pt- or Pd-based chemicals. These sensors form the basis of a range of portable explosimeter instruments.

5.3.1.1   MSA Model 5 explosimeter.   This portable measuring instrument, manufactured by MSA (Britain) Ltd, utilises sensors that operate at elevated temperatures. The sensors, which are specially adapted for use with leaded petrol vapours (BASEEFA approval SFA3007 no. Ex 70058 for group II gases and vapours), are calibrated in heptane (C_7H_{16}) and are resistant to poisoning by tetraethyl lead.

5.3.1.2   MSA Model 2E explosimeter.   Model 2E explosimeters incorporate sensing coils that run at lower temperatures and, as such, they are more suited to *go/no go* testing of sites such as manholes, sewers, chemical refineries and plants and paint factories (Certificate of Intrinsic Safety 1S 3329 for gas groups 2a, 2c, 2d and 2e). They should not be used for (i) O_2-enriched atmospheres including oxyacetylene and oxyhydrogen and (ii) O_2-deficient atmospheres.

It should be noted that this type of sensing element is susceptible to silicon compounds such as silicones, silicates and silanes, and if such poisoning is suspected then re-calibration must be carried out before resuming operational mode.

5.3.2   Catalytic oxidation pellistors

These pell(et res)istors or pellistors were established several decades ago (Baker, 1962) as a robust form of filament sensor that has a relatively large surface area for depositing Pd catalyst. The basic design was improved upon by others (Firth and Guest, 1970; Watson, 1970).

5.3.2.1   Pellistor manufacture.   A detailed and comprehensive description of commercial pellistor manufacture is given elsewhere (Jones, 1987). It should be emphasised that these pellistors have been stalwarts to the gas-sensing industry and continue to be the basis of many gas-measuring instruments.

Briefly, coils of 1 mm diameter and 1 mm length are wound using 50 μm diameter Pt wire and successively dipped into aluminium nitrate solution until they become embedded in a small core of refractory α-alumina; the basic geometry is shown later in the Figure 5.9. To produce a balanced instrument, individual pellistors are electrically matched for a range of temperatures (Figure 5.4) and selected for closeness of physical size, shape and surface texture prior to pairing. Catalytic oxidation pellistors are then

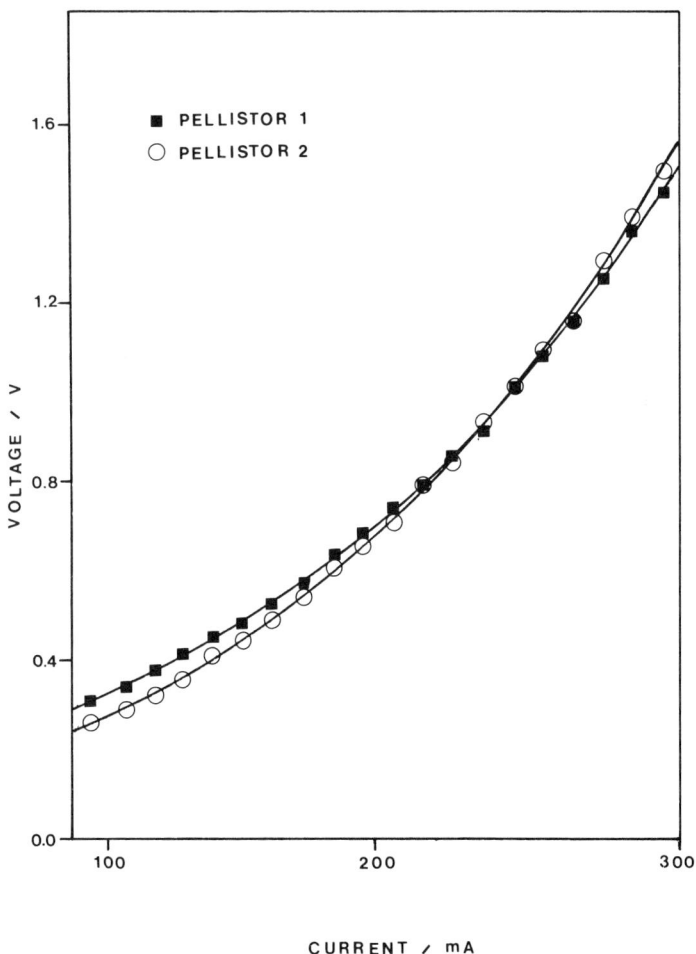

Figure 5.4 Electrical characteristics of two similar pellistors.

coated with a Pt- or Pd-based catalyst while the compensating pellistors have any catalytic tendency suppressed by dipping them in boiling potash solution. Modern pellistors have low power requirements because of their relatively low masses and operating temperatures (400–500°C) (Curry, 1990). The brown or black coloration of the catalytic surface causes the radiative emissivity to differ from that of the white compensating sensor and so to ensure that the bridge is properly compensated, a trimming resistor is usually placed in parallel with the catalytic pellistor. The bridge voltage can be ramped to ensure a good output signal stability as a function of ramping voltage (Figure 5.5).

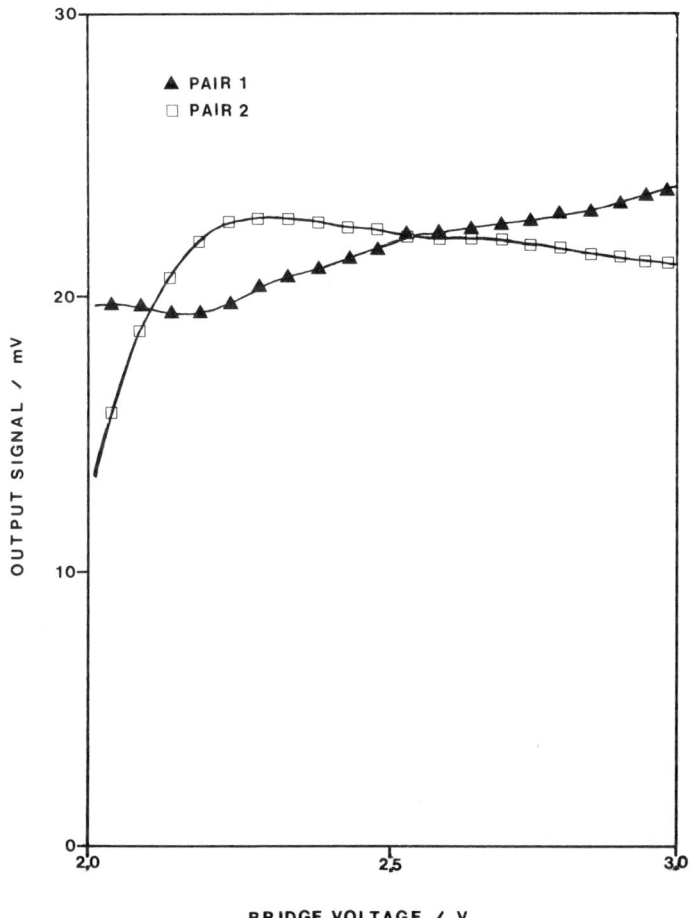

Figure 5.5 Wheatstone bridge operating plateaus.

5.3.2.2 CiTipeLs. City Technology Ltd produces a series of pellistor types, some of which are mounted separately in metal can enclosures (Figure 5.6) while others reside in combined flameproof enclosures (Figure 5.7). This particular manufacturer also produces new series-4 miniature CiTipeLs which have both pellistors mounted in a stainless steel case with integral sinter (Figure 5.8) to enable the product to meet European and North American flameproof approval standards.

CiTipeLs 50N and 90N are low-power devices that enable battery operated instruments to operate for longer periods of time before replacement batteries are required.

The general purpose sensor CiTipeL 200N can withstand low con-

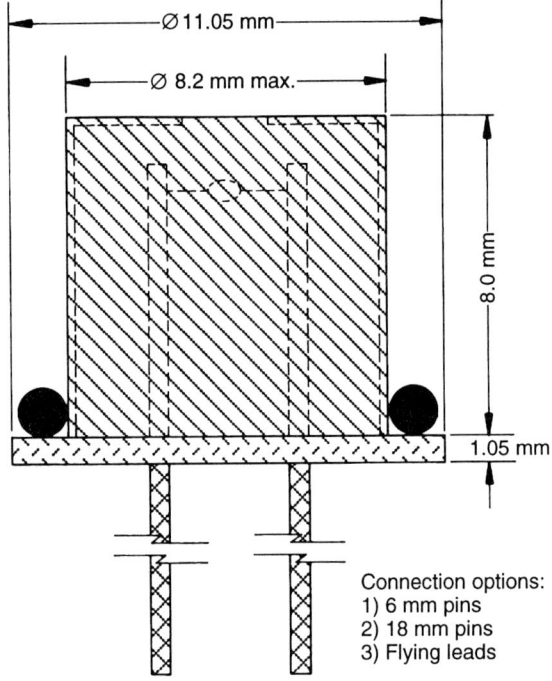

Figure 5.6 Typical pellistor header (courtesy of City Technology Ltd).

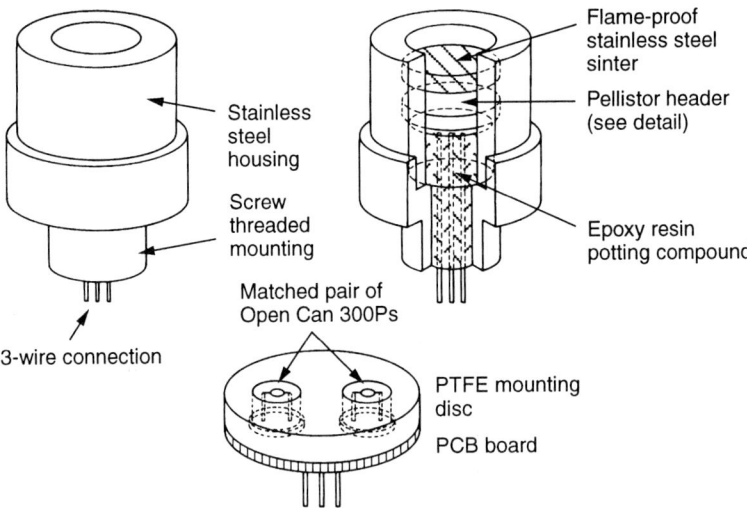

Figure 5.7 Customised pellistor pair headers (courtesy of City Technology Ltd).

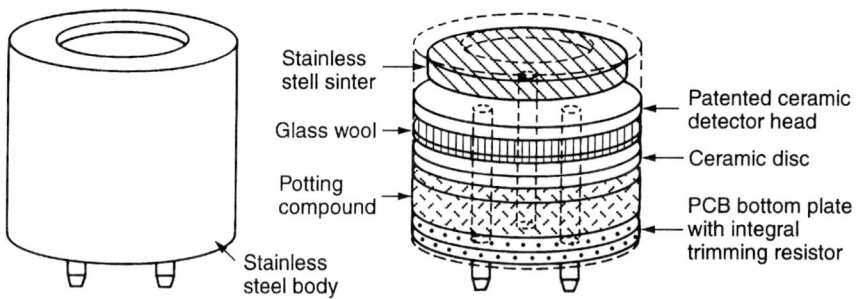

Figure 5.8 Ceramic type pellistor pair headers (courtesy of City Technology Ltd).

centrations of silicone oil vapours but is not specifically poison resistant. All of these devices are used by instrument manufacturers world-wide.

5.3.2.3 Resistance to chemical poisoning. The susceptibility of catalytic oxidation pellistors, especially to virulent silicon-based poisons such as hexamethyldisiloxane (HMDS) has been well studied (Gentry and Jones, 1978; Gentry and Walsh, 1984). The problem has been partly overcome by fabricating porous catalytic pellistors based on a γ-alumina structure (Dabill *et al.*,1982). These devices have a greater size compared with the non-porous variety. In a comprehensive review of the subject of catalyst poisoning (Gentry and Walsh, 1987), it was noted that while the resistance to poisoning depended on (i) the nature of the poison, (ii) the fuel and (iii) the physical and chemical *form* of the noble metal catalyst, it was essentially independent of the actual noble metal catalyst used (Pd, Pt or Rh). In the case of methane, poisons such as silicones, lead- and phosphorus-containing vapours produce an irreversible change. These effects are particularly important since contaminated instruments produce a decreasing output reading so that the device fails to danger mode.

The CiTipeL 300P device is a poison-resistant sensor (Figure 5.9) that can be used in environments where catalyst poisons have a continuous presence. The CiTipeL 4P range of sensors contain the catalytic and compensating sensors within a single enclosure that is surrounded by a stainless steel flameproof mesh. Such devices exhibit reasonable resistance to poisoning as well as low power consumption and meet North American and European flameproof approval standards.

The extensive range of CiTipeLs marketed by City Technology Ltd is summarised in Table 5.5.

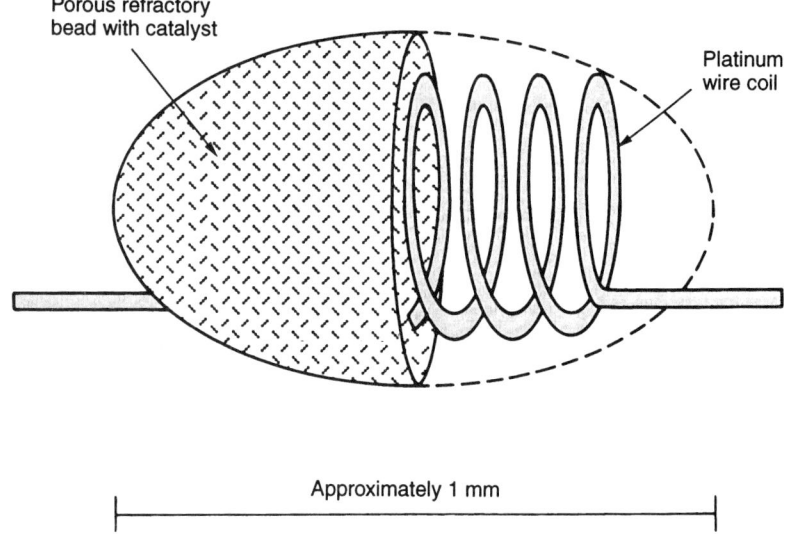

Figure 5.9 Catalytic pellistor architecture (courtesy of City Technology Ltd).

Table 5.5 CiTipeL specifications[a]

Sensor type	Supply voltage (V)	Detector operating current (mA)	Output (mV/% methane)	Poison resistance	T90 response time(s)
50N	4.25 ± 0.1	< 55	43 ± 5	H2S	< 15
90N	3.5 ± 0.1	75	35 ± 5	H2S	< 15
200N	2.0 ± 0.1	180	30 ± 5	H2S	< 15
300P	2.0 ± 0.1	280	13 ± 2 or 10 ± 2	Most poisons	< 15
4P-50	4.25	60	44 ± 10	H2S	< 15
4P-90	3.3	75	34 ± 7	H2S	< 15
4P-100	3	90	30 ± 5	Silicones	< 15
4P-200	2.3	200	17 ± 3	Silicones	< 15

[a]Courtesy of City Technology Ltd

5.3.2.4 MSA model L combustible gas indicator. The MSA model L combustible gas indicator (CGI) is a typical instrument. It is produced by MSA (Britain) Ltd and has a proven track record in flammable gas detection. It can monitor flammable gases in the 0–100% LEL range with measuring resolutions of 0.1% LEL in the range 0–9.9% LEL and 1% in the range 10–100% LEL. Flashback arrestors, dust filters and a pump/hand aspirator are standard features. The sensors have an in-built electronic auto-zeroing feature. MSA CGIs are BASEEFA certified and electronics intrinsically safe to BS5501 Parts 1 and 7. They are designed to perform in accordance with the requirements of BS6020:1981

5.3.2.5 Recent developments in pellistor technology. Two main disadvantages of pellistors are that they (i) consume reasonable amounts of electrical power and (ii) suffer to some extent from a lack of selectivity. The electrical power consumption may possibly be reduced by minaturisation and devices consuming only 50 mW (Gall, 1993) and 100 mW (Krebs and Grisel, 1993) have been reported.

Gall fabricated an $850\,\mu m^2$ device on a silicon substrate that contained a Pt meander pattern. The Pt acted as the catalyst, heater and sensor and was thermally insulated from the silicon by a thin layer of silicon nitride. An integrated array was constructed where each sensor within the array was operated at a different temperature, with one sensor using an Ir catalyst. The effect of the different catalyst was to alter the activation energy of oxidation and to change the rate of reaction. Using pattern recognition techniques, the array could distinguish between methane, pentane, trichloroethylene, methanol, ethanol and CO.

Krebs and Grisel (1993) fabricated a $6\,mm^2$ device on a silicon substrate using a $0.3\,\mu m$ Pt meander as the heater and temperature sensor but with a Pd catalyst. At an operating temperature of 400°C, a sensitivity of 13 mV/% methane was achieved in comparison with 9 mV/% for most devices.

One major difficulty with pellistor devices is that when two or more combustible gases are present these give rise to different temperature increases. Hence the signal from 1% methane is higher than for 1% xylene. In general, the resistance change for a given percentage of gas decreases as the molecular weight of the gas increases. This obviously presents problems in calibration and leads to inaccurate measurements if two or more gases are present. To overcome this problem, the potential applied across the heater wire may be altered during the course of the measurement thereby altering the temperature of the catalyst (Iredale, 1990). Readings can then be resolved for the individual gases or integrated over time so that the signal produced is approximately the same for unit amounts of many gases.

5.4 Thermal conductivity pellistors

Pairs of electrically matched compensating (catalytically suppressed) pellistors are extensively used in the gas industry to measure the presence of volume leakage of gas. This is often necessary since, for flammable gases, catalytic pellistors record low readings in safe conditions (very small leakage) and also in extremely dangerous conditions (high leakage). The reason for this ambiguity is straightforward – insufficient fuel and

Table 5.6 Thermal conductivities of some gases

Gas	Thermal conductivity $(mJ\,s^{-1}\,m^{-1}\,K^{-1})$
Air	2.4
Oxygen	2.3
Hydrogen	13.2
Argon	1.6
Carbon dioxide	1.45
Methane	2.7
Helium	14.3
Chlorine	0.7

insufficient O_2, respectively. The principle of operation is based on the fact that different gases have different thermal conductivities (K), as illustrated in Table 5.6. To utilise this property of gases, it is necessary to keep one pellistor in a air reference cell while the other is exposed to the gas. Depending on the gas in question, the reading obtained from the Wheatstone bridge configuration will either be small or very small, since the resistive changes are themselves relatively small compared with the typical values induced by catalytic oxidation.

5.4.1   An inexpensive gas alarm for CO_2 leakage

This particular gas has very many applications, as shown in Table 5.7. Recalling the physiological effects on humans listed in Table 5.3, it is obvious that a relatively inexpensive gas alarm would find a ready market. The K values of methane and air are similar and yet the principle of detection based on their differences is effective, as mentioned above. It is obvious, therefore, that it will be even more effective for gases which have quite different K values, such as air and CO_2. A simple gas alarm for CO_2, based on this principle, has been shown to be reasonably effective

Table 5.7 Uses of CO_2

Raw material for the manufacture of chemical products
Glasshouse crop yield and date enhancement
Solvent for solvent extraction processes
Binding agent for silicate moulds and cores
Inert gas for blanketing and purging fires
Tobacco processing
Pipe freezing
Carbonation of beers and soft drinks
Refrigerator and freezer coolant
Inert atmospheres for steel welding
Heat-transfer medium for advanced gas-cooled reactor power stations

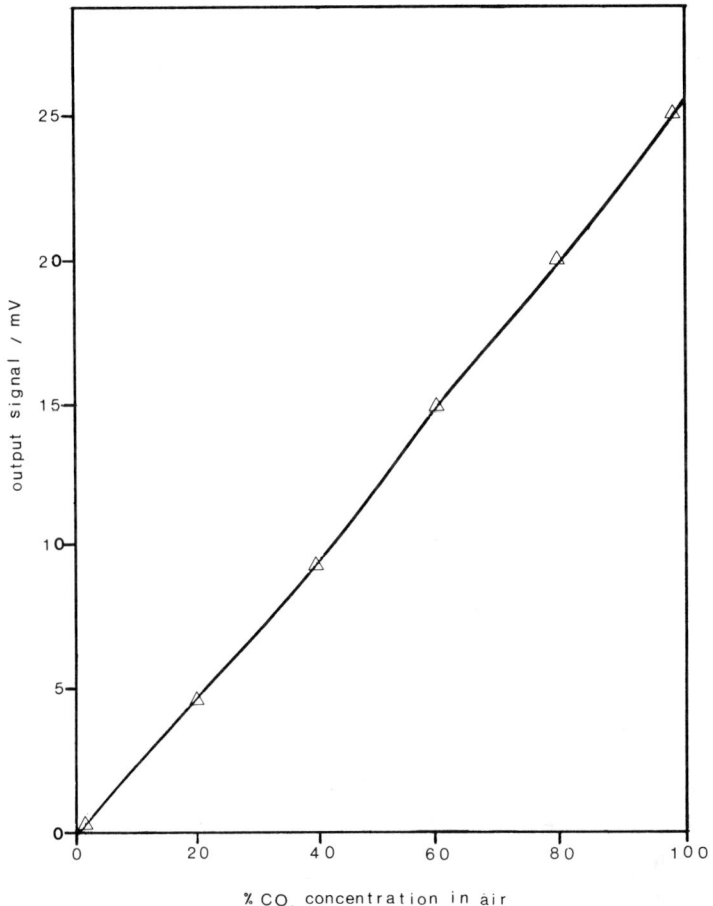

Figure 5.10 Linearity of thermal conductivity pellistor pairs as a function of CO_2 concentration.

(Campbell, 1993). The system displays good linearity (Figure 5.10), lack of hysteresis (Figure 5.11), good repeatability (Figure 5.12), reasonably small baseline excursion and electronic drift (Figure 5.13) and good response (T_{90}) times at all concentrations (Figure 5.14). The simple circuit, which contains a comparator for altering the preset alarm threshold, is shown with LED indicator, latched alarm level and buzzer (Figure 5.15).

5.4.2 MSA model LV combustible gas indicator

This is similar to the model L mentioned in section 5.3.2.4 but has the additional thermal conductivity range for methane of 0–100% v/v with

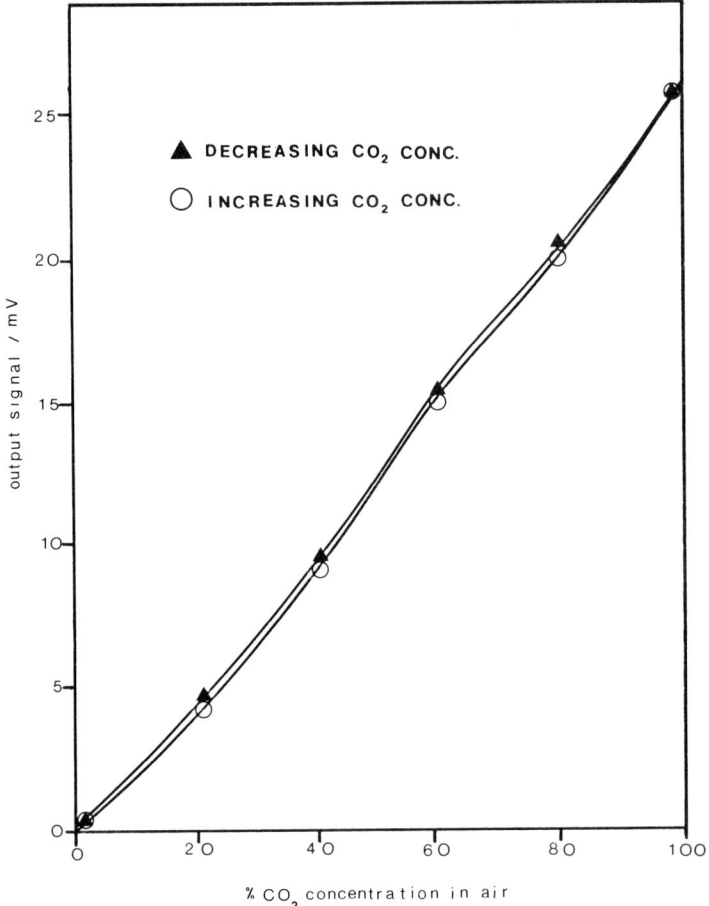

Figure 5.11 Hysteresis in thermal conductivity pellistors pairs as a function of CO_2 concentration.

a resolution of 1% v/v. They have the same high specifications as the model L.

5.5 Taguchi sintered semiconductor sensors

5.5.1 Principle of operation

The Taguchi gas sensor (TGS) is a bulk semiconductor device that is based mainly on SnO_2 (Watson and Tanner, 1974). It is sensitive to alcohols, acetone, n-hexane, benzene, isobutane, propane, methane, CO, NH_3, H_2S,

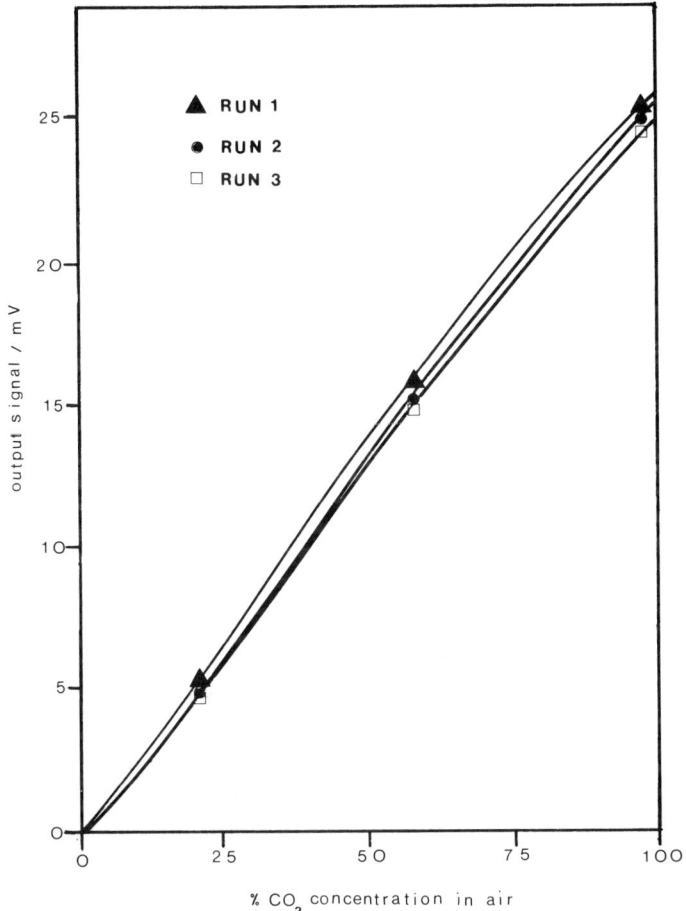

Figure 5.12 Repeatability of thermal conductivity pellistor pairs as a function of CO_2 concentration.

H_2, etc. Even though the TGS and other metal oxide devices are sensitive to most combustible gases in the ppm range, the non-linearity of their response, long-term drift and poor selectivity have hindered their use in sensitive applications. The steady-state and transient behaviour of these devices have been extensively studied and found to be dependent on a number of factors (Clifford, 1983). The sublinear response to combustible or reducing gases is the most salient feature of the behaviour. The resistance, R, of a gas sensor is related to the partial pressure, p, of a combustible gas by:

$$R = R_0 p^{-\beta} \tag{5.1}$$

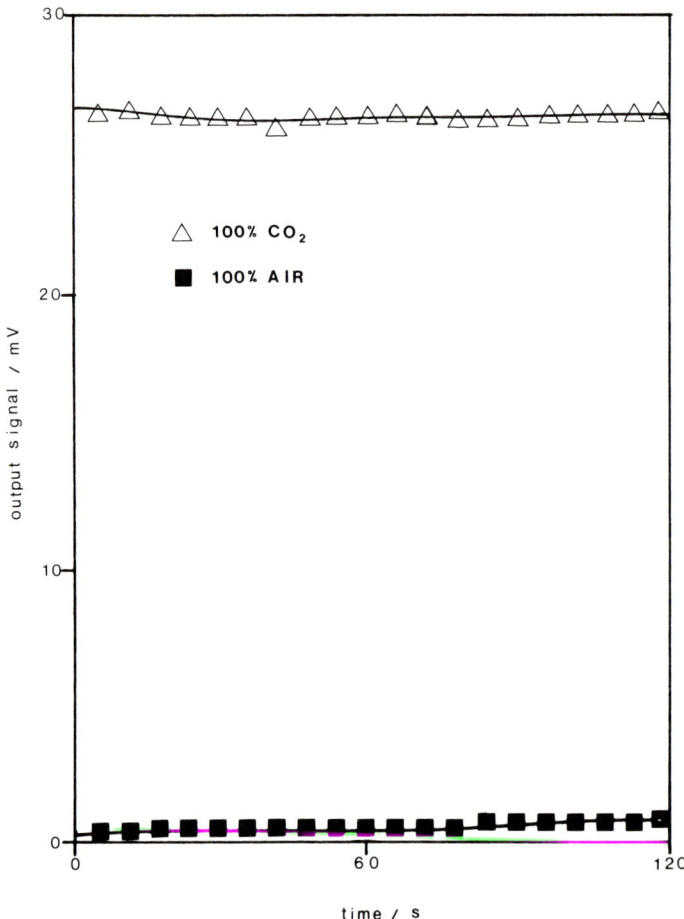

Figure 5.13 Electronic drift at 0% and 100% CO_2 (v/v).

where β is typically between 0 and 1 and is proportional to the temperature (Clifford and Tuma, 1983). The transient gas response is characterised by Elovich type kinetics for changes in the sensor barrier potential and by exponential relaxation and non-linear drift for changes in conduction. Interference and response enhancement phenomena are also seen from gases detected simultaneously and have been studied for CO and H_2S using the TGS. The conduction response of the devices is mediated by a barrier potential or activation energy for conduction, which changes with gas concentration. Finally, the response is also dependent upon sample preparation and composition.

There have been many mechanisms proposed to describe the processes by

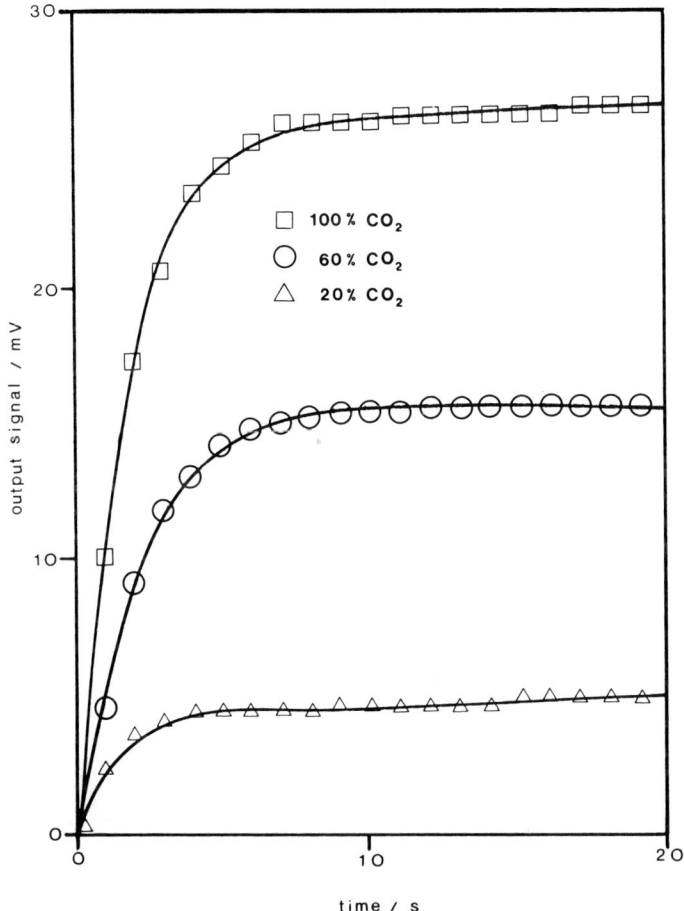

Figure 5.14 Response curves as a function of CO_2 concentration.

which the conductivity of a semiconductor is affected by ambient gases. It is generally assumed that for a combustible gas or a reducing agent to influence the conductivity of an n-type semiconductor it must participate in one of the three following reactions and for each reaction various models have been proposed.

5.5.1.1 Oxygen ionosorbate reaction. The reducing gases react with oxygen ionosorbate to yield a product which is a strong enough reducing agent to release an electron into the conduction band and the product gas subsequently desorbs. The conductivity increases as a result of the extra electron in the conduction band and perhaps as a result of a change in the

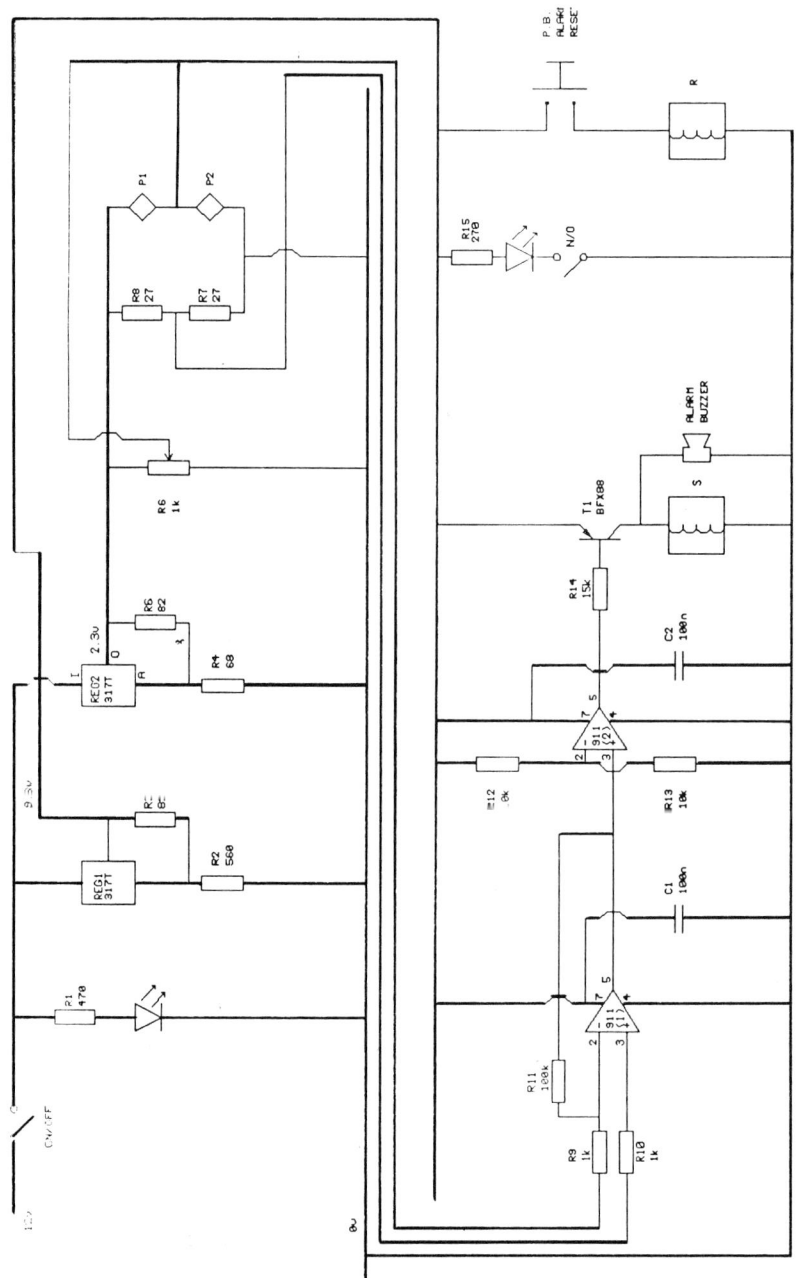

Figure 5.15 Simple CO_2 gas alarm circuit.

mobility (Clifford and Tuma, 1983).

$$G_{gas} + O_{surf}^- \rightarrow P_{gas} + e_{bulk}^- \tag{5.2}$$

5.5.1.2 Electron donation mechanism. The reducing or combustible gas acts as an electronic donor (i.e. as an occupied surface state). When the gas is absorbed it injects an electron into the conduction band of the semiconductor and increases the conductivity (Clifford and Tuma, 1983).

$$G_{gas} \Leftrightarrow G_{surf}^+ + e_{bulk}^- \tag{5.3}$$

5.5.1.3 Defect reaction. The combustible gas participates in a defect reaction by way of an oxidation–reduction process that affects the bulk stoichiometry of the semiconductor. As the dominant electronic donor is a stoichiometric defect, the bulk conduction band electron concentration reflects the reducing gas concentration (Clifford and Tuma, 1983).

$$G_{gas} + MO_{2bulk} \Leftrightarrow P_{gas} + (MO)_{bulk}^{2+} + 2e_{bulk}^- \tag{5.4}$$

Each of these reactions has been used as a basis of gas sensor operation. In particular, the reaction of combustible gases with pre-adsorbed oxygen ions is that most frequently cited as being responsible for the response of the TGS type device. A further model was suggested whereby the combustible gases do not directly interact electronically with the semiconductor but with physically adsorbed O_2 (Clifford, 1983). These non-equilibrium reactions are catalysed by the metal oxide surface, resulting in a steady-state concentration of oxygen physisorbate that is far less than would be the case if no combustible gas were present. The oxygen physisorbate concentration and the bulk electron concentration determines the amount of O_2 adsorbed in a charged state as ionosorbate. Consequently, the presence of combustible gas decreases the amount of oxygen ionosorbate and the conductivity of the metal oxide. In this model, only O_2 interacts with the electronic structure of the semiconductor, so contributions to the conductivity by reducing gases are mediated by the reactive desorption of physically adsorbed O_2.

Commercially available TGS devices are bulk semiconductor sensors comprised mainly of SnO_2. In the Taguchi No. 308, two Pt/Ir alloy electrodes are encapsulated in sintered SnO_2 and the sensor head is contained in a flame-proof cover to prevent ignition of the test gas. The circuit diagram is shown in Figure 5.16 and the variation in resistance is measured indirectly as a change in voltage appearing across the load resistor, R_L. In air, the current passing through the sensor and R_L is steady but in the presence of a combustible gas such as H_2 or CO, the sensor resistance decreases in proportion to the gas concentration. The voltage

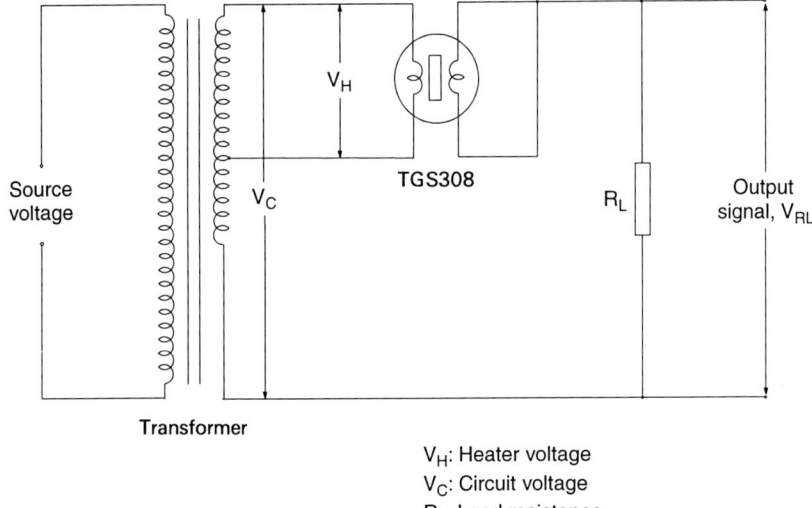

Transformer

V_H: Heater voltage
V_C: Circuit voltage
R_L: Load resistance

Figure 5.16 TGS 308 gas sensor circuit. V_H, heater voltage; V_c, circuit voltage; R_L, load resistance.

change across R_L is the same when V_C and V_H are supplied from AC or DC sources.

One problem with many TGS type devices is the internal heating produced by the circuit current. The increased current, resulting from decreased sensor resistance caused by the presence of the combustible gas, produces Joule heating within the sensor element. In air the sensor receives heat only from the heater coil and so its temperature is constant. In the presence of combustible gases, the sensor is heated both by the coil and by internal heating effects, causing the temperature to increase. This means that saturation of the sensor does not occur when the TGS is exposed to certain gases. In these circumstances, the sensor continues to respond to increased gas concentrations and the sensitivity curves show a hysteresis when supply voltage changes occur.

5.5.2   Diffusion-reaction model

The electrical conductivity of thick porous SnO_2 gas detectors is diffusion limited since the reaction rate is fast compared with the diffusion of gas. A diffusion-reaction model was developed (Gardner, 1989) to describe the immobilisation of the gas species in terms of specific sites at grain boundaries with the semiconductor. These have the effect of changing the band structure of the semiconductor. The model can be used to characterise transient and steady-state responses.

5.5.3   Thermal cycling of TGS heating filaments

Several years ago, some work was carried out to try to increase the sensitivity to reducing gases by thermally cycling the heater voltage (Sears et al., 1989). It was found that the conductance change in the lower temperature section of the cycle varied at a rate faster than linear for low gas concentrations and, since the conductance of the sensor in air is very low, this meant that the sensitivity to reducing gas was very high. Some recent research has shown that triangular wave excitations of the heating coils can result in greatly increased sensitivities to methane and propane (Campbell, 1996).

5.6   Recent developments

5.6.1   Electronic noses

Electronic noses, similar to the one developed by Neotronics, are sensor systems based on conducting polymers such as poly(pyrrole) deposited onto a silicon substrate. The principles of operation are broadly similar to the SnO_2-based devices, with conductivity changes being monitored on exposure to the test gases. The basic poly(pyrrole) is a p-type semiconductor with the charge carriers being either polarons or bipolarons depending on the doping level. This means that the polymers respond to oxidising and reducing gases in a similar manner to the response of inorganic semiconductors. The action of the vapours on these polymer films is described in more detail in section 5.9. The polymers are grown across two electrodes separated by a $10\,\mu m$ gap to enable resistance measurements to be made. Each type of sensor responds to a vapour in a slightly different manner because of the different dopants in the polymers and response times are approximately 60 s. Readings are taken at a specific point in time and the output of each sensor displayed on a bar chart or polar plot. The device consists of 12 individual sensors that monitor gas or vapour samples and the resultant signals are compared with a reference signal and the differences recorded.

Although the primary application for these sensors has been in the food and drink industry, the use of these devices in much broader applications, such as environmental monitoring, is imminent. A recent version of the 'nose', based on arrays of piezoelectric sensors, has also been developed recently (Byfield et al., 1995).

5.6.2   Metal oxide sensors

Metal oxide sensors are usually based on SnO_2 although Zn and W oxides have been investigated (Madou and Morrison, 1989). These sensors are based on the changing electrical conductivity of the metal oxide in different ambient atmospheres. Normally for a p-type material, the conductivity is decreased by an oxidising gas and increased by a reducing gas (and conversely for n-type materials). These sensors are usually thin film or single crystal devices. The single crystal devices have not proved to be very successful because the conductivity changes involved are too small. Thin film devices with SnO_2 are available commercially for the detection of methane, H_2, CO and H_2S, although they need to be operated at temperatures of around 573 K (Madou and Morrison, 1989). The non-specificity of thin film metal oxide sensors can be overcome to some extent (i) by controlling the film microstructure and (ii) by the use of incorporated catalysts. When these devices are operated at lower temperatures, they often lose their sensitivity to many hydrocarbons and become sensitive to relative humidity.

5.6.3   Single crystal thin film sensors

In single crystal sensors there are no grain boundaries to block the current flow along the sample and so the conductance is controlled by surface effects (and therefore by an ambient gas). Adsorbed O_2 is usually dominant, extracting electrons from an n-type semiconducting oxide. The surface conductance, σ_S is given by:

$$\sigma_s = N_s q \mu_n \frac{W}{L} \qquad (5.5)$$

where N_s is the density of charged surface states (N_s could be the density of O_2 adsorbed), q is the electronic charge, μ_n is the carrier mobility, W is the width of the sample and L the distance between the contacts (Madou and Morrison, 1989).
The bulk conductance, σ, is given by:

$$\sigma = N_D q \mu_n \frac{Wd}{L} \qquad (5.6)$$

where N_D is the donor density in the bulk and d the sample thickness. From equations 5.5 and 5.6 then:

$$\frac{\sigma_s}{\sigma} = \frac{N_s}{dN_D} \qquad (5.7)$$

so the sensitivity is greatest when d is small. Assuming N_s to be $10^{12}\,cm^{-3}$ and N_D to be $10^{17}\,cm^{-3}$ then d needs to be approximately $10^{-5}\,cm$ for σ_s to be of the order of σ and permit high sensitivity. Such thin samples of single crystal semiconducting metal oxides are not easy to prepare and so the majority of the work to date has been with sputtered or evaporated thin film devices.

5.6.4 Thin film metal oxide sensors

Thin film metal oxide sensors are usually prepared by (i) reactive sputtering, (ii) evaporation followed by oxidation, (iii) chemical vapour deposition or (iv) spray pyrolysis. Mechanisms suggested for the sensitivity of thin films to gases are either an idealised thin film model of the single crystal mechanism or a model that assumes the grain boundaries are affected by ambient gases.

A model was suggested that relates the sensitivity of thin films of ZnO to the diameter of spherical grains (Leary et al., 1982). This model assumes that there is no intergranular resistance, so any resistance changes arise from a loss of conduction electrons to adsorbed O_2 and that this process can occur at the surface of every grain. If the grains are below 100 nm the electrons are fully depleted by the adsorbed O_2, causing the resistance to be high. Grains sizes between 10^2 and 10^3 nm are in a transition region while with larger grains the resistivity of the bulk grains is measured.

Most thin film sensors are still in the development phase (e.g. Williams and Cole, 1995), but sensors for H_2S have been commercially available for a number of years. The films are usually WO_3 or SnO_2 sputtered films with sensitivity down to 0.1 ppm. The mechanisms are not fully understood, but the presence of H_2S at the surface causes the conductivity to increase, even though H_2S is not known as a strong reducing agent. An ion-exchange mechanism has been suggested in which the surface layer of the oxide converts to the sulphide in the presence of H_2S. As with all metal oxide-based sensors, there are problems with specificity. Au is often incorporated into H_2S sensors as it seems to act as a good catalyst for sulphide sensing (Madou and Morrison, 1989).

5.6.5 Incorporation of catalysts

In order to improve the sensing capabilities of both bulk and thin film metal oxide sensors, various catalysts have been incorporated into the sensor. The effects of these materials have been studied in detail (Okayama et al., 1983; Yamazoe et al., 1983; Duh et al., 1989), although the precise mechanisms of the interaction are not clear. Duh et al. studied the effects of a sintered SnO_2 device doped with ThO_2, MgO and $PdCl_2$. During the sintering process, the $PdCl_2$ was converted into PdO and this catalysed the oxidation of CO to CO_2 at temperatures of 300°C. However, it was found that the PdO tended to be reduced to Pd and the sensitivity deteriorated accordingly. A post-oxidation treatment at 600°C proved effective in retaining the PdO, and high sensitivity without degradation was obtained. The incorporation of Pt and Sb into SnO_2 enabled the detection of 100 ppm CO at room temperature (Okayama et al., 1983), although the response time was greater than for operation at elevated temperature.

The addition of Ag to SnO_2 was found to enhance the sensitivity to H_2 by a factor of five at temperatures of $250°C$, and this improvement was found to be much greater than that seen with sensors doped with Pd (Yamazoe *et al.*, 1983). This increase in sensitivity may be explained by an electronic interaction between the Ag and SnO_2, in contrast to the spillover effects normally used to explain the effects of Pd doped devices.

5.7   Zirconia devices

These devices have been developed over a period of time for the measurement of O_2 by the partial pressure (Benammar and Maskell, 1991), by concentration (Ioannou and Maskell, 1991) and in the presence of combustion gases (Benammar and Maskell, 1993).

5.8   Organic conductors

Since the earliest studies of electrical conductivity in organic conductors, it has been realised that electron-accepting gases may influence conductivity via surface (and possibly bulk) charge-transfer interactions that facilitate the generation of positive charge-carrying holes. If the majority carriers are holes, then exposure to an electron-accepting gas, such as NO_X, would be expected to increase the carrier concentration and, therefore, the conductivity. If the majority carriers are negative, then the holes produced by the electron-accepting gas will recombine with the original charge carriers, thus reducing the conductivity. Analogous effects (with the carrier signs reversed) are expected on exposure to electron-donating gases.

These effects were originally used to determine the sign of the majority carriers in molecular crystals. More recently, they have been studied quantitatively with the goal of developing solid-state gas sensors (Wright, 1989). Exposures of metal phthalocyanines to electron-accepting gases such as NO_2 cause conductivity increases by factors of up to 10^8. Ideally, single crystal samples with good quality surfaces should be used for a clear interpretation of these conductivity changes. Experiments on single crystals using an earthed guard ring (see Figure 5.17) show that in most cases the conductivity changes are confined to the surface layers with these changes vanishing when the guard ring is earthed (Wright, 1989).

For compressed pellets and films, diffusion along internal surfaces can complicate the discrimination between surface and bulk effects and it becomes important to distinguish effects caused by adsorbed gases reducing interparticle resistance from genuine electronic effects.

The effects of gases on the magnitude, rate and reversibility of conductivity changes have been studied for a wide range of molecular crystals

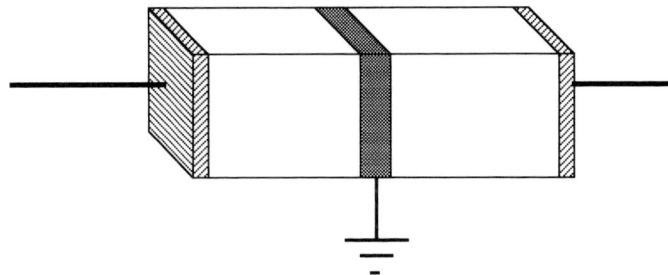

Figure 5.17 Earthed guard ring.

(Grate *et al.*, 1988; Leznoff and Lever, 1989; Bates, 1993). The strength of the surface charge-transfer interaction is the single most important factor controlling these properties. If the interactions are too weak, then the effect on the activation energy for carrier generation is small, whereas if they are too strong and localised, the charge carriers are coulombically bound to the adsorption sites and the conductivity changes are small.

The rate of conductivity change and the rate of its reversal are optimised when the surface charge-transfer interactions are weak; although this must be balanced with the retention of conductivity effects. Whilst it is obvious that the reversal of these effects will be faster for more weakly bound gases, the faster rise in conductivity for weaker surface charge-transfer inter-actions is less apparent, until it is appreciated that adsorption of a gas is primarily limited kinetically by the ease of the displacement of adsorbed O_2 from the surface. Under normal atmospheric conditions, the surface may be assumed to be completely covered with an adsorbed layer comprising mainly oxygen species. (It has been shown that a monolayer of O_2 forms on a clean phthalocyanine surface in less than 1 min on exposure to 10^4 Pa O_2 (Wright, 1989).) Weaker surface charge-transfer interactions weaken the binding of O_2 to the surface and, therefore, enhance the rate of its displacement by stronger electron acceptors.

The effects of gases on the electrical conductivity of molecular crystals provide the basis of a sensitive device for detecting electron-accepting gases in the atmosphere, although misleading conclusions on the scope of such devices can easily be drawn unless careful examination is made of all the variables influencing the performance. Operating temperatures can be varied in order to identify the temperature that optimises the response rate, reversibility and selectivity, while retaining reasonable sensitivity and device stability. Initial generation of active sites, such as by thermally induced crystallisation of amorphous films, may lead to apparently low sensitivity and lack of reproducibility unless careful thermal ageing is carried out. In assessing response speed and preparing calibration curves for these sensors, adequate time for the attainment of equilibrium surface coverage must be allowed, recognising that chemisorption kinetics follow the Elovich equation:

$$\frac{d\theta}{dt} = ae^{-b\theta} \qquad (5.8)$$

It can be seen from equation 5.8 that the fractional surface coverage (θ), and hence the conductivity, varies linearly with log (time) rather than with linear time. Calibration must also be carried out under realistic conditions; exposure to low concentrations of electron-accepting gases in an inert carrier gas removes the effect of O_2 competition and is not, therefore, a realistic indicator of atmospheric sensing ability. Furthermore, organic solids that respond reversibly to low concentrations of various gases often show irreversible damage if tested in the presence of much higher concentrations. The effects of potentially interfering gases must be examined as a function of temperature and concentration; for example, water vapour adsorbs to many organic solid surfaces and tends to reduce sensitivity, particularly at moderately low temperatures.

5.8.1   Metal phthalocyanine sensors

Metal phthalocyanine (Pc) materials were among the first to be considered as potential gas sensors (Van Oirsschot, 1972). These sensors are usually prepared as thin films on an interdigitated electrode structure. Phthalocyanines are generally weakly conducting, $\sigma = 10^{12}\,S\,cm^{-1}$ (Leznoff and Lever, 1989) and so voltages of between 10 and 700 V are required to measure the conductivity on conventional electrodes. As well as such a high power requirement, problems arise with electrochemical degradation at the Pc–electrode interface. The development of microtechnology has meant electrode spacings of 1 μm can be used and this substantially reduces the voltages required.

The effect of O_2 on the conductivity of Cu–Pc films was observed in 1948 when the presence of O_2 was found to increase the conductivity and lower the activation energy of conduction. The effects of O_2, NO, NO_2 and NH_2 on sublimed films of metal-free Pc and on Fe–, Co–, Ni–, Cu– and Zn–Pc were later studied by Kaufhod and Hauffe (1965). They observed that the nitrogen oxides caused the largest conductivity increases and reductions in activation energy with a similar, but less pronounced effect for O_2. These changes were reversed on exposure to NH_3, but it appeared the NH_3 interacted chemically with the Pc crystals (Kaufhod and Hauffe, 1965). The interaction of NO_2 has one of the largest effects on the conductivity and so has been studied extensively. The effect of NO_2 has been examined for metal-free and for Mn–, Co–, Ni–, Cu–, Zn– and Pb–Pc (Leznoff and Lever, 1989). All the Pc crystals studied showed surface conductivity

Table 5.8 Activation energy of conduction of phthalocyanine single crystals in vacuum and NO_2

Phthalocyanine	E_{act} vacuum (eV)	E_{act} NO_2 saturated (eV)
MnPc	0.37	0.10
CoPc	0.60	0.22
NiPc	0.74	0.10
CuPc	0.79	0.17
ZnPc	0.69	0.17
PbPc	0.58	0.28
H_2Pc	0.85	0.15

From Leznoff and Leder, (1989)

increases from 10^{-15} to 10^{-7} S cm^{-1} and the changes in activation energy are listed in Table 5.8. These changes were found to be irreversible on evacuation for the Pb–, Mn– and Co–Pc, while the conductivity of the metal-free, Ni–, Cu–, and Zn–Pc decreased significantly. Heating under vacuum at 150°C returned the conductivity of the metal-free, Ni–, Cu– and Zn–Pc to within a factor of five of the initial level, whereas the Pb–, Mn– and Co–Pc required heating at 250°C for 12 h for the conductivity to fall to this level. Exposing the Pc crystals to NH_3 restored the conductivity to its initial level in all cases.

The conductivity increases were explained by the formation of a charge-transfer complex between the Pc donor and the NO_2 acceptor, with the increase in charge carriers arising from the holes produced in the Pc matrix. The Pb–, Mn– and Co–Pc all show a less reversible interaction with the NO_2 than the metal-free, Ni–, Cu– and Zn–Pc because Pb, Mn and Co can adopt oxidation states greater than +2, thereby enhancing the donor power of the Pc and allowing it to form stronger donor–acceptor complexes.

As NO_2 is a strong electron acceptor, the electron accepted from the Pc is delocalised over the planar NO_2 structure. This delocalisation means the coulombic force between the opposite charges are weakened and so charge carrier movement is facilitated. Exposure to BF_3, a strong σ-electron acceptor, causes conductivity increases in Cu– and Ni–Pc, although the effects are not as marked as with NO_2. Although BF_3 forms a stronger charge-transfer complex with Pc than does NO_2, the electron is more localised in a σ-orbital. The resulting coulombic forces are much stronger than in the NO_2–Pc complex, thereby retarding the movement of the charge-carrying holes. In both the BF_3 and the NO_2 cases, the effects are fully reversed by exposure to NH_3; this is because the NH_3 acts as a competing donor and displaces the Pc in the donor-acceptor interaction.

The conductivity response of sublimed films is different from that of single crystals: the response is not completely reversed by exposure to electron donors such as NH_3 or H_2S and the conductivity dependence on NO_2 concentration differs between different Pc films. Sublimed films are likely to be either amorphous or polycrystalline, so defects and grain boundaries will affect absorption sites and charge-carrier transport. The morphology and crystal structure of sublimed films can be changed from either the amorphous phase or the α-Pc (monoclinic) phase to the β-Pc (triclinic) phase by thermal annealing at temperatures greater than 300°C for 60 min. The conductivity and gas response characteristics are also affected by this structural change (the response times are typically less than 1 min for the β-Pc compared with 5 min for the amorphous and α-Pc). However, the triclinic structure is not stable over long periods of time and so periodic re-annealing is required.

Phthalocyanines can also be laid down as Langmuir–Blodgett films and these films display significantly different behaviour to the sublimed films, with conductivity *increases* reported for the exposure of tetra-cumylphenoxy-substituted Cu–Pc to NH_3 (Leznoff and Leder, 1989). An asymmetrically substituted trismethylene isopropylamine Pc LB film showed an enhanced response and recovery time in comparison with the unsubstituted film. This peripheral substitution apparently places the Pc in a matrix where absorption and desorption of the gases is easier.

Evidently the film morphology plays an important role in the film's electrical response to gas exposures. The morphology must accommodate both the charge-transfer interaction and charge-carrier transport. These mechanisms are not well understood, but it has been suggested that the charge-transfer interaction is easier if the face of the Pc ring rather than the edge is available for an electron transfer. Polycrystalline Pc films are composed of crystallites of columnar stacking of cofacially oriented Pc rings. This feature, coupled with partial oxidation, facilitates charge-carrier transport through the film. Films of peripherally substituted Pc are believed to exist as oligomeric aggregates with short-range order (Leznoff and Leder, 1989). Such a morphology would mean a greater number of Pc ring faces would be available for charge-transfer interactions as well as having a more porous microstructure for gas permeation. However, the lack of long-range order means lower conductivities are observed upon exposure to oxidising gases. This has been observed in LB films of tetracumylphenoxy Pc where the conductivity increased by four orders of magnitude on exposure to iodine, compared with the increase of ten orders of magnitude observed for unsubstituted Pc LB films (Leznoff and Leder, 1989).

The modification of the Pc ring by the substitution of peripheral groups to vary the vapour sensitivity and selectivity of the films has promising prospects for the development of sensor devices. These peripheral groups can modulate the vapour response by affecting the film morphology or by an inductive effect on the Pc ring, which would affect the strength of the charge-transfer interaction.

5.8.2   Other materials

In addition to metal Pc a number of other organic conductors have been considered as potential gas sensor materials, including the nickel salt of tetrabutylammonium bis(1,3-dithiol-2-thione-4,5-dithiolate), $Bu_4NNi(dmit)_2$, for the detection of SO_2 and NO (Bates, 1993) and bis((diethylamino)dithiobenzil)nickel, BDN, for the detection of hydrazine (Grate et al., 1988). Metal complexes of sulphur ligands have been of interest for many years because of their applications in analytical chemistry, catalysis and their relevance to bio-organic systems, although more recent work has been concerned with the solid-state properties of these materials (Cassoux et al., 1991).

$Bu_4NNi(dmit)_2$ based devices were fabricated on an interdigitated electrode structure. Any changes in the resistance of an electrodeposited film of $Bu_4NNi(dmit)_2$ in SO_2 and NO were then recorded (Bates et al., 1995, 1996a). The resistance of the films increased by 40% (on a baseline of $150\,\Omega$) when exposed to SO_2, and this response was almost completely reversible by flushing with argon. Further testing of the device showed increases of 0.7% to 1000 ppm SO_2. In a practical situation, typical SO_2 concentrations that need to be measured are in the 1–100 ppm range for trace analysis, although for flue gas analysis the desirable detection range is from 100 to 2000 ppm. The response time of the devices was typically 180 s compared with 50 s for the pellistor and metal oxide devices.

The devices were also sensitive to NO with increases of a factor of 4.5 observed on exposure to 1% NO in argon. This increase was found to be only partially reversible on flushing the chamber with argon and on evacuation. No response to CO or CH_4 was observed. The fact that NO (a powerful oxidising agent) and SO_2 (a reducing agent) both caused an increase in the film resistance was unexpected, as the action of these gases would normally be expected to change the resistance in opposite directions, as described in earlier. This was explained by assuming the gases acted on the high conductivity ($\sigma = 10\,S\,cm^{-1}$) mixed valence $(Bu_4N)_{0.29}Ni(dmit)_2$ species and, in the case of SO_2, reduced it to the low conductivity ($\sigma = 10^{-8}\,S\,cm^{-1}$) monoanion, $Bu_4NNi(dmit)_2$. In the case of NO, it was proposed that the gas oxidised the mixed valence species to the low conductivity, ($\sigma = 10^{-3}\,S\,cm^{-1}$) neutral species $Ni(dmit)_2$. Plots of resistance versus log (time) were straight lines, indicating that the absorption kinetics followed the Elovich isotherm (equation 5.8). The surface morphology of the deposited films was found to greatly influence the sensitivity of the devices (Bates et al., 1996b).

5.9   Conducting polymer sensors

Investigations into the effects of various gases on the electrical and optical properties of poly(pyrroles) (Walmsley et al., 1991; Weimar et al., 1990;

Miasik *et al.*, 1986; Gustafsson and Lundstrom, 1987; Bartlett *et al.*, 1990), poly(thiophenes) (Foot *et al.*, 1988; Honawa *et al.*, 1989; Assadi *et al.*, 1990), poly(paraphenylene), poly(acetylene) (Chaing, 1977; Park, 1979; Yoshino 1985; Yoshino and Gu, 1986) and poly(phenylacetylene) (Hermans, 1984) have also been carried out. The bulk of this work has been with strong oxidising or reducing gases such as NO_x and NH_3, although it has been shown that poly(aniline) is sensitive to acetone and methanol vapours, and poly(pyrrole) is sensitive to methanol, toluene and ether vapours (Bartlett and Ling Chung, 1989; Bartlett *et al.*, 1989; Slater *et al.*, 1992; 1993).

Yoshino *et al.* (1985) examined the effects of NO, N_2O, NH_3 and CO on poly(thiophene)/BF_4. Exposure of the films to NO caused a conductivity increase of over two orders of magnitude, while only a slight increase was observed for N_2O. No increase at all was observed on exposure of the films to CO, whereas a conductivity decrease was observed for NH_3. It was suggested that doping with an electron acceptor such as NO leads to the formation of positive holes in the poly(thiophene) chain and an increase in conductivity was consequently observed. The process of charge-carrier transfer was viewed as an electron transfer from the valence band of the polymer to the unoccupied electron states of the acceptor. In the case of CO, the large negative electron affinity $(-1.5\,eV)$ prevents the formation of a donor–acceptor complex, whereas the electron affinity of the NO $(+0.024\,eV)$ allows the formation of a donor–acceptor complex and, therefore, increasing the conductivity.

Hanawa *et al.* (1989) observed similar effects for poly(thiophene)/ClO_4^- exposed to NO_2 and NH_3 and also observed that the conductivity fell on exposure to H_2S. In all cases, the conductivity change was irreversible on flushing with N_2 and upon evacuation. It was also noted that conductivity increases were greater if the films were electrochemically reduced prior to NO_2 exposure, indicating that the conductivity changes were probably caused by chemical doping. The poly(thiophene)/ClO_4^- films were also exposed to methane, C_3H_8, C_3H_6, CO and SO_2 but no response was observed.

Miasik *et al.* (1986) exposed poly(pyrrole) to NH_3, H_2S and NO_2 and found the conductivity changes were reversible for short exposure times and also demonstrated the dependence of the conductivity decrease on NH_3 concentration. Foot *et al.* (1988) suggested a model to describe why the NH_3 process was irreversible for poly(thiophene)/BF_4^- and not for poly(pyrrole)/BF_4^-. During the exposure of poly(thiophene) films it was noted that N_2 gas was evolved during the undoping process, and it was suggested that this came from the decomposition of the NH_3 :

$$8NH_3 \rightarrow 6NH_4^+ + N_2 + 6e^- \qquad (5.9)$$

The electrons produced in this scheme would compensate the mobile holes on the conjugated polymer chain and the NH_4^+ would complex with the

BF_4^-; a crystalline NH_4BF_4 phase was detected by X-ray diffraction in the NH_3-exposed films. In the case of poly(pyrrole), the undoping process proceeded via the deprotonation of the pyrrole nitrogen to form NH_4^+. As no NH_4BF_4 was observed, it was suggested that a 'hydrogen bonding bridge' connected the two nitrogen atoms and so immobilised the NH_4^+. At higher NH_3 concentrations, the process was irreversible, suggesting the NH_4^+ was mobilised; this was supported by the presence of NH_4BF_4 in the film after exposure to high NH_3 concentrations.

Gustafsson and Lundstrom (1987) exposed poly(pyrrole)/ClO_4^- to NH_3 and found the conductivity changes to be reversible, although in the presence of water vapour the changes were irreversible. They suggested a similar model to that of Foot et al. involving deprotonation of the pyrrole nitrogen, which created an extra electron on the nitrogen that would compensate the charge-carrying hole. In the presence of water vapour, they suggested a mechanism whereby the NH_3 reacts with the water and ClO_4^- to form NH_4ClO_4 and OH^- species. The OH^- would then interact with the conjugated chain to form carbonyl groups and so decrease the conductivity irreversibly.

In addition to the action of NO_x and NH_3 on the p-type nature of poly(pyrrole), vapours such as alcohols have been shown to have a solvent type action on the polymer, causing it to swell (Slater et al., 1992). These dimensional changes were accompanied by a change in the conductivity and this effect was related to the polymer thickness – thicker layers did not swell uniformly, suggesting that the penetration of the vapour was limited. Slater et al. (1993) used these phenomena by preparing a device based on a polymer-coated array of conduction bands that allowed differential conductivity measurements to probe relative resistances of zones of the polymer. The aim was to obtain broadband, overlapping conductivity responses corresponding with different zones of the same polymer to allow the response of a partially swollen surface zone to be compared with an unswollen zone. These devices were prepared with a number of substituted pyrroles and a variety of possible mechanisms for the interaction of the gas were briefly outlined. When exposed to methanol, the poly(pyrrole) films swelled, but this swelling appeared to be incomplete and different relative responses were observed between electrode pairs. Using data reduction techniques, the devices were able to distinguish between methanol, ethanol and propanol as well as between four different alcoholic drinks.

Other recent work by Pearce et al. (1993) involved the construction of devices that mimicked the mammalian olfactory system by using an array of non-specific, chemical sensors with data acquisition and processing software. The sensors were fabricated using poly(pyrrole) films prepared with a number of different counter ions deposited across the gap between two thin Au electrodes. These devices were able to distinguish between various commercial beers and lagers using simple changes in film resistance.

5.10. Electrochemical cells for gas sensing

Electrochemical cells, in general, are dealt with in detail in Volume 1, Chapter 4. This discussion will, therefore, be limited to specific types of cell that are suited to the measurement of gases by partial pressure and by concentration.

5.10.1 Clark cells

In the early days of cell development, the available instrumentation was large scale and, as such, tended to be laboratory based and certainly could not be considered as portable. Much of the present cell technology owes its existence to pioneering work carried out in the 1950s (Clark, 1956). The Clark cell was the first practical cell for measuring the partial pressure of gaseous oxygen (pO_2) and consisted of concentric electrodes, a suitable electrolyte and a gas-permeable, non-porous membrane, as shown in Figure 5.18. The membrane is stretched across the end of the container walls to be flush with the electrodes, thereby trapping the electrolyte as a thin liquid film. The chemical reactions that occur at the electrodes are:

central cathode: $\qquad O_2 + 2H_2O + 4e^- = 4OH^-$ $\qquad\qquad$ (5.10)

cylindrical anode: $\qquad 2OH^- + Metal = Metal\,O + H_2O + 2e^-$ $\quad$ (5.11)

Whenever a suitable potential is applied to the electrodes, the cathode reaction takes place. By ramping the applied voltage, a polarogram is obtained for air, as shown in the upper trace of Figure 5.19. Too low a potential causes no reaction to occur while too high a potential causes dissociation of OH^- to produce gaseous H_2. The lower trace in Figure 5.19 is that which would be obtained for pure N_2. For single voltage point

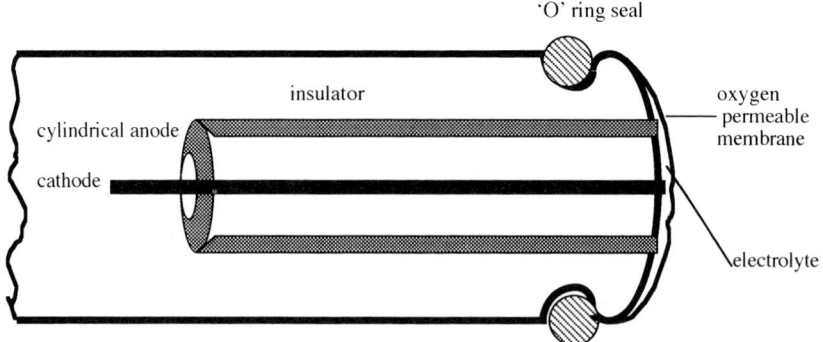

Figure 5.18 Cross-section through a Clark type cell.

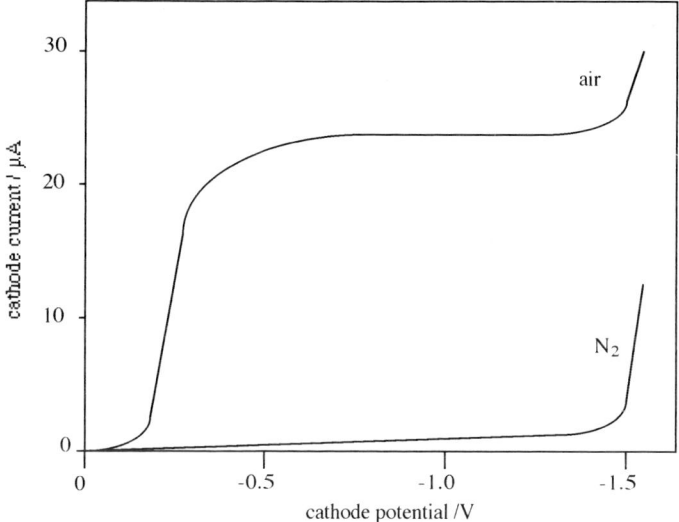

Figure 5.19 Polarographs for a Clark cell in air and in pure N_2.

operation, it would be appropriate to set the cell voltage to say -1 V, which is within the plateau region of cell response.

Two conditions should be satisfied for proper cathode reaction to take place: (i) the rate of reaction should be low (achievable by diffusion through a membrane) and (ii) the cathode should be a suitable metal, such as Ag. The drawbacks encountered with this cell structure are the result of membrane tensile weakness and the fact that the gaseous oxygen has not only to pass through the membrane but also through a liquid electrolyte film, very much a limiting factor when fast response times to changing conditions may be necessary.

5.10.2   Metallised membrane cells

As a consequence of the slow response times exhibited by Clark cells, much effort was expended in finding a suitable method for solving the problem of O_2 transport through the electrolyte film. A metallised form of membrane was proposed for two electrode pO_2 cells (Bergmann and Windle, 1972) and three electrode CO cassettes (Bergmann, 1975). The electrode could be fabricated by evaporating a relatively thick, gas-permeable layer of Ag onto very thin ($\sim 10\,\mu m$) PTFE film. It was found that the adhesion of Ag was greatly assisted by initially evaporating a very thin layer of Au onto the PTFE membrane. The metallised membrane formed the cathode and a low current density Ag coil formed the anode. The KCl electrolyte did, however, have a tendency to leak from the perspex and PTFE cell bodies, which greatly reduced their lifetimes. Conducting electrolyte gels do not tend to leak but do tend to shrink within the cell body, which, of course, causes other problems.

5.10.3   CiTiceLs for O₂ measurements

City Technology Ltd achieved a breakthrough in this technology in the late 1970s with a radical new approach to the structure of O_2 measuring cells. This technology gave rise to a series of cell that are suited to various gases and conditions.

5.10.3.1   Standard O₂ CiTiceLs.   All oxygen CiTiceLs are of the self-powered, diffusion-limited, metal–air battery type and consist of an air cathode, an anode and electrolyte, as shown schematically in Figure 5.20. By monitoring the potential difference across the load resistor, a signal is

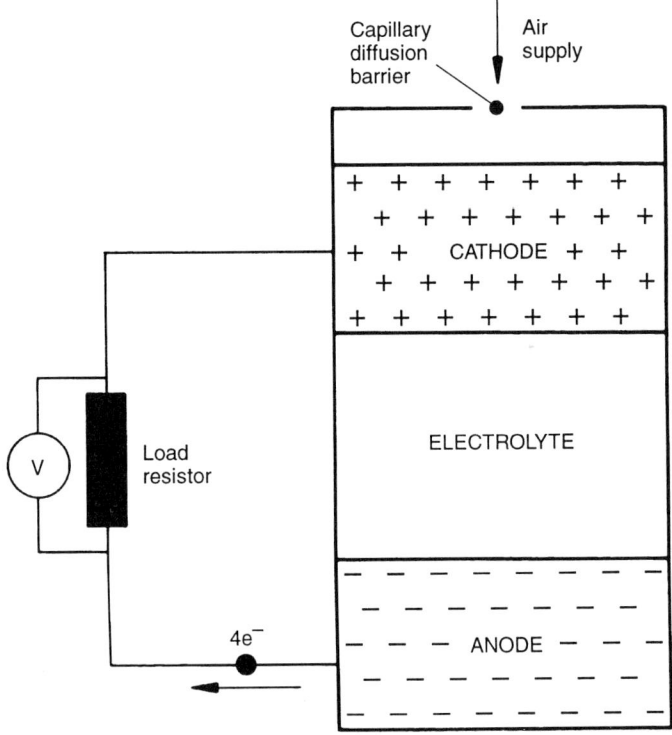

Figure 5.20 Standard CiTiceL for measuring O_2 (courtesy of City Technology Ltd).

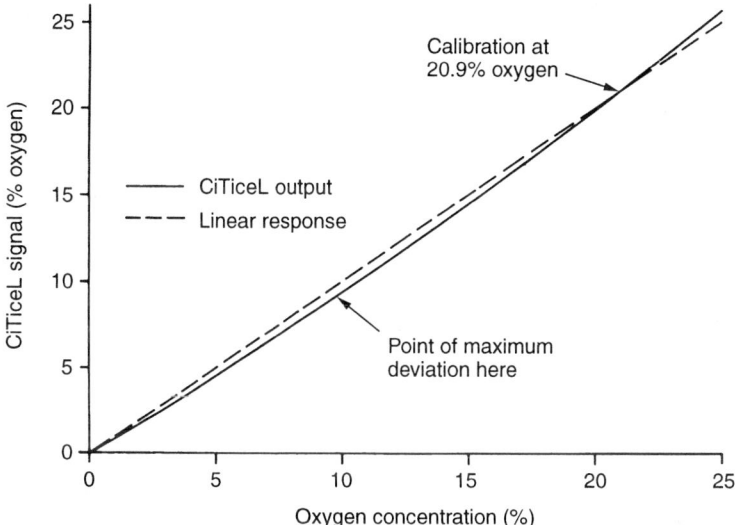

Figure 5.21 CiTiceL output as a function of O_2 concentration (courtesy of City Technology Ltd).

obtained that is very close to being linear with the concentration of O_2 as indicated in Figure 5.21.

5.10.3.2 The AO2 CiTiceL for pO_2 measurements. This cell has been specifically designed for the measurement of oxygen in automotive exhaust gases. It measures pO_2 rather than concentration by volume and has a resolution of 0.1 ppm and a T_{90} response time of less than 20 s. It is linear in the range 0–100% O_2. It utilises an industry standard screw-threaded plastic housing and Molex connector.

5.10.3.3 The 2FO CiTiceL. This O_2 sensor is ideally suited to monitoring O_2 in flue gases. Its resolution and T_{90} response time are similar to the AO2 cell. The 2FO features a plastic housing, bayonet mounting system and gas tight interface to sample drawing instruments.

5.10.4 Toxic gas cells

The increasing concentrations of toxic gases such as NO_X, CO_X, SO_X and volatile organic compounds (VOCs) in our environment have a profound impact on us all. The technology now exists for routine measurement of many of these toxic compounds and electrochemical cells, in particular, have an important role to play. Several sensors and complete instruments will now be mentioned.

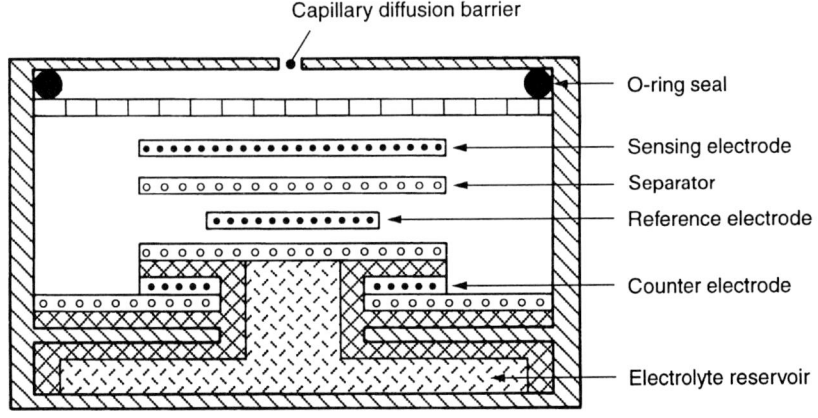

Figure 5.22 Schematic diagram of the CiTiceL for toxic gases (courtesy of City Technology Ltd).

5.10.4.1 CiTiceL toxic gas cells. The micro fuel cells designed by City Technology Ltd are three electrode cells which are now rapidly becoming the industry standard. The CiTiceL toxic gas cell structure is shown schematically in Figure 5.22. The standard circuit which maintains the reference electrode potential at a constant value is shown in Figure 5.23. The selectivity of these cells is to some extent dependent upon removing other gases that have cross-sensitivities. The simple schematic sampling layout shown in Figure 5.24 illustrates the precautions that should be designed into any particular application involving a variety of specific sensing cells.

5.10.4.2 Draeger polytron 2. The polytron 2 is a smart sensor that carries sensor-specific data within an EEPROM and which is temperature compensated using digital processing. Up to seven polytron 2 gas-sensing cells can be configured using the HART interface (see Volume 1, Chapter 8). This enables the analogue (4–20 mA) signals to be exchanged with digital information. The smart polytron 2 system has several desirable features, including (i) sensor self-test, (ii) free configuration of measuring ranges, (iii) pre-calibrated sensors, and (iv) computer-aided calibration. In addition, Draeger, in keeping with other manufacturers, claim long-life, high sensitivities, low cross-sensitivities and low influence of changing ambient conditions. The system can apparently be customised to measure a range of over 60 toxic gases.

5.10.4.3 MSA Gasgard 2. The instrument shown in Figure 5.25 is the versatile MSA Gasgard 2 gas detector. It can accommodate up to four sensors for O_2 (0–25% with 0.1% resolution), combustible gases (0–100% LEL with 1% LEL resolution) and toxic gases (0–999 ppm CO with 1 ppm resolution

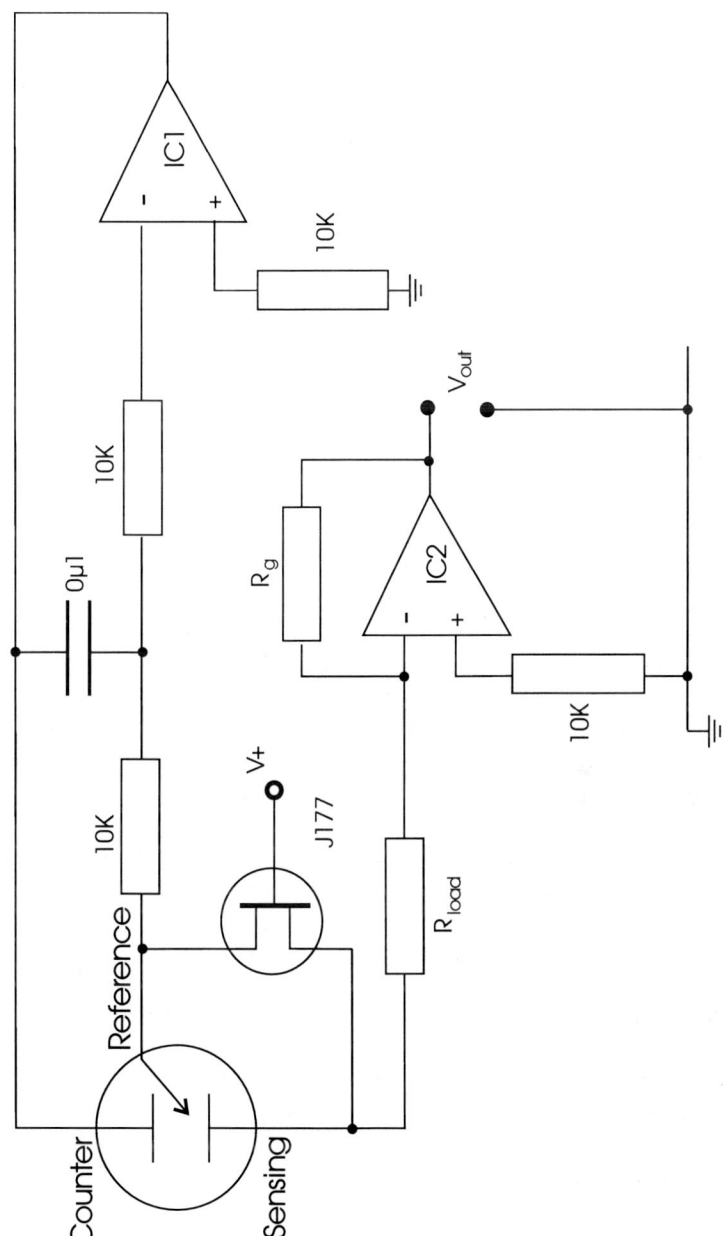

Figure 5.23 Circuit diagram for the CiTiceL toxic gas cell (courtesy of City Technology Ltd).

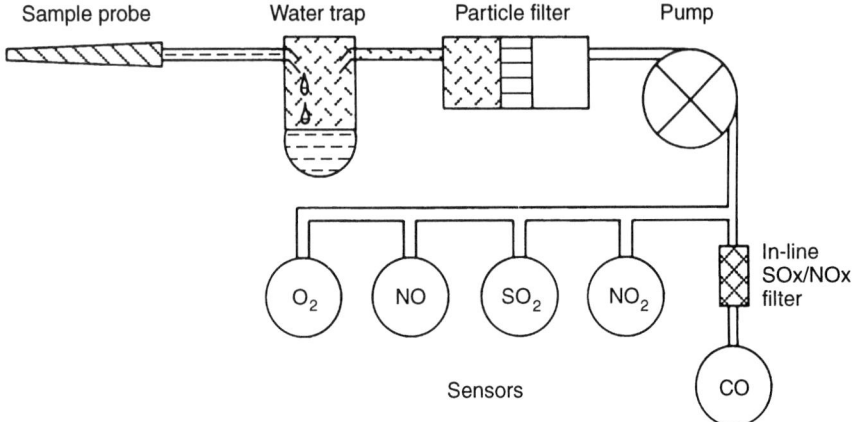

Figure 5.24 Simplified gas sampling and cleaning system (courtesy of City Technology Ltd).

Figure 5.25 MSA Gasgard portable monitor (courtesy of MSA (Britain) Ltd).

and 0–99 ppm for all other gases with 0.1 ppm resolution for 0–9.9 ppm and 1 ppm resolution for 10–99 ppm). Toxic gases include H_2S, SO_2, HCl, NO_2, NO, Cl and HCN.

5.11   Gas analysers

This section deals with instruments that are intended to measure specific gaseous species which are present in trace concentrations.

5.11.1   Introduction

The nature of gas analysis requires that the methods currently used for the monitoring of atmospheric gases are discussed. The regular and reliable monitoring of the atmosphere has only begun relatively recently and new legislation in many countries has led to the search for new and inexpensive methods for accurately monitoring gases such as NO_x, SO_2, CO, N_2O, NH_4, O_3 and methane. Numerous analytical methods such as mass spectrometry, gas chromatography and titration have been used to detect trace gases, but spectroscopic methods, especially using lasers, can give *in situ*, highly sensitive, real-time measurements (see Volume 1, Chapter 3). There are a number of problems that need to be overcome when monitoring gases in the atmosphere. The diverse range of concentrations of atmospheric trace gases means that sensitivity is the most important requirement of a monitoring system. The average concentration of methane in the troposphere is 2 ppm compared with a OH^- concentration of approximately 10^{-3} parts per trillion (Fehér and Martin, 1991). In addition to this large sensitivity range, many applications require high temporal and spacial resolution, particularly for flux and distance measurements. The ability to detect several different species, either simultaneously or successively, without interference from other species is often desirable. Although no single system covers all these requirements, spectroscopic techniques, particularly those based on lasers, are very versatile.

5.11.2   Principles of operation

Spectroscopic monitoring techniques use the UV and visible regions of the electromagnetic spectrum to detect electronic transitions and the IR region to detect vibration–rotation transitions. Some of the emitted radiation is absorbed at certain characteristic wavelengths, causing a decrease in the intensity of the beam at that wavelength. The ratio I/I_0, where I and I_0 are the transmitted and initial beam intensities, respectively, is related to the concentration, C, of the absorbing gas by the Beer–Lambert–Bouguer law:

$$\frac{I(\lambda)}{I_0(\lambda)} = e^{-A(\lambda)} \tag{5.12}$$

and

$$A(\lambda) = \alpha_m(\lambda)CL \tag{5.13}$$

so that:

$$\ln \frac{I_0}{I} = \alpha_m(\lambda)CL \qquad (5.14)$$

where $A(\lambda)$ is the absorbance at wavelength λ, $\alpha_m(\lambda)$ is the molecular absorption coefficient and L is the radiation pathlength through the gas. From equation 5.13, it can be seen that the absorbance is proportional to CL and the proportionality constant, $\alpha_m(\lambda)$, enables the gas to be identified as it is particular to each species. It must be noted that the absorption features are to some extent both temperature and pressure (i.e. altitude) dependent.

Electronic transitions usually have the advantage of large absorption coefficients, but not all molecules have transitions in this spectral region. However, nearly all molecules have vibrational transitions, although they are intrinsically weaker. For molecular absorption methods, the sensitivity can be increased by increasing the pathlength, although this approach is limited by the attenuation of the light beam; pathlengths of 5 km have been used (Plane and Nien, 1992).

A commonly used technique that utilises UV or visible spectra is differential optical absorption spectroscopy (DOAS), while the most common IR technique is Fourier transform IR Spectroscopy (FTIR). These and other absorption methods will now be described together with interferometry.

5.11.3 Differential optical absorption spectroscopy

While the vast majority of molecules absorb in the IR region of the electromagnetic spectrum, there are a few with weak absorption spectra, others that absorb in the same region in which CO_2 and water vapour absorb strongly and some, such as homonuclear diatomics, that have no IR absorption at all. Many of these materials do absorb in the UV or visible region. The DOAS technique, developed in the late 1970s (Platt and Perner, 1979; 1980), measures the concentrations of trace species in the atmosphere by recording their differential absorption of a UV–visible light source over a path length of several kilometres. A spectrometer is used in conjunction with a remotely positioned UV–visible light source, such as a xenon arc lamp, to define a long pathlength. Differential rather than absolute absorption spectra are measured, because most atmospheric species are present continuously in the atmosphere and so an unattenuated atmospheric spectrum cannot be obtained to act as a reference (NO_3 is an exception to this). In order to obtain the differential spectra, two laser beams are required: one whose frequency is tuned to an absorption feature of the particular gas while the other is tuned to be slightly off the feature to provide a reference beam. This enables a differential measurement to be

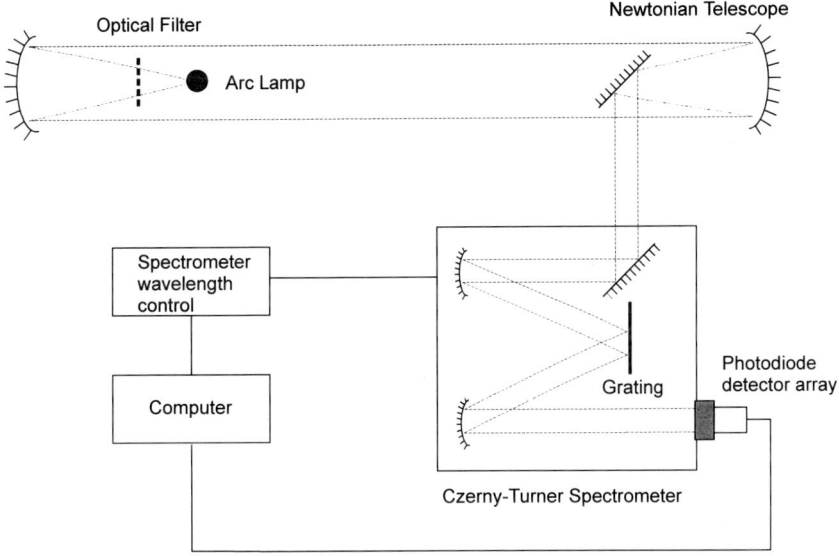

Figure 5.26 Schematic diagram of the DOAS system.

made. The DOAS technique has several advantages. If the instrument can scan the electromagnetic spectrum from 300 to 700 nm, there is the potential for real-time measurements of a large number of different species with a single instrument. Comparative tests between the commercially available OPSIS DOAS system and conventional fixed point monitoring methods showed a high correlation between the two techniques with the DOAS system showing extremely low detection limits (Stevens *et al.*, 1993). A schematic of a typical DOAS system is shown in Figure 5.26.

5.11.3.1 Draeger Polytron IREX gas analyser.   This instrument is an IR point gas detector that is based on well proved IR absorption principles. It employs state-of-the-art analogue and microprocessor technologies to give high reliability, stability and accuracy in the measurement of hydrocarbon gases and vapours. The analyser has a broadband IR LED source and the optical layout is illustrated in Figure 5.27. The LED beams that exit the sample chamber are routed to the detectors via a 50:50 beam splitter. The photodetectors shown in this Figure have in-built interference filters so that one detector measures the attenuated signal at an absorption wavelength while the other detector monitors the IR source at a non-absorbing wavelength to ensure that the long-term stability of the instrument is maintained despite possible changes in ambient conditions.

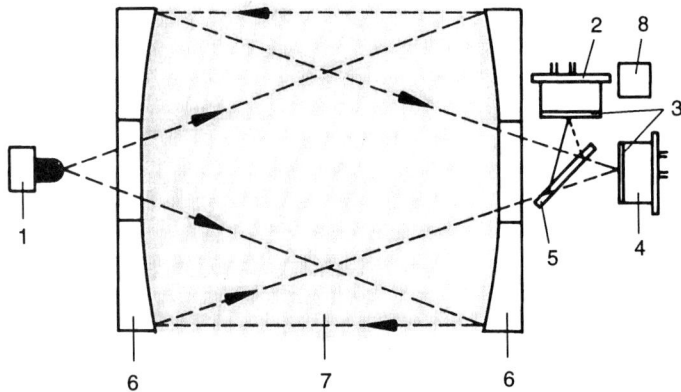

Figure 5.27 Optical arrangement for the Polytron IREX system. 1, IR light source; 2, reference detector; 3, interference filter; 4, measuring detector; 5, beam splitter; 6, mirror; 7, sample gas; 8, temperature sensor. (courtesy of Draeger Ltd).

Table 5.9 IREX detectable gases and vapours[a]

	Detection limits	
	Minimum	Maximum
Acetone	0–100% LEL	0–100% LEL
Iso-butane	0–2000 ppm	0–100% LEL
n-Butane	0–2000 ppm	0–100% LEL
Cyclohexane	0–2000 ppm	0–100% LEL
Cyclopentane	0–2000 ppm	0–100% LEL
Dimethylether	0–2000 ppm	0–100% LEL
Ethanoic acid	0–5000 ppm	0–100% LEL
Ethane	0–5000 ppm	0–100% LEL
Ethyl acetate	0–10 000 ppm	0–100% LEL
Ethylene	0–80% LEL	0–100% LEL
n-Hexane	0–2000 ppm	0–100% LEL
n-Heptane	0–2000 ppm	0–100% LEL
Methane	0–20% LEL	0–100 v/v
Methanol	0–2000 ppm	0–100% LEL
Nonane	0–2000 ppm	0–100% LEL
Octane	0–2000 ppm	0–100% LEL
Pentane	0–2000 ppm	0–100% LEL
Propane	0–1000 ppm	0–100% LEL
Iso-propanol	0–5000 ppm	0–100% LEL
n-Propanol	0–5000 ppm	0–100% LEL
Toluene	0–100% LEL	0–100% LEL

[a]Courtesy of Draeger Ltd. Values for other hydrocarbon gases and vapours are available on request from Draeger Ltd

When no gas is present, the inbuilt ambient temperature compensation ensures that a zero reading is always obtained. The selectivity and maximum sensitivity of the instrument derive from the selection of appropriate interference filters, whose bandpass maximises the absorption by the selected hydrocarbon (see Table 5.9). The presence of any absorbing gas or vapour between two heated curved mirrors will cause the sample signal intensity to be less than that of the reference beam so that a resultant signal is produced which is proportional to the concentration of the gas present in the sample. The advantages include (i) very fast response, (ii) fail-safe operation, (iii) use in inert atmospheres, (iv) immune to poisons, (v) extensive periods between maintenance, (vi) no moving parts and (vii) ranges above hydrocarbon LELs.

Typical full-scale deflection (FSD) specifications include:

methane	0–20% LEL FSD up to 0–100% LEL FSD and also 0–100% v/v FSD
propane/butane	0–1000 ppm FSD up to 0–100% LEL FSD
ethylene	0–80% LEL FSD up to 0–100% LEL FSD.

5.11.3.2 Chopped ratiometric analyser.   In this method (Figure 5.28), the beams are split with equal intensity using a mechanical chopper so that each beam passes through either the sample or the gas reference cell. The exit beams are dumped into compartments of a chamber separated by a flexible diaphragm. The differential absorption of these beams, owing to the presence of sample gas, causes the diaphragm to change the capacitance of a pick-up microphone so that the resulting signal has an amplitude related

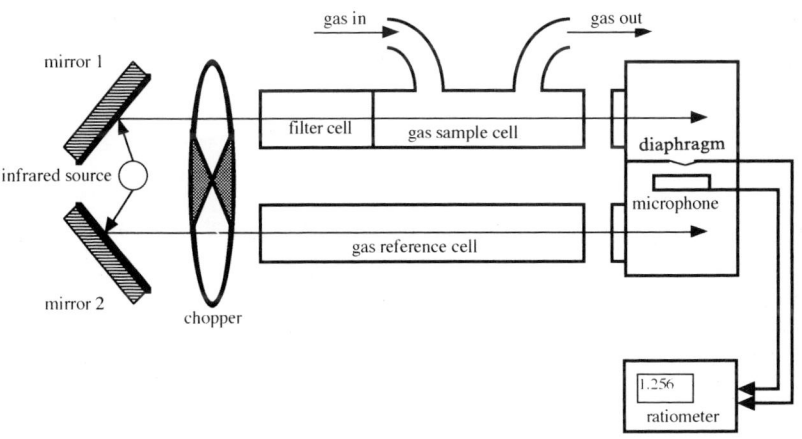

Figure 5.28 Chopped ratiometric gas analyser.

to the concentration of the gas and a frequency equal to twice that of the chopper.

5.11.4   Fourier transform IR spectroscopy

Most toxic gases covered under the 1990 Clean Air Act Amendments in the United States (Lee, 1991) have absorption spectra in the two atmospheric IR window regions, i.e. $750–1200\,cm^{-1}$ and $2400–3000\,cm^{-1}$. A Fourier transform IR (FTIR) analyser covering these regions has the distinct advantage of simultaneously measuring large numbers of gaseous species that have absorption features in these regions (see Volume 2, Chapter 5). The measurement of IR active vibrational and rotational transitions are measured using a very broad spectral coverage, with all wavelengths detected simultaneously.

Path-averaged concentration levels can be detected using FTIR from the absorption coefficient of a particular species. These calculations are greatly simplified by using a differential system, i.e. by using two frequencies, as described earlier. In addition, it is assumed that spectral interference by other species is absent, although this condition is not always satisfied, especially if the line overlaps an atmospheric water vapour line or a CO_2 line.

The main disadvantage with IR systems, even with using long pathlengths and multipass cells, is that the sensitivity is poor because of the low intensity of IR light sources, although the spectral resolution can be as high as $0.004\,cm^{-1}$ for some instruments (Fehér and Martin, 1991). It is more usual, however, to use systems with a spectral resolution of $0.5–2\,cm^{-1}$ so that smaller, transportable instruments can be used to make more rapid measurements. A schematic diagram of a typical FTIR spectrometer is shown in Figure 5.29.

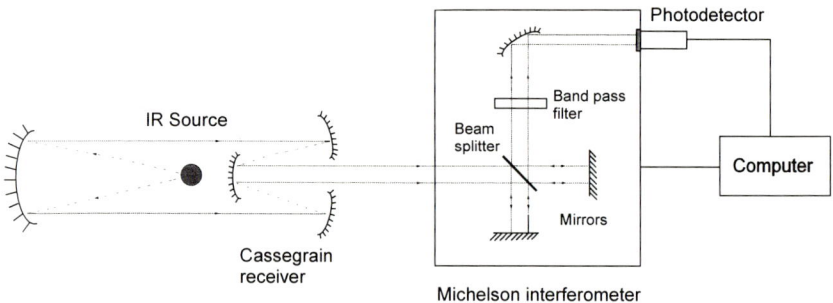

Figure 5.29 Schematic diagram of the FTIR system.

5.11.5 Differential absorption LIDAR

Light detection and ranging (LIDAR) systems require two visible or IR lasers whose wavelengths are different. Both wavelengths are back-scattered by a layer of the atmosphere and detected in a receiver located by the transmitter. As in DOAS, a differential system is used with one frequency tuned to an absorption feature while the second is tuned slightly off the absorption feature to provide a reference beam. The signal is stronger for the wavelength that is not absorbed and this provides a differential spectrum. The non-absorbed beam calibrates the backscatter of the LIDAR system in molecular (Rayleigh) and aerosol (Mie) scattering and for any slowly varying molecular absorption, such as water vapour continuum. The Beer–Lambert–Bouguer law is used to determine the concentration profile of the analyte species from the absorbances (logarithms of the signal ratios) at the two laser frequencies as a function of distance. Spatial information is obtained by pulsing the laser and measuring the time delay of the backscattered signal. The depth of the scattering layer is $c\Delta t$, where c is the speed of light and Δt the laser pulse duration. The spatial resolution is roughly twice the depth of the scatter layer, i.e. $2c\Delta t$, as two regions must be sampled to make a range resolved measurement (Grant et al., 1992). The differential LIDAR technique (DIAL) has been in use for measuring atmospheric trace constituents since the mid-1960s and has been most successful using the UV or visible spectral regions for the detection of O_3, NO, NO_3 and Cl_2 (Grant et al., 1992). More recently, mobile DIAL systems have been developed for monitoring industrial pollutants (Ednar et al., 1995). LIDAR and DIAL techniques are discussed more thoroughly in Volume 1, Chapter 3.

5.11.6 Laser-induced fluorescence

Molecules are electronically excited by laser radiation and when the molecule relaxes to its ground state the energy is emitted in the form of light that corresponds to certain energy level differences in the particular molecule. The detected fluorescence is characteristic of the species. Laser-induced fluorescence (LIF) is one of the most sensitive optical monitoring methods, with concentrations as low as 4.5 ppm NO and 10 ppt NO_2 detectable. However, LIF is not widely applicable because a different excitation scheme is needed for each species and many polyatomic molecules tend to photodissociate rather than fluoresce.

5.11.7 Chemiluminescent techniques

Chemiluminescent gas analyser systems tend to be used for the detection of NO, NO_2 and NO_X. The chemiluminescent reaction occurs only with NO,

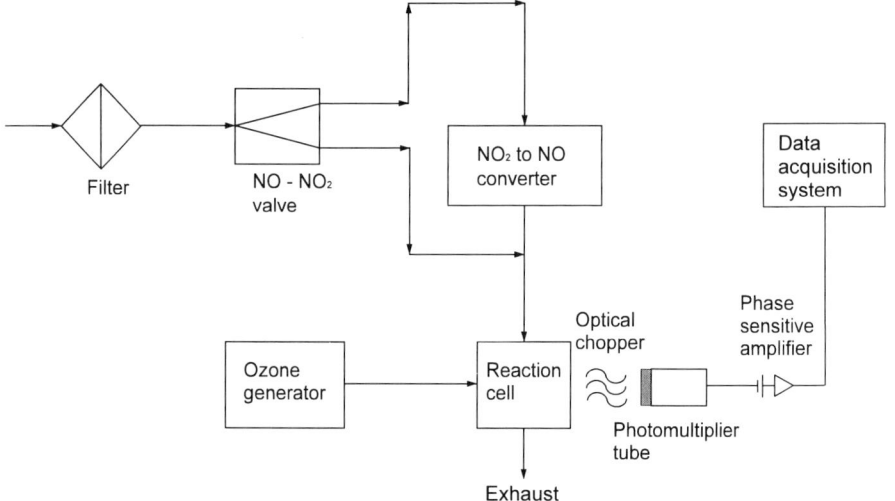

Figure 5.30 Schematic diagram of the chemiluminescent gas analyser.

so measurements of NO_2 and NO_3 are not possible directly, although by converting the NO_2 and NO_3 to NO a value for the total oxides of nitrogen can be obtained. Most commercially available systems provide NO_2 converter systems. The intensity of the chemiluminescence emitted from the reaction of NO with O_3 is measured to give an NO concentration. The O_3 is generated *in situ* from dry air. The reaction of NO with O_3 is shown below:

$$NO + O_3 \rightarrow NO_2 + O_2 \qquad (5.15)$$

Some of the NO_2 produced is in an excited state and as this excited NO_2 relaxes to its ground state, light is emitted, which is detected by a thermally cooled photomultiplier tube. The reaction chamber is normally operated under a vacuum in order to enhance the reaction and to reduce water vapour and CO_2 quenching. Detection limits can be as low as 0.4 ppb with response times of about 30 s. A schematic of a typical chemiluminescent system is shown in Figure 5.30.

5.11.8   Jamin type interferometer

The principle of operation of Jamin type interferometers relies on the fact that identical optical pathlengths can be altered using refractive index. This is accomplished by causing the source beam to split into two double pass beams (Figure 5.31), one of which passes through an air reference cell while the other passes through a sample cell. The system is arranged so that if the sample gas is air then the detector receives a minimum signal resulting from

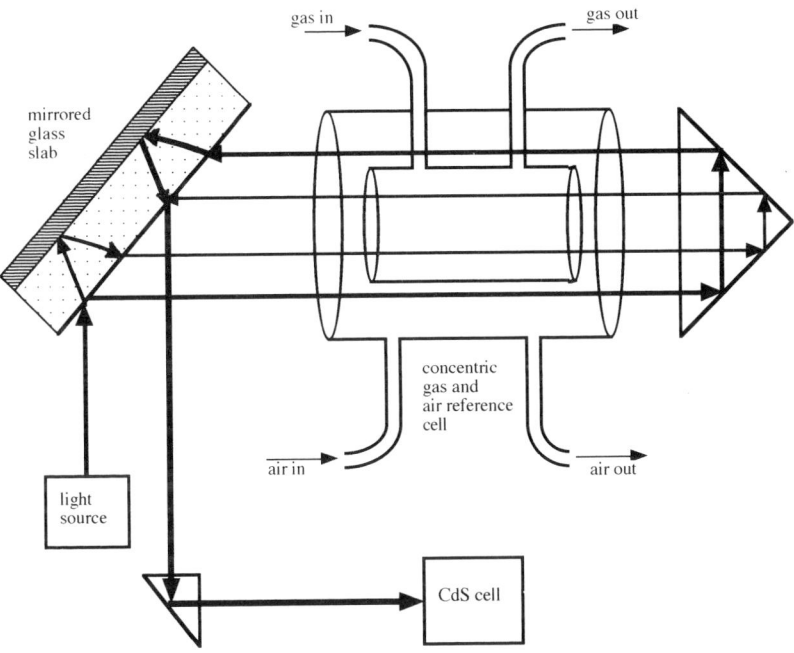

Figure 5.31 Jamin type interferometer.

the suitable phasing of the interfering beams. Introducing an air sample that contains an impurity of different refractive index to air causes the velocity of propagation in the sample arm to change; this results in a change of phase between the interfering beams at the detector, i.e. the signal increases.

5.12 Conclusions

The future market for gas sensor devices is likely to be very large because of the current emphasis on environmental protection and monitoring of pollution. The two main types of solid-state gas sensor in use today have been briefly outlined and the advantages and disadvantages discussed. The main problem with both the pellistor and the Taguchi gas sensor is their lack of selectivity between gases. The development of electronic nose-type devices has been discussed and it is likely that these types of device will be more common over the next five years. The trends towards miniaturisation will continue, with lower power requirements being a main objective of future development.

The drive for improving the selectivity of gas sensor devices will continue with work to improve catalysts in both thin film and bulk SnO_2, and the use of organic conductors will be extended. Perhaps the most important developments will be in the integration of neural networks with sensor devices. By learning the complex odour patterns of a particular analyte, the neural network can then use this information to identify a sample subsequently presented to it, effectively mimicking the mammalian olfactory system. It seems we have come full circle to the canary in the coal mine.

References

Assadi, A., Gustafsson, G., Willer, M., Svensson, C. and Inganas, O. (1990) Determination of the field effect mobility of poly(3-hexylthiophene) upon exposure to NH_3 gas. *Synthetic Metals*, **37**, 123–130.

Baker, A.R. (1962) Improvements in or relating to electrically heatable filaments,. *UK Patent Specification 892530*.

Baker, A.R. and Firth J.G. (1969) The estimation of firedamp – applications and limitations of the pellistor. *Mining Engineer*, **128** (No. 100, Part 4), 237–244.

Bartlett, P.N. and LingChung, S.K. (1989) Conducting polymer gas sensors. Part III: results for four different polymers and five different vapours. *Sensors and Actuators*, **20**, 287–292.

Bartlett, P.N., Archer, P.B.M. and LingChung, S.K. (1989) Conducting polymer gas sensors Part I: fabrication and characterisation. *Sensors and Actuators*, **19**, 125–140.

Bartlett, P.N., Gardner, J.W. and Whitaker, R.G. (1990) Electrochemical deposition of conducting polymers onto electronic substrates for sensor applications. *Sensors and Actuators*, **A21A23**, 911–914.

Bates, J.R. (1993) An investigation into the use of novel organic materials in gas sensor devices. PhD Thesis, University of Northumbria.

Bates, J.R., Kathirgamanathan, P. and Miles, R.W. (1995) Fabrication of a device for the detection of gases using conductivity changes of electrodeposited $Bu_4NNi(dmit)_2$ thin films. *Electronics Letters*, **31** (15), 1225–1227.

Bates, J.R., Miles, R.W. and Kathirgamanathan, P. (1996a) Modification of the room temperature conductivity of electrodeposited $Bu_4NNi(dmit)_2$ in ambient gases. *Synthetic Metals*, **76**, 313–315.

Bates, J.R., Kathirgamanathan, P. and Miles, R.W. (1996b) The influence of the electro-deposition parameters on the morphology of organo-transition metal complexes for thin film gas sensor applications. *Thin Solid Films* (in press).

Benammar, M. and Maskell, W.C. (1991) Measurement of oxygen partial pressure using fully-sealed zirconia devices operated in tracking mode. In *Sensors, Technology, Systems and Applications*, (ed. K.T.V. Grattan) pp. 113–119, Adam Hilger, Bristol.

Benammar, M. and Maskell, W.C. (1993) A novel miniature zirconia gas sensor with pseudo-reference. Part 2: combined amperometric and potentiometric operation for monitoring combustion systems. In *Sensors VI, Technology, Systems and Applications*, (ed. K.T.V. Grattan and A.T. Augousti) pp. 21–26. IOP, Bristol.

Bergman, I. (1975) Electrochemical carbon monoxide sensors based on the metallised membrane electrode. *Annals of Occupational Hygiene*, **18**, 53–62.

Bergman, I. and Windle, D.A. (1972) Instruments based on polarographic sensors for the detection, recording and warning of atmospheric oxygen deficiency and the presence of pollutants such as carbon monoxide. *Annals of Occupational Hygiene*, **15**, 329–337.

Byfield, M.P., May, L.P., Wunsche, L.F. and Vuilleumier, C.R. (1995) Development and applications of an electronic nose based on arrays of piezoelectric sensors. In *Sensors and their Applications*, (ed. A.T. Augousti) pp. 52–58. IOP, Bristol..

Campbell, M. (1993) Low cost carbon dioxide monitor. *International Journal of Electronics*, **75**, 777–780.

Campbell, M. (1996) unpublished data.

Cassoux, P., Valade, L., Kobayashi, H., Clark, R. and Underhill, A.E. (1991) Molecular metals and superconductors derived from metal complexes of 1,3-dithiol-2-thione-4,5-dithiolate (dmit). *Co-ordination Chemistry Reviews*, **110**, 115–160.

Chaing, C.K., Fincher, C.R., Park, Y.W., Heeger, A.J., Shirakawa, H., Louis, E.J., Gau, S.C. and MacDiarmid, A.G. (1977) Electrical conductivity in doped polyacetylene. *Physics Review Letters*, **39**,1098–1099.

Clark, J.C. (1956) Monitoring and control of blood and tissue oxygen tension, *Transactions of the American Society for Artificial Organs*, **2**, p. 41.

Clifford, P.K. (1983) Homogeneous semiconducting gas sensors: a comprehensive model. *Analytical Chemistry Symposium, Series 17 (Chemical Sensors)*, **17**, 135–146.

Clifford, P.K. and Tuma, D.T. (1983) Characteristics of semiconductor gas sensors. Part I. *Sensors and Actuators*, **3**, 233–254.

Curry, R. (1990) Gas alarms – some truths and realities. *Gas and Engineering Management*, 66–70.

Dabill, D.W., Gentry, S.J., Hurst, N.W., Jones, A. and Walsh, P.T. (1982), *UK Patent Specification 2083630*.

Duh, J.G., Jou, J.W. and Chiou, B.S. (1989) Catalytic and gas sensing characteristics in Pd doped SnO_2. *Journal of the Electrochemical Society*, **136**, 2740–2747.

Ednar, H., Ragnarson, P. and Wallinder, E. (1995) Industrial emission control using LIDAR techniques. *Environmental Science and Technology*, **29**, 330–337.

Fehér, M. and Martin, P.A. (1991) Something in the air? *Chemistry in Britain*, 883–886.

Firth, J.G. (1966) Catalytic oxidation of methane on palladium–gold alloys, *Transactions of the Faraday Society*, **62**, 2566–2576.

Firth, J.G. and Guest, A. (1970) Improvements in or relating to electrically heatable filaments. *UK Patent Specification 1184184*.

Firth, J.G., Jones, A. and Jones, T.A. (1973) The principles of the detection of flammable atmospheres by catalytic devices. *Combustion and Flame*, **21**, 303–311.

Foot, P., Ritchie, T. and Mohammad, F. (1988) Mechanisms of chemical undoping of conducting polymers by ammonia. *Journal of the Chemical Society: Chemical Communications*, 1536–1537.

Gall, M. (1993) The Si–planar–pellistor array: a detection unit for combustible gases. *Sensors and Actuators B*, **15–16**, 260–264.

Gardner, J.W. (1989) A diffusion-reaction model of electrical conduction in tin oxide gas sensors. *Semiconductor Science and Technology*, **4**, 345–350.

Gentry, S.J. and Jones, A. (1978), *Journal of Applied Chemistry and Biotechnology*, **28**, 727.

Gentry, S.J. and Walsh, P.T. (1984) *Sensors and Actuators*, **5**, 239.

Gentry, S.J. and Walsh, P.T. (1987) The theory of poisoning catalytic flammable gas-sensing elements. In *Solid State Gas Sensors*, Ch. 3 (ed. P.T. Mosley and B.C. Tofield) pp. 32–50. Adam Hilger, Bristol.

Grant, W.B., Kagann, R.H. and McClenny, W.A. (1992) Optical remote measurement of toxic gases. *Journal of the Air and Waste Management Association*, **54**(1), 18–30.

Grate, J.W., Rose-Pehrsson, S. and Barger, W.R. (1988) Langmuir–Blodgett films of a nickel dithiolene complex on chemical microsensors for the detection of hydrazine. *Langmuir*, **4**(6), 1293–1301.

Gustafsson, G. and Lundstrom, I. (1987) The effect of ammonia on the physical properties of polypyrrole. *Synthetic Metals*, **21**, 203–208.

Hanawa, T., Kuwabata, S., Hashimoto, H. and Yoneyama, H. (1989) Gas sensitivities of electropolymerised polythiophene films. *Synthetic Metals*, **30**, 173–181.

Hermans, E.C.M. (1984) CO, CO_2, CH_4 and H_2O sensing by polymer covered interdigitated electrode structures. *Sensors and Actuators*, **5**, 181–186.

Iredale, P.J. (1990) Combustible gas detection. *International Patent Application: WO 91/06849*.

Ioannou, A.S. and Maskell, W.C. (1991) A screen printed amperometric zirconia oxygen gas sensor. In *Sensors, Technology, Systems and Applications*, (ed. K.T.V. Grattan) pp. 157–161. Adam Hilger, Bristol.

Jones, E. (1987) The Pellistor Catalytic Gas Detector. In *Solid State Gas Sensors*, Ch. 2, (ed. P.T. Mosley and B.C. Tofield) pp. 17–31. Adam Hilger, Bristol.

Kaufhod, J. and Hauffe, K. (1965) Über das Leitfähigkeitsverhalte verschiedener Phthalocyanine im Vakuum und unter dem einfluß von Gasen. *Berichte Bunfengesellschaft für Physikalische Chemie*, **69**, 168–174.

Krebs, P. and Grisel, A. (1993) A low power integrated catalytic gas sensor. *Sensors and Actuators B*, **13-14**, 155–158.

Leary, D.J., Barnes, J.O. and Jordan, A.G. (1982) Calculation of carrier concentration in polycrstalline films as a function of surface acceptor state density: application for ZnO gas sensors. *Journal of the Electrochemical Society*, **129**, 1382–1387.

Lee, B. (1991) Highlights of the Clean Air Act Amendments of 1990. *Journal of the Air and Waste Management Association*, **41**, 16–25.

Leznoff, C.C. and Lever, A.B.P. (eds) (1989) *Phthalocyanines: Properties and Applications*. VCH, London.

Madou, M.J. and Morrison, S.J. (1989) *Chemical Sensing with Solid State Devices*. Academic Press, London.

Miasik, J.J., Hooper, A. and Tofield, B.C. (1986) Conducting polymer gas sensors. *Journal of the Chemical Society, Faraday Transactions 1*, **26**, 1117–1126.

van Oirschot, G.J., van Leeuwen, D. and Medema, J. (1972) The effect of gases on the conductive properties of organic semiconductors. *Journal of Electroanalytical Chemistry*, **37**, 373–385.

Okayama, Y., Fukaya, H., Kojima, K., Terasawa, Y. and Handa, T. (1983) Characteristics of CO gas sensor of Pt and Sb dispersed SnO_2 ceramics. *Analytical Chemistry Symposium, Series 17 (Chemical Sensors)*, **17**, 29–34.

Park, Y.W., Druy, M.A., Chaing, C.K., MacDiarmid, A.G., Heeger, A.J., Shirakawa, H. and Ikeda, S. (1979) Anisotropic electrical conductivity of partially oriented polyacetylene. *Journal of Polymer Science, Polymer Letters Edition*, **17**, 195–201.

Pearce, T.C., Gardner, J.W., Friel, S., Bartlett, P.N. and Blair, N. (1993) Electronic nose for monitoring the flavour of beers. *Analyst*, **118**, 371–377.

Plane, J.M.C. and Nien, C.F. (1992) Differential optical absorption spectrometer for measuring atmospheric trace gases. *Review of Scientific Instrumentation*, **63**, 1867–1876.

Platt, U. and Perner, D. (1979) Detection of nitrous acid in the atmosphere by differential optical absorption. *Geophysical Research Letters*, **6**, 917–919.

Platt, U. and Perner D. (1980) Direct measurement of atmospheric CH_2O, HNO_2, O_3, NO_2 and SO_2 by differential optical absorption in the near UV. *Journal of Geophysical Research*, **85**(C12), 7453–7458.

Sears,W.M., Colbow, K. and Consadora, F. (1989) General characteristics of thermally cycled tin oxide gas sensors. *Semiconductor Science and Technology*, **4**, 351–359.

Slater, J.M., Watt, E.J., Freeman, W.J., May, I.P. and Weir, D.J. (1992) Gas and vapour detection with poly(pyrrole) gas sensors. *Analyst*, **117**, 1265–1270.

Slater, J.M., Paynter, J. and Watt, E.J. (1993) Multi-layer conducting polymer gas sensor arrays for olfactory sensing. *Analyst*, **118**, 379–384.

Stevens, R.K., Drago, R.J. and Mamane, Y. (1993) A long path differential optical absorption spectrometer and EPA approved fixed point methods intercomparison. *Atmospheric Environment*, **27B**(2), 231–236.

Walmsley, A.D., Haswell, S.J. and Metcalfe, E. (1991) Methodology for the selection of suitable sensors for incorporation into a gas sensor array. *Analytica Chimica Acta*, **242**, 31–36.

Watson, J. and Tanner, D. (1974) Applications of the Taguchi gas sensor to alarms for inflammable gases, *The Radio and Electronic Engineer*, **44**, 85–91.

Watson, V.L. (1970) Improvements in or relating to combustible gas detectors. *UK Patent Specification 1387412*.

Weimar, U., Schierbaum, K.D., Gopel, W. and Kowalkowski, R. (1990) Pattern recognition methods for gas mixture analysis: application to sensor arrays based upon SnO_2. *Sensors and Actuators*, **1**, 93–96.

Williams, G. and Coles, S.V. (1995) Thin tin dioxide films for combustible gas sensors. In *Sensors and their Applications*, (ed. A.T. Augousti) pp. 69–74. IOP, Bristol.

Wright, J.D. (1989) *Molecular Crystals*. Cambridge University Press, Cambridge.

Yamazoe, N., Kurokawa, K. and Seiyama, T. (1983) Catalytic sensitisation of SnO_2 sensor. Analytical Chemistry Symposium, *Series 17 (Chemical Sensors)*, **17**, 35–40.

Yoshino, K. and Gu, H.B. (1986) Effect of ammonium gas on electrical property of conducting polymers. *Japanese Journal of Applied Physics*, **25**(7), 1064–1068.

Yoshino, K., Nalwa, H.S., Rabe, J.G. and Schmidt, W.F. (1985) The influence of nitric oxide, nitrous oxide and carbon monoxide on the electrical conductivity of polythiophene and polyacetylene. *Polymer Communications*, **26**, 103–104.

6 Piezoelectric sensors

M.J. HEPHER and D. REILLY

6.1 Introduction

Piezoelectric crystals offer the potential of real-time or near-time measurement of a vast number of measurands: these include temperature, pressure, magnetic field strength and analytes in both the liquid and gaseous phase. Moreover, their high sensitivity of response to very small changes in nano mass at their surface offers the potential of extraordinary low detection limits for analytes. Consequently, piezoelectric crystals (PZs) have received increasing attention since the mid-1970s, particularly in laboratories in the USA, UK and Japan. The majority of publications have presented the virtues of highly specific applications under very controlled conditions; there are very few instances where such devices have been taken into an environment outside the laboratory. A major feature of PZ or micromass chemical sensors is the reaction of specific coatings applied to their surface: these coatings hold the key to future developments. Of particular interest in environmental and health and safety studies is the measurement of dusts and toxic gases in the atmosphere and toxic chemical species in natural waterways.

6.2 Piezoelectric crystal theory

The piezoelectric effect (from *piezen*, to press) was discovered in 1880 by the Curie brothers. They found that if pressure is applied to opposite faces of a slice of α-quartz then an electrical potential is developed across them (direct effect). The phenomenon of piezoelectricity is observed in certain natural crystals (e.g. quartz, rochelle salt, tourmaline), in ceramics (e.g. lead zirconate, lead titanate, barium titanate) and in polymeric material (e.g. polyvinylidene fluoride, polyvinylidene chloride).

The converse piezoelectric effect, predicted and shown by Lippmann, was that an elastic deformation of piezoelectric materials (all crystals possessing a structure lacking a centre of symmetry) occurs when the material is subjected to an electric field. By alternating the electric field, the piezoelectric material is set into oscillation, the amplitude of which will be maximum when the frequency of the applied field is equal to the resonant frequencies of the piezoelectric material. The alternating electric field applied to the crystal can be obtained by incorporating it into a simple

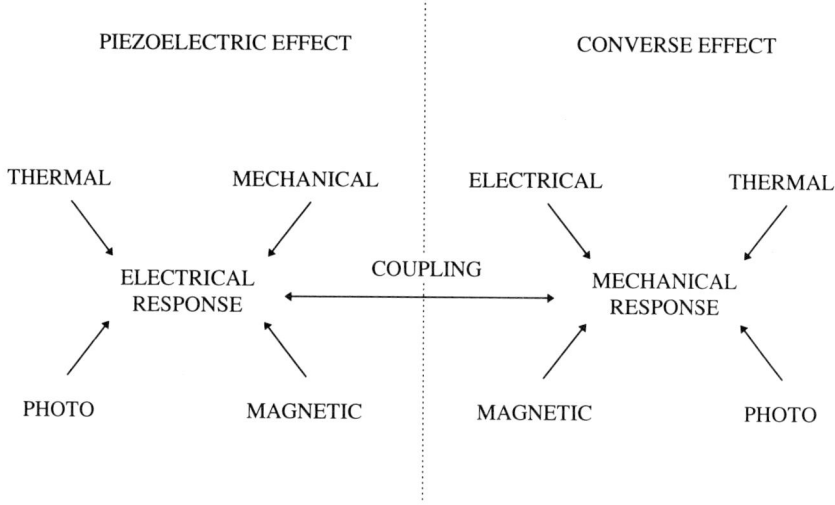

Figure 6.1 Piezoelectric direct and converse effects and influencing factors.

oscillator circuit. The mode (shape) of oscillation occurring in natural piezoelectric materials depends on the angle of cut relative to the x, y, z crystal axes and in ferroelectric materials depends on the orientation of the electrodes with respect to the application of the poling voltage. Common cuts for quartz crystals are x, y, AT and BT as reviewed by Cooke (1979). The 'AT cut' α-quartz crystal is the one most studied during the early days of sensor development; this consists of a slice of quartz cut at 35° 19′ between the x–y axes and parallel to the z axis, which provides crystals that oscillate in thickness shear mode in the megahertz range. Figure 6.1 illustrates these concepts and the combination of direct/converse effect utilised in chemical sensors. Also illustrated are other energy processes that can affect the resonant frequency of a PZ.

The resonant frequency of a device depends on:

- the piezoelectric material
- the dimensions of the crystal
- the media surrounding the material (i.e. the material at the surfaces of the PZ)

Furthermore, the addition or removal of relatively thin layers at the surfaces of the crystal reduce or increase the frequency, respectively. It is this phenomenon that is exploited in the use of PZs as mass sensors.

Piezoelectric elements can be characterised by their electrical impedance/

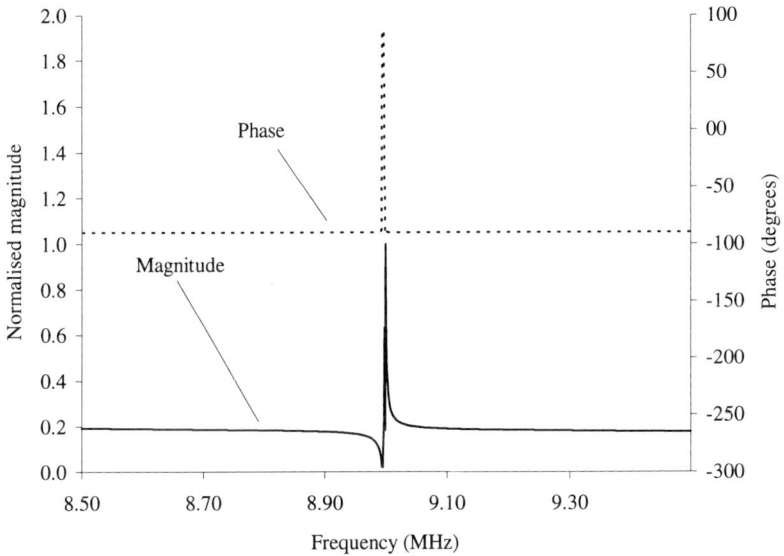

Figure 6.2 Impedance characteristics of a (9 MHz) quartz crystal.

admittance (Hayward, 1984); an example of the impedance of a quartz crystal is shown in Figure 6.2.

Piezoelectric elements are essentially capacitive, as may be observed by the general decreasing trend in the magnitude response and the $-90°$ phase value. However, at frequencies where the electromechanical behaviour of the device dominates, a rapid decrease from the characteristic capacitance line is obvious; this is then followed by a rapid increase in the magnitude value, which then decays away and the response rejoins the capacitive trend. It should also be noted that within this region the phase response is 90°, i.e. the device is now inductive. When the device is connected into an oscillator circuit, it controls the operating frequency as a result of the rapid change in the impedance of the device with respect to other components in the circuit; that is the circuit is forced to oscillate at a frequency between the frequencies corresponding to the minima and maxima in Figure 6.2. The first occurrence of this resonant frequency is termed the fundamental resonant frequency. The process repeats at odd numbers of this fundamental frequency; these are termed the harmonics of the piezoelectric.

The impedance of a piezoelectric device may be determined following the method proposed by Martin and Sigelmann (1975), which involves deriving the Thevenin equivalent circuits for the device. The analysis is based on the following relationships:

$$F_1 = \frac{Z_0}{j}\left[\frac{v_2}{\sin(kl)} + \frac{v_1}{\tan(kl)}\right] + \frac{hI_3}{j\omega} \qquad (6.1)$$

$$F_2 = \frac{Z_0}{j}\left[\frac{v_2}{\tan(kl)} + \frac{v_1}{\sin(kl)}\right] + \frac{hI_3}{j\omega} \qquad (6.2)$$

$$V = \frac{h}{j\omega}(v_2 + v_1) + \beta_{33}\frac{lI_3}{j\omega S} \qquad (6.3)$$

where a number of parameters are defined in Figure 6.3 and

$$Z_0 = S\rho V_t \qquad k = \frac{\omega}{v_t} \qquad \omega = 2\pi f \qquad (6.4)$$

v_t is the velocity of sound propagation through the piezoelectric material; h is the piezoelectric stress constant; β is the reciprocal of dielectric constant; l the thickness of the sensor; and S the surface area of the sensor.

Using the above, the electrical impedance, Z_{eeq}, for a piezoelectric material is given by:

$$Z_{eeq} = \frac{1}{j\omega C_0} + \frac{f^2}{ac - b^2}(2b + c + a) \qquad (6.5)$$

where

$$a = -\left(Z_1 + \frac{Z_0}{j\tan(kl)}\right), b = \frac{Z_0}{j\sin(kl)}, c = -\left(Z_2 + \frac{Z_0}{j\tan(kl)}\right), f = \frac{h}{j\omega} \qquad (6.6)$$

The above theory can be further developed to include a layer in contact with the device (D. Reilly and G. Brown, unpublished data).

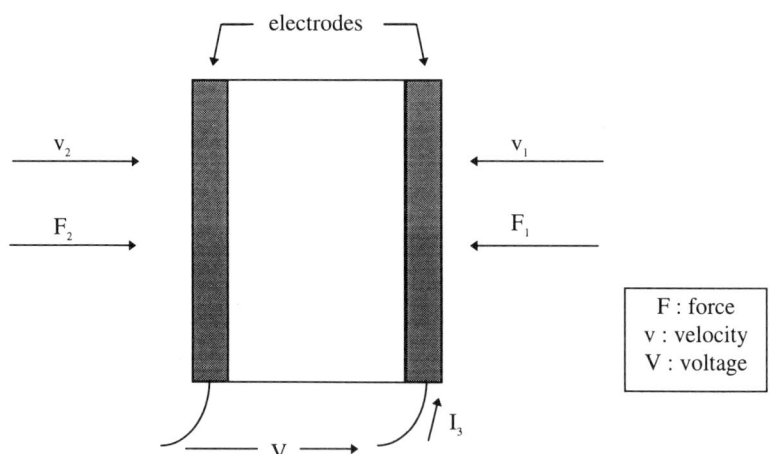

Figure 6.3 Variable definitions for impedance model: F, force; v, velocity; V, voltage.

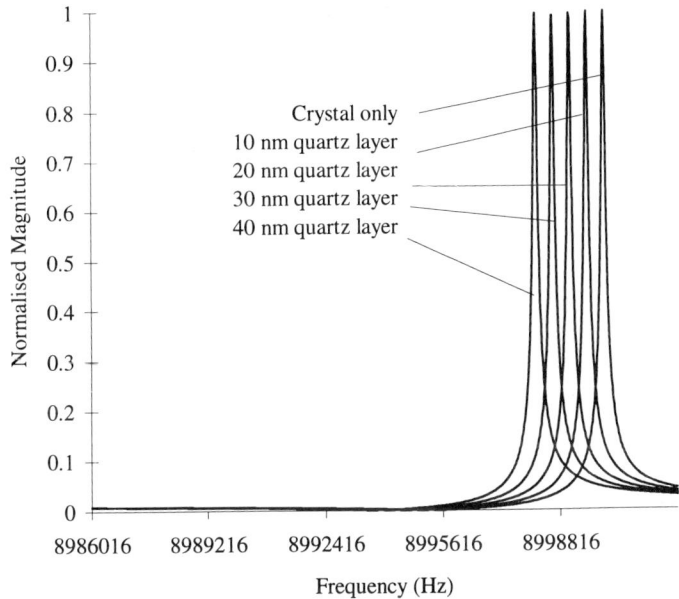

Figure 6.4 Impedance characteristics (magnitude only) for a 9 MHz quartz crystal with additional quartz layers in steps of 10 nm.

Equation 6.5 was used to generate the impedance characteristics shown in Figure 6.2, for an AT-cut quartz crystal device of dimensions 0.185 mm thick and 5 mm radius. Figure 6.4 illustrates the impedance characteristics (magnitude only is shown; however, the phase characteristics also shift by a corresponding amount) of the same device for several layer additions of quartz in steps of 10 nm. As may be observed, there is a reduction in the resonant frequency as the layer size increase, i.e. as the added mass increases.

Figure 6.5 indicates the change in the piezoelectric crystal response for the addition of copper, again in steps of 10 nm layers. As may be observed, the change in the resonant frequency per 10 nm is larger than that which occurred with quartz. However, consider the impedance characteristics for the addition of equal masses of quartz and copper, i.e. different thicknesses, as shown in Figure 6.6. For these cases the change in resonant frequency is equivalent. This is the basis for the use of the piezoelectric effect within mass balances; that is the change in resonant frequency for a given added mass is equal to the change that would occur for the equivalent mass addition of quartz. The relationship, which was first proposed by Sauerbrey (1959), may be described by the

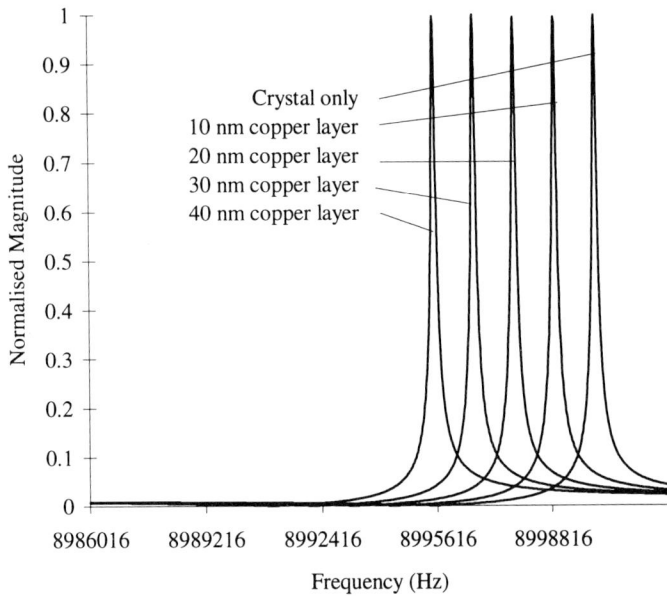

Figure 6.5 Impedance characteristics (magnitude only) for a 9 MHz quartz crystal with additional layers of copper in 10 nm steps.

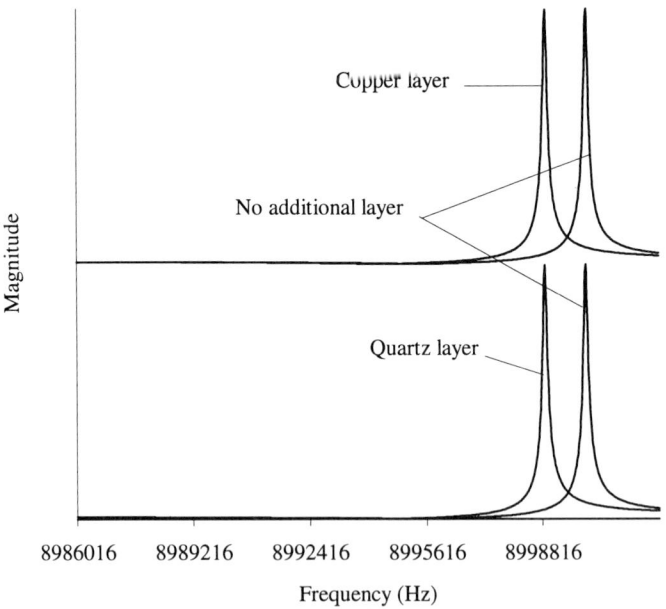

Figure 6.6 Impedance characteristics (magnitude only) for a 9MHz quartz crystal for additional layers of quartz and copper of equivalent mass.

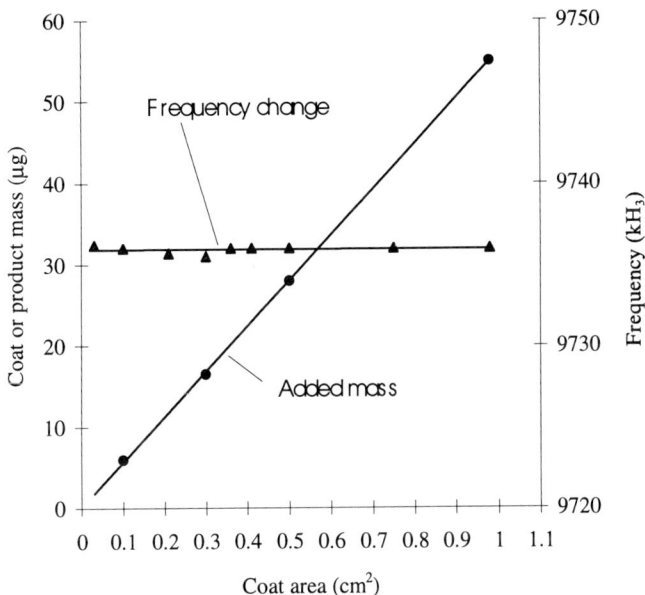

Figure 6.7 Theoretical sensitivity of a 9 MHz crystal for constant coat mass over varying areas.

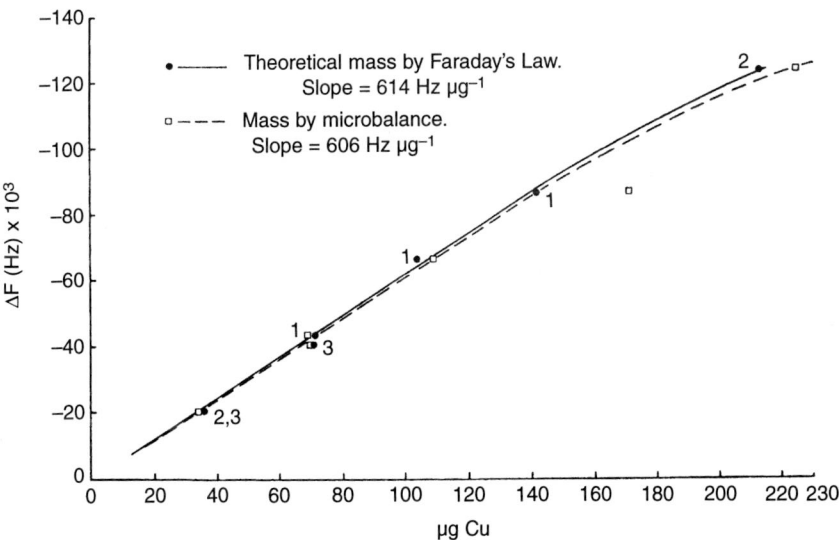

Figure 6.8 Quartz crystal response for evenly distributed added mass: ●—●, theoretical mass by Faraday's Law (slope 614 Hz μg⁻¹); □– –□, mass by microbalance measurement (slope 606 Hz μg⁻¹).

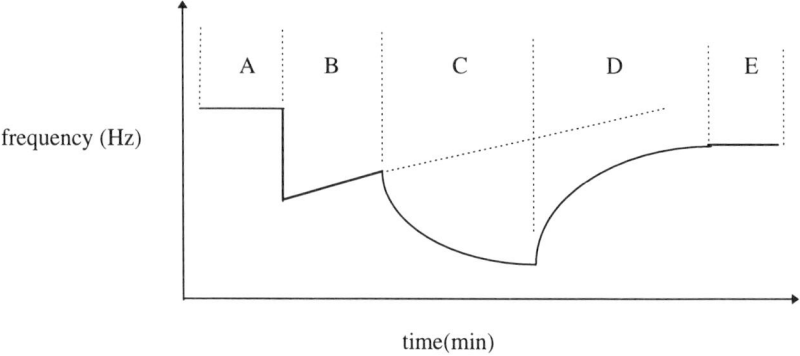

frequency (Hz)

time(min)

Figure 6.9 Real-time sensor response, where A is the resonant frequency of uncoated crystal; B, the new frequency owing to the mass of applied coating, drift is also indicated because of coating bleed; C is the sorption reaction progress curve between one gas analyte and the coating; and D is the desorption reaction as analyte is switch out of line from the gas cell (note that either the reaction is not completely reversible or there has been a humidity increase in the gas line); E is the humidity effect often observed as a change in baseline.

following equation:

$$\Delta f = -\Delta m N_1 \qquad (6.7)$$

where Δf is the change in frequency: Δm, the change in mass and N_1 a constant:

$$N_1 = \frac{f_0}{A \rho t}$$

where f_0 is the fundamental resonant frequency of the crystal; A is the area at the face of the crystal (usually $\sim 1\,cm^2$ for mass changes over the whole crystal face or $0.28\,cm^2$ for mass changes over a typical gold electrode area); ρ is the density of piezoelectric material (quartz is $2.65\,g\,cm^{-3}$); and t is the crystal thickness (typically $0.18\,mm$ for a $9\,MHz$ crystal).

Therefore, for a $9\,MHz$ crystal experiencing a mass change over the electrode area, a $1\,Hz$ frequency change is observed for a mass change of $1.6\,ng$. Figure 6.7 shows the theoretical sensitivity of such a crystal for constant coat mass over varying areas. Experimental data shown in Figure 6.8 demonstrate the linear response of the quartz crystal for evenly distributed added mass (copper by electrodeposition) at one surface. The figure also demonstrates a close agreement between changes in frequency predicted via the impedance simulations and the experimental data. Thus, the Sauerbrey equation is verified under very controlled mass addition conditions, as is the proposed model. When applied to real-time sensors the response may be depicted as illustrated in Figure 6.9.

6.3 Instrumentation

In order to measure the changes occurring in the resonant frequency of the piezoelectric device as a result of changes in mass at its surface, the PZ is usually incorporated within an oscillator circuit. The PZ within these circuits controls the frequency of the oscillations produced. A variety of oscillator circuits may be utilised; some of which are presented in Figure 6.10. Frerking (1978) presents an excellent summary of these and also provides an indication of the most suitable circuit for a particular

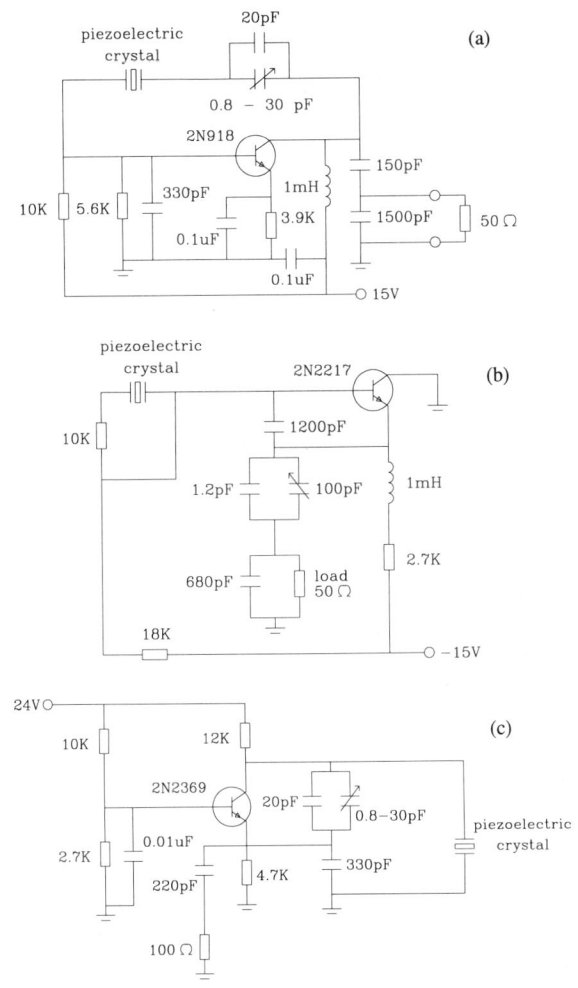

Figure 6.10 Oscillators circuits for operation with a 9 MHz crystal: (a) Pierce oscillator, (b) Colpitt's oscillator and (c) Clapp oscillator.

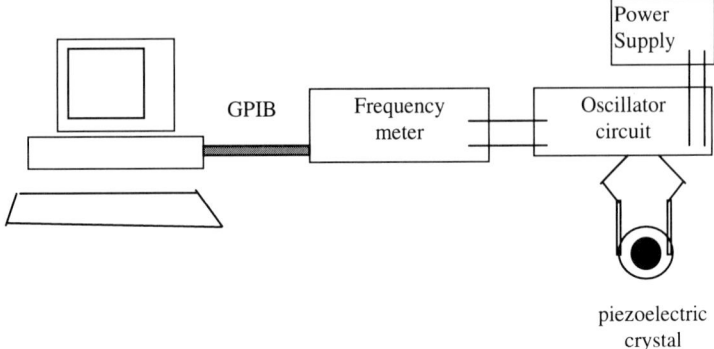

Figure 6.11 Example of an experimental configuration.

application. The PZ element forms the sensor, which must be mounted in a suitable cell that both protects the fragile PZ and allows delivery of the analyte material to the PZ in the most appropriate manner.

The frequency of the signal output from the oscillator circuit may be measured using a frequency counter. These are available with accuracies of 0.1 Hz for measurements in the megahertz range, hence they are extremely accurate. The meter can be connected to a PC via a suitable link, e.g. GPIB, and the system can then operate as a real-time mass balance. A schematic of a possible system is illustrated in Figure 6.11.

Alternatively, the frequency change can be measured by using a reference oscillator (i.e. with a similar crystal unaffected by the measurand) and an analogue multiplier circuit mixer. The output of the circuit will be a signal with a carrier frequency, at twice the crystal resonance, modulated at the frequency difference. Consequently, the use of a low pass filter will remove the high-frequency content of the signal and leave only the difference frequency. This may then be connected to a frequency-to-voltage circuit. Hence, the change in frequency may be converted into a change in voltage. The voltage value may then be measured via a meter connected to a PC, similar to the first case, or may be directly displayed on an LCD display module incorporating analogue-to-digital conversion.

6.4   Gas analysis

If a coating that selectively sorbs gaseous species is applied to the crystal, then the crystal will act as a monitoring device with its sensitivity dependent on coating/analyte area, temperature, and molar reaction ratios and its response time dependent on reaction kinetics, which are affected by gas flow, rate of presentation of analyte gas sample to the coating, temperature and analyte concentration (Cheney and Homolya, 1975). According to the Sauerbrey equation, sensitivity is inversely proportional to the coating area (Figure 6.7) (Earp, 1966), but sensitivity is also related to the mass of adsorbate (i.e. coating) provided we are considering total movement of analyte molecules through the coating and not simply monolayer effects (McBain, 1932; Bickerman, 1970). At atmospheric pressure, the adsorption data can be represented by Frendlich's equation:

$$\frac{\Delta m}{m_c} = kc^n \tag{6.8}$$

where Δm is the weight of substance adsorbed (g); m_c the mass of adsorbate (coating); c the concentration of analyte gas (ppm (v)) and k and n are constants that depend on the temperature.

Therefore, it should be possible to use this relationship in terms of detector response. If

$$\Delta m = -\frac{\Delta f}{N_1} \text{ (Sauerbrey)}$$

then

$$-\Delta f = N_1 m_c kc^n \tag{6.9}$$

Response Δf should increase with coating mass, the limit occurring when approaching the crystal oscillation 'damping-out' mass. The relationship also allows standardisation of results between different coating masses used over the same area. For the two parameters of area and mass of coating, there should be optimum values to achieve maximum sensitivity. In theoretical terms, it should be possible to calculate the optimal area and mass values and test these experimentally. However, the picture becomes complex once it is realised that many of the best coatings available are for modelling experiments with selected analytes: viscous gas chromatography stationary-phase (GCSP) materials. These coatings radially migrate (Cooke, 1979) using the impinger cell system; therefore, the mass distribution at the crystal surface changes, altering sensitivity.

By performing theoretical calculations to demonstrate Δf for various distributions areas of material at the crystal surface with m_c/A constant, it is apparent that Δf remains constant (Figure 6.12). Thus, we obtain the same response for the range of masses distributed over different areas. This is important because it means one should be able to keep the response time of a detector to a minimum while still maintaining the same sensitivity by simply using the smallest area of coating for a given m_c/A value. For a liquid coating used in a gas cell this will depend on its bleed characteristics (volatility) and its radial migration (dependent on the coating viscosity and

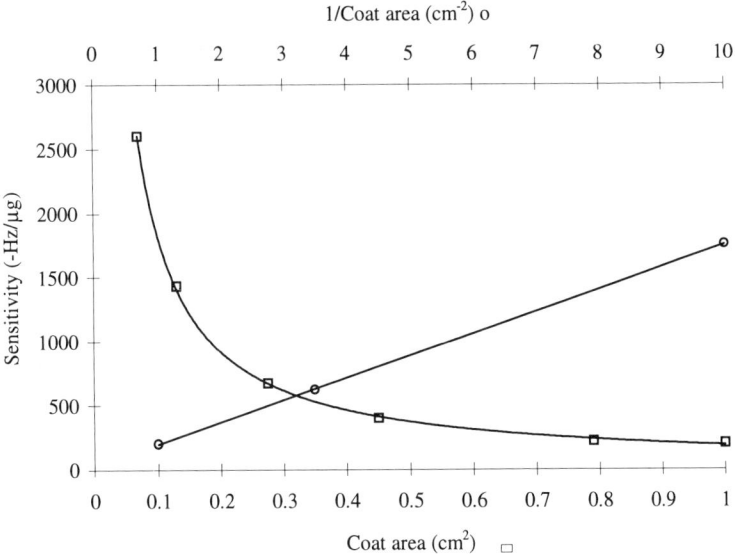

Figure 6.12 Frequency change versus distribution area.

gas jet strike velocity). For a solid, it may be possible to achieve greater mass/area coatings than with the liquids, though the overall loading factor causing damping out of oscillation, i.e. m_c/A should still remain constant; such factors as the particle size, density and elasticity properties of the solid will determine this.

Coating materials that are stable, free from ageing under atmospheric conditions (i.e. oxidation, hydrolysis), are reactive with specific gases and have good reaction reversibility properties are the best approach to dynamic real time trace gas analytical systems.

King (1964; 1965; 1969) first used a coated piezoelectric crystal for gas chromatography in 1964. He developed a detector sensitive to 0.1 ppm (V) water. Coatings investigated for adsorption of water were molecular sieves, hygroscopic polymer, silica gel, polar liquids, gold, nickel and aluminium. Since 1964, there has been a commercially available water detector based on a hygroscopic polymer-coated crystal and having a detection level at parts per million, a 30 s response time and long lifetime (Crawford, 1964). Gjessing *et al.* (1967) developed a humidity detector consisting of an SiO_x film evaporated on a crystal.

Guilbault and Lopez-Roman (1971) investigated crystal response to temperature and found increased frequency response with increased temperature, particularly in the range 100–200°C. Between 25 and 40°C the effect of temperature was very small. This work was completed prior to their investigation of the use of the following coating materials for the detection of sulphur dioxide: sodium tetrachloromercurate, apiexon, silicone, SE-30, silicone QF-1, Carbowax 20M and Versamid 900. A detection limit of 5 ppm (V) sulphur dioxide was claimed using a sodium tetrachloromercurate coating, which had been applied to the crystal by spraying. During 1973, two groups of researchers reported their findings for work on sulphur dioxide detectors at the same time; these groups were Janghorbani and Freund (1973) and Frenchette and Flashing (1973). The former group investigated the following GCSP coatings: Carbowax 400, 20M; dinonyl phthalate; polyphenyl ether; β, β'-oxidipropionitrile; SAIB (sucrose acetate–butyrate); triethanolamine, Amine 220 and squalane. The latter group investigated styrene–dimethyl-aminopropyl maleimide 1:1 co-polymer and quote 0.1 ppm (V) detection limit for sulphur dioxide using a syringe sampling technique. Karmarkar and Guilbault (1974) developed a glass impinger cell and used it with crystals coated with p-toluidine, Amine 220, triethanolamine, Quadrol and Armeen 25 for detection of sulphur dioxide down to 1.0 ppb (V) limit of detection. Karmarker *et al.* (1975) investigated the use of a hydrophobic membrane to eliminate moisture interference in sulphur dioxide determinations using a Quadrol-coated crystal. Triethanolamine was thoroughly tested as a potential coating for the detection of sulphur dioxide down to a limit of 25 ppm (V) by Cheney and Homolya (1975). These researchers showed response time to equilibrium decreases as gas concentration increases, using a flow cell (non-impinger type) and nitrogen carrier gas. Typically 25 ppm (V) sulphur dioxide had a sensor response (t_{eq}) of 13 min and Δf of -1962.5 Hz, with a flow rate of $100 cm^3 min^{-1}$, at a temperature of 25°C. Sorption of nitric oxide, nitrogen dioxide, carbon dioxide and oxygen was also tested using a triethanolamine coating. Nitrogen dioxide was found to be the only one of these gases to give a non-reversible reaction. Karmarker *et al.* (1976) reported on the use of a portable device for measurement of sulphur dioxide from automobile exhausts and stack gases. The device incorporated a glass impinger cell and a Quadrol-coated PZ. It operated on two ranges with a detection limit of 20–50 ppm (V) sulphur dioxide on range 1 and 300 ppm (V) on range 2. Street (1976) investigated many inorganic solid coating materials for use as sorbants for sulphur dioxide and nitrogen dioxide. In particular nickel II hydroxide, sodium tetrachloromercurate, silver oxide/silver metavanadate, the thermal decomposition product of silver permanganate and manganese dioxide. He developed a fluidised bed technique for applying solids in powder form to the crystal surface, using polyvinyl alcohol as a binder (20% success rate production). The organic stationary phase Carbowax 20M was also studied as a coating material for sulphur dioxide. Also considered were temperature effects, baseline autocorrection using a double crystal system (i.e. a reference and sample line). Edmonds (1976) studied organic vapours sorbed on GCSP coatings using a diffusion cell. He obtained frequency response data for ethylbenzene, o-xylene, acetone, hexane, cyclohexane and choloroform using the following coatings: Pluronic L64, Carbowax 20M and squalene. Multiplexing was thus possible using the simultaneous equation method (SEM) or signal ratio method (SRM) (Ostojic 1974). These data treatment techniques rely on the different degree of partition that each coating material has for the analyte and could be used in a field-based monitor with today's computer technology. In 1976, Cheney *et al.*, reported results obtained for the detection of sulphur dioxide using an ethylene–dinitrotetraethanol-coated crystal. A greater sensitivity to sulphur dioxide was obtained using a centre-coated 9 MHz crystal (340 Hz) than with a whole-area coated crystal (260 Hz); however, this is only true if the masses of both coatings are equal and bulk effect sorption occurs rather than surface layer effects.

Nitrogen dioxide and ammonia were found to be sorbed by Ucon-H-90,000 and Ucon-LB-300X (Karmakar and Guilbault, 1975); nitrogen carrier gas was used. They used IR spectroscopy to analyse product formation and claimed nitrogen dioxide and ammonia had good sensitivity on the coatings after first pre-reacting and coating with nitrogen dioxide, i.e. conditioning the coating.

Other substances analysed by piezoelectric sorption monitors, including pesticides, hydrogen sulphide, hydrogen chloride, aromatic hydrocarbons and mercury vapour, are given in a review of applications (Hlavay and Guilbault 1977). Later research includes detection of toluene diisocyanate, (Alder and Isaac, 1981).

The application of PZ sensors for monitoring the acid gases sulphur dioxide and nitrogen dioxide have been studied in relationship to acid rain and they are of interest in agriculture.

None of the above piezoelectric monitors claim to have reached the stage of development required for an accurate, high-precision, reliable, long-term, real-time dynamic continuous sampling system. Many have used inert carrier gas and a system of grab sampling of analyte gas, which is injected into the carrier gas. Therefore, these systems are really discontinuous samplers. Those who have attempted to achieve a real-time dynamic system with air as the carrier gas have run into problems of non-reversible oxidation or hydrolysis of the coating materials. Therefore, there is some scope for synthesis of coatings that resist these changes and react reversibly with specific gases/vapours. To this end, the concept of enzyme coatings and anchor polymeric coatings is considered. Figures 6.13 and 6.14 illustrate a number of coating types and routes which may be used for sulphur dioxide and nitrogen dioxide, respectively.

An alternative approach is to develop a system that incorporates filters which are presaturated with the analyte gas of interest as an integral part of the detector and use available GCSP coating materials. This was briefly investigated by Cooke (1979) using hydrophobic membranes for a sulphur

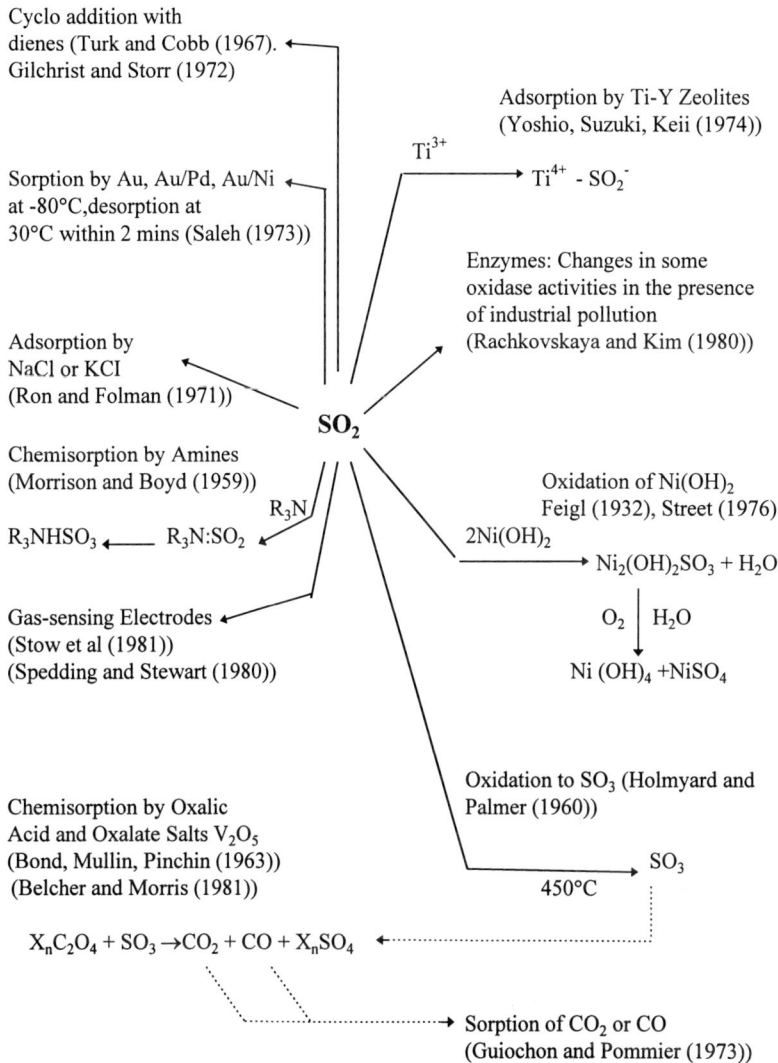

Figure 6.13 Selected reactions/sorptions of sulphur dioxide.

dioxide/triethanolamine analyte using a coating system. If one considers the ageing effect he monitored, using the piezoelectric sorption monitor in an equilibrium shift mode of operation (ESMO), they are dramatic, rendering the coating to be almost unreactive after several sorption/ desorption cycles. At 25–30°C using the ESMO, there is considerable detector dead time (i.e. prior to the reaction progress curve showing

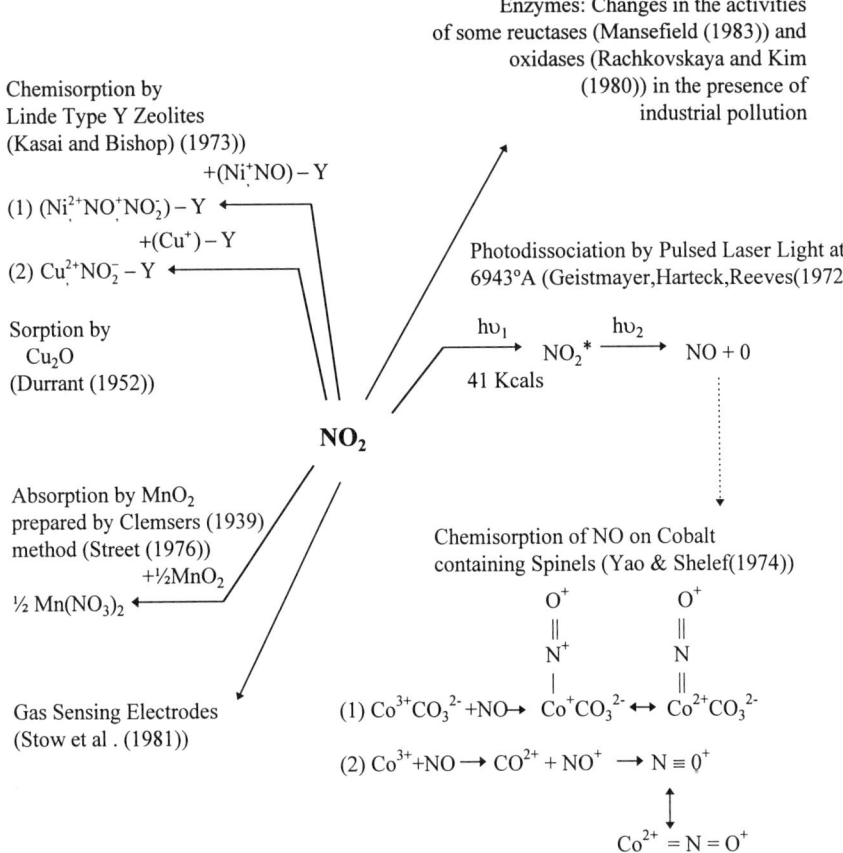

Figure 6.14 Selected reactions/sorptions of nitrogen dioxide.

equilibrium has been attained). An alternative mode of response proposed and investigated by Hepher (1984) is that of an initial rate of reaction mode of operation (IRRMO) or peak height obtained in a finite time.

The feasibility of temperature programming for rapid sorption/ desorption cycling was studied using the model system sulphur dioxide/ triethanolamine. A Peltier heater/cooler device was used in conjunction with a thermistor temperature controller unit to achieve temperatures in the 0 to 60°C range (Edmonds *et al.*, 1986). This was for later application to the solid coatings, which generally require either thermal desorption or a photostimulated desorption/reversible reaction. The model gas/solid system nitrogen dioxide and manganese dioxide was used to demonstrate this (Edmonds *et al.*, 1988). Inorganic metal oxides and hydroxides generally have a critical humidity at which they dissolve, oxidise or combine with other groups. The nitrogen dioxide/manganese dioxide system was studied in detail for a range of humidities, as was the proposed product manganese nitrate (Street, 1976).

Recent work by Hepher (1994) has shown that solvent fumes from lacquers may be monitored using Carbowax 20M coated 'AT-cut' 5 MHz quartz PZs. The work was towards development of sensor arrays for pattern recognition of mixtures of fumes from stack emissions, automobile exhausts and such processes as welding and spraying of paints and lacquers.

6.5 Piezoelectric aerosol sensors

Aerosols may be defined as liquid or solid particulate matter capable of dispersion through a gaseous atmosphere. Sprays and mists are aerosols of defined spherical size and arise from the liquid phase. Dust arises from the solid phase and can be divided into living, i.e. biological microorganisms (bacteria and fungi), and non-living, i.e. solid particulate matter including irregular shaped material and fibrous material (e.g. cement, coal, plaster, asbestos, glass and cotton). Some dusts exist in both irregular and fibrous form, e.g. carbon. The particulate size of the aerosol is of particular interest in health and safety and also in the clean rooms used in the electronic silicon chip industry. Figure 6.15 illustrates typical common particles and their size ranges. The respirable range is $< 7.5\,\mu m$ diameter and it is this range which is of most concern as a hazard to health or as a contaminant in certain delicate micro-electronics processes. Of recent concern are submicron particulates from both diesel and petrol engines. Currently these may be counted and displayed in the 10 to 1000 nm diameter range using TSI cloud chamber systems, but these are extremely costly and may eventually be competed with by PZ systems. PZs have been studied by a number of workers as a possible transducer device for real-time dust measurement over larger particulate ranges, i.e. $1-10\,\mu m$, though the tighter the range and the smaller the particle the better.

The major difference between the mechanisms operating at the PZ surface for dusts as opposed to gas or liquid analytes is related to the size and binding capacity of the analyte. With chemical species in the gaseous or the liquid phase and with a suitable chemical coating on the PZ, sorption and desorption or reversible chemical reactions can occur. However, with dusts this is not possible and the best options for causing the particles to adhere to the PZ surface appear to be the use of viscous oil coatings or electrostatic precipitation. Wilson (1992) studied the use of silicon oil coatings as a method of modelling dust distribution at the PZ surface for a series of sensor cell designs. Wilson and other workers have investigated dust charging techniques and this approach seems to offer the best hope for near real-time nanogram dust sensors of the future.

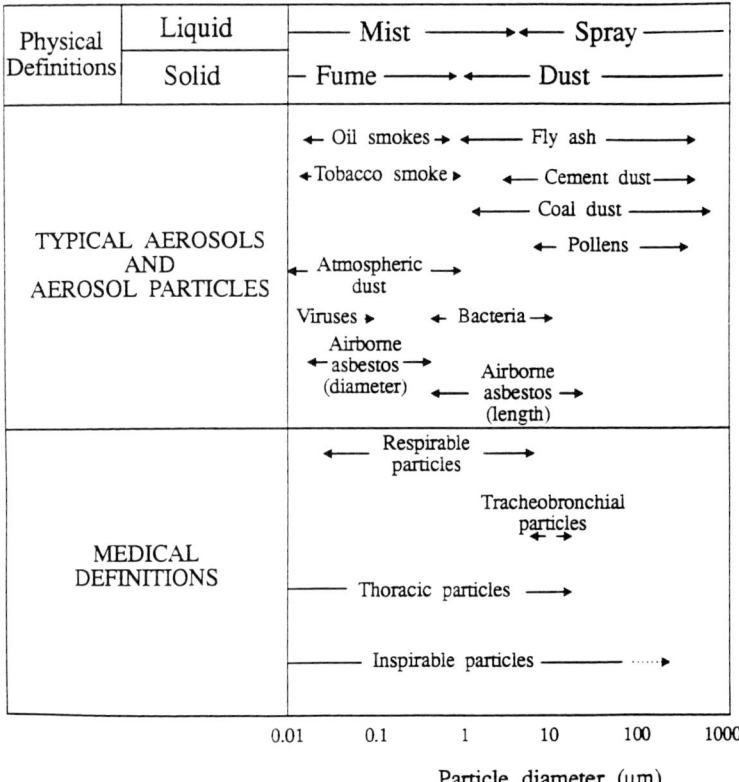

Figure 6.15 Common particles and their size ranges (Vincent, 1989).

The principal problems observed with 9 and 5 MHz quartz PZs are lack of dynamic linear range and deviation from Sauerbrey theory. These problems are associated with particle size range and lack of particle binding at the PZ surface. Lu and Lewis (1972) observed that as a deposit becomes thicker a point is reached when additional mass to the crystal produces less than the theoretical response. This can be caused by deviation from the monolayer requirement of Sauerbrey theory, non-binding of dust or dust bounce effects at the PZ surface. The response, therefore, becomes curvi-linear, and these factors explain departure from linear response. Olin (1971) suggested this point is reached for deposits equivalent to 1.5–6.0 $\mu g\,mm^{-2}$ for unspecified ambient and pollution source aerosols.

Temperature and humidity changes can also cause errors in dust measurement (Daley and Lundgren 1975).

Dust direction stage
electrostatic precipitator

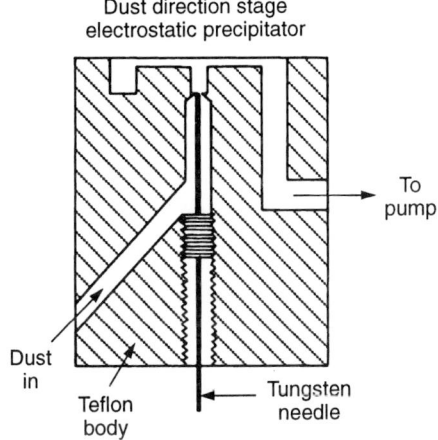

To
pump

Dust
in

Teflon
body

Tungsten
needle

Figure 6.16 Electrostatic precipitator dust sensor.

In the electrostatic precipitation method (Figure 6.16), the point-to-plane distance between a high-voltage electrode and the PZ sensor surface is critical. The force, F, applied to a particle within an electric field is a function of the number of charges (n) acquired and the field intensity, E, i.e.

$$F = neE \tag{6.10}$$

The particle mobility (B) within an electric field may be quantified using equation 6.11, which may be simplified to equation 6.12:

$$V = \frac{neEC}{3\pi\eta d} \tag{6.11}$$

where V is the terminal drift velocity, η is the viscosity of air and C is the Cunningham slip factor.

On simplification, let

$$B = \frac{C}{3\pi\eta d}$$

then

$$V = neEB$$

$$B = \frac{V}{neE} \tag{6.12}$$

$$Bne = \frac{V}{E}$$

Therefore, the particle mobility is a function of drift velocity and the intensity of the electric field, i.e. Z ; therefore, $Z = neB$. For a singly charged particle, particle mobility decreases as particle size increases. The electrostatic drift velocity of a charged particle can be much higher than the gravitational or inertial velocities and, therefore, electrostatic precipitation is favoured over these methods. White (1963) produced equation 6.13 to describe the number of charges n on particle of diameter d in an electric field:

$$n = \frac{3\varepsilon}{\varepsilon + 2} \frac{Ed_p^2}{4e} \frac{\pi N_i e Z_i t}{\pi N_i e Z_i t + 1} \qquad (6.13)$$

where E is the applied electric field; ε, the dielectric constant of the particle; Z_i the mobility of the species; e, the electron charge; N_i, the species concentration and t the charging time.

Equation 6.13 suggests a saturation charge or limited charge is reached with sufficient charging time. This normally occurs within a very short period.

In a point-to-plane electrostatic precipitator using a positive corona discharge, free primary electrons are drawn to the anode electrode creating electron-positive pairs by impact ionisation in the glow region (Lowe and Lucas, 1953). Outside this region, positive ions, aerosol molecules and neutral air molecules exist. The positive ions migrate towards the receiving electrode, interacting with the aerosol particles, charging them positively; thus, the positively charged aerosol accelerates towards the PZ surface, which may be grounded or negatively charged.

The corona along a wire electrode has distinct characteristics: for example, a positive corona is smooth and uniform whereas a negative corona shows as a series of localised glow points that can move along the wire. The overall effect on particle aerosol charge is that a negative corona imparts negative charge to dusts while a positive corona imparts positive charge. Figure 6.17 illustrates the apparent effect of different dust charges on the deposition over a PZ sensor surface (Wilson, Hepher and Reilly, 1994). Luminescent dust tracking has also been attempted in an experimental electrostatic dust cell (Hepher and Wilson, 1995). Other features about coronas include the fact that negative coronas are produced at lower applied voltages and yield higher corona current at any applied voltage. No current flows until a minimum voltage to ionise the air molecules in air is approached. Current increases rapidly with applied voltage until one of two limiting conditions is reached: (i) spark over resulting from localised field intensity becomes large, i.e. this can occur with dusts causing tracking; or (ii) over potential causes expansion of the glow region to a generalised corona to fill the complete air gap between the electrodes, resulting in non-charging of the particles. Figure 6.18

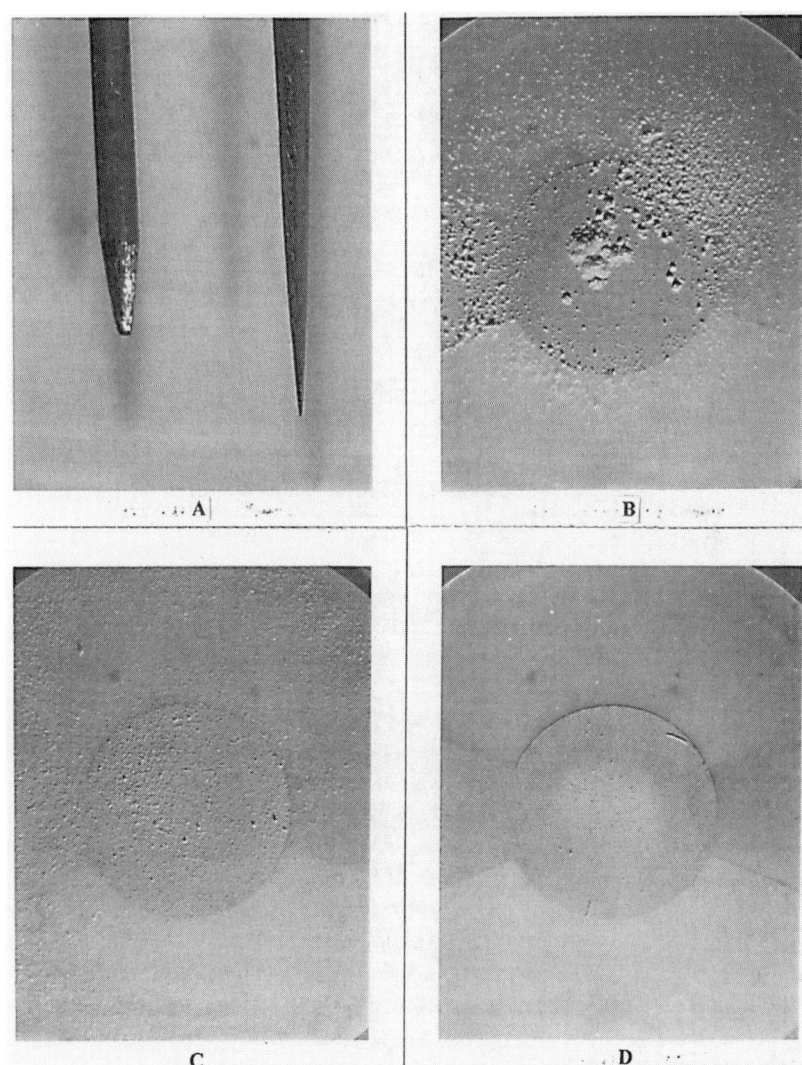

Figure 6.17 Dust charge effect on deposition over sensor surface. (a) Tungsten electrodes used in electrostatic precipitator. (b) No electrostatic precipitation. (c) Electrostatic precipitation using positive corona (point-to-plane distance 9.5 mm, 5000 V). (d) Electrostatic precipitation using negative corona (point-to-plane distance 9.5 mm, 5000 V).

illustrates the voltage–current relationship and the various regions of operation.

Brown *et al.* (1951) reported no measurable difference in the collection efficiency between positive and negative corona systems; however, this fact

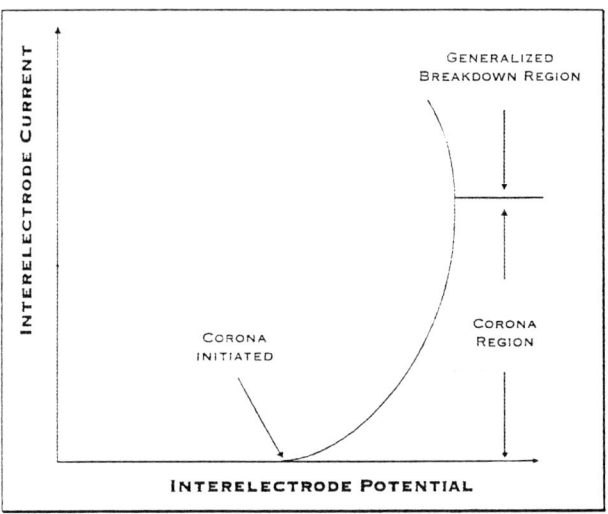

Figure 6.18 Typical voltage–current curve for a point-to-plane electrode system used in particle focusing and precipitation. Note the rapid current rise in the corona region prior to corona breakdown and arcing.

seems to be disputed in terms of the PZ response linking back to dust distribution at the PZ surface, as indicated earlier in Figure 6.17. Cheng *et al.* (1981) indicated that the precipitation efficiency is a function of particle size and flow rates; one way to minimise these effects is to keep flow rates low and use particle discriminators to monitor only respirable particles, i.e. $< 7.5\,\mu m$ diameter. Larger particles and particle ranges contribute towards departure from pseudomonolayer effects. Adhesion (Penney, 1962) is dependent on the kind of dust layer formed, i.e. inherent dipole characteristics of particles affect adhesive properties of the species. For most effective electrostatic precipitation, the particles should have some electrical conductivity otherwise non-conducting particles within the corona zone can form an insulating barrier that reduces the corona current and thus the charging capacity. Humidity can aid particle conductivity through water vapour absorption on the surface; high humidity can lead to electrical breakdown at lower voltages than the corona discharge has been set for with a particular point-to-plane distance and cell configuration. White (1963) reported that high-voltage industrial precipitators produce parts per million levels of ozone; other workers who have investigated these effects are Silverman and Dennis (1956), Beadle *et al.* (1954) and Machala (1965). A problem area in dust measurement using PZ/electrostatic precipitator sensors may arise from ultrafine particles, $< 0.01\,\mu m$ diameter, because these are difficult to charge and no account is taken of the initial mechanism involved for these sizes of material, though Mercer (1957) and Liu and Yeh (1967) have attempted explanations.

Current thinking on PZ technology for dust monitoring includes disposable 'one shot' PZ quartz crystals coated with silicon oil to be used for each half of an 8 h working shift, development of porous PVDF dust filter trap sensor elements and electrostatic coating technology for quartz PZ dust sensor elements.

Wilson (1996) reported > 95% collection efficiency for Arizona road dust using field charging with a corona discharge and a 5–7 kV tungsten electrode with a 5–10 mm point-to-plane distance. In this work, gold electrodes were used on the PZ sensing element and they tended to erode, the amount of erosion depending on the abrasive nature of the dust type examined. Also there were memory effects that appeared to be caused by dust occlusion in the micropore gold electrode surface. Scanning tunnelling microscopy has been used to examine these surface effects and indicate the surface topography of such a gold electrode on the PZ surface (Figure 6.19). Finally such experimental systems are prone to 'burn out' caused by arcing, this is particularly likely through electrical tracking via dust to the PZ surface. When this occurs, there is a surge in electrical current; therefore, current limiters are required on these devices to prevent premature sensor element destruction. In a working experimental electrostatic precipitation piezoelectric system, the charging chamber to needle tip distance also affects the sensitivity, as illustrated in Figure 6.20.

6.6 Piezoelectric crystal liquid sensors

Prior to 1980, there were few publications on applications of PZs in the analysis of specific analytes in liquids, and in fact early reported work seems to have concentrated on their uses in organic solvents. The main problems seem to have been: (i) an inability to prevent damping out, i.e. complete detuning of the integrated oscillator PZ system; and (ii) a lack of knowledge on production of suitable immobilised surface-active coatings for these sensors. Essentially, these are problems in two very distinct disciplines, i.e. electronics and chemistry. However, there has been a gradual merging of the operational development areas of electronics and chemistry since the very early work described on gas and humidity sensors.

Nomura and Hattori (1980) have pioneered the use of silver-plated PZs for micromolar concentrations of cyanide in solution. They used a 9 MHz 'AT-cut' quartz crystal having 5.0 mm diameter gold electrodes and an International Kits oscillator. The gold electrodes were plated with silver and subsequently exposed and found to respond to potassium cyanide concentrations in the range 2.6–260 ppb cyanide ions. The system was not a dynamic real-time system and did not function under liquid; rather it

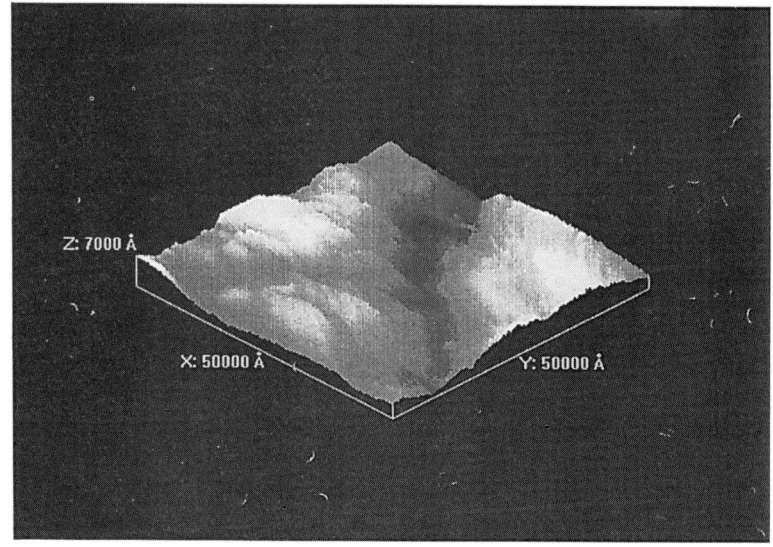

(a)

Height (Angstroms)

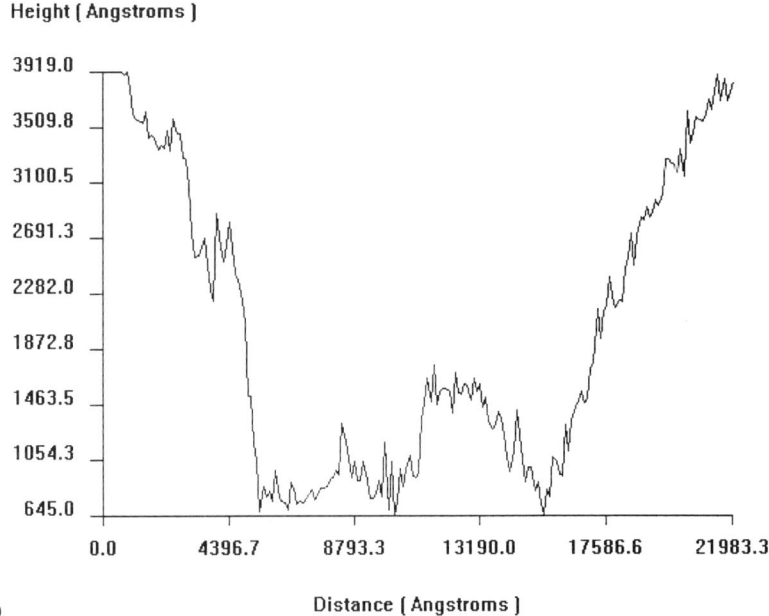

(b)   Distance (Angstroms)

Figure 6.19 (a) Surface topography of a gold electrode on a quartz crystal. (b) Sample cross-section of the channel region present in (a). This illustrates that pores are large enough to occlude dust particles at the lower end of the respirable region ($< 3\,\mu m$) and cause sensor memory effects, i.e. the channel depth is $\approx 0.33\,\mu m$, channel width tapering is from $\approx 2.0\,\mu m$ at top to $\approx 1.0\,\mu m$ at the bottom in this case.

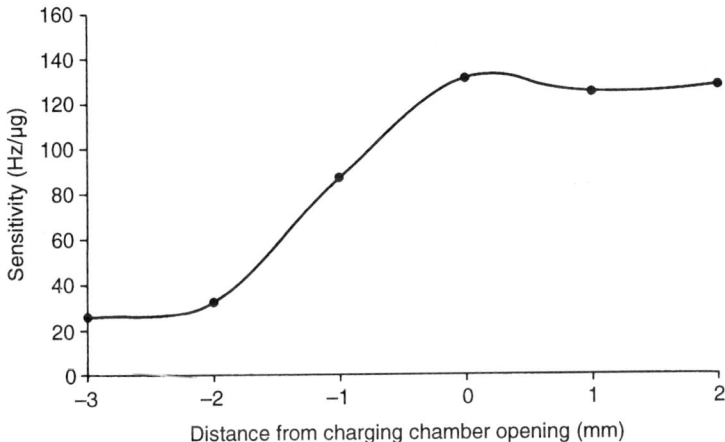

Figure 6.20 Apparent electrostatic dust sensor sensitivity in relationship to the charging chamber to central electrode needle tip distance.

was a discrete wet/dry lift and lay system. It should be noted that these systems, which may be considered electrochemical in nature, have restricted working lifetimes; however, reactivation is often possible via electro-chemical stripping methods. These systems are, therefore, cyclic systems that can be made rapid by employing a bank of sensors and multiplexors, they are alternative and more primitive systems than the immobilised phase-reversible reaction systems being pursued by other workers, but they have much to credit them as real working systems. Thus the crystal frequency is only measured when the crystal is dry; however, it does pose the question, could such a system be automated to give near real-time sensors? This might be achieved via a simple rotating arm (Figure 6.21)

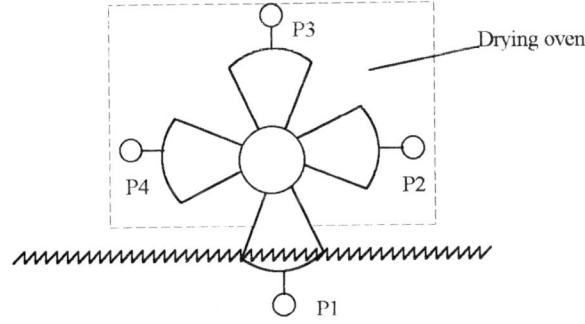

Figure 6.21 Proposed rotating arm device for multiplexed liquid sensor near-real-time system. P1 is the sampling position under the liquid surface; P2 and P3 are drying positions; and P4 is the data reading position.

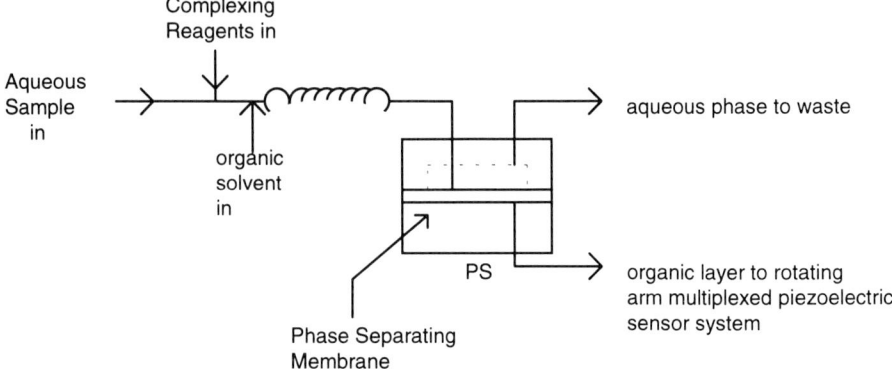

Figure 6.22 Phase separator automated lead quinolate measurement using a rotating PZ sensor head.

device allowing a sequential sampling, drying and data reading programme to operate.

Nomura and Hattori (1980) also claimed lack of interference from a number of anions and cations (nitrate, percholate, carbonate, sulphate, aluminium, chromium(III), iron(III) and ammonium ions). Iodide, thio-sulphate and sulphide interfered through silver salt formations. Silver(I), mercury(II) cobalt(II) and nickel(II) interfered because they formed stable complexes. Lead(II), cadium(II), maganese(II) copper(II) and zinc(II) could be prevented from interfering by addition of EDTA (0.005 M final solution strength). Also in 1980, Nomura and Minemura reported a far more interesting finding on the operation of a PZ that has one of its electrodes immersed in solution; this appears to offer a route to solving some of the electrical problems associated with obtaining a stable oscillator drive system for a liquid load at the surface of a crystal. Nomura and Minatsu (1982) also appear to have successfully analysed trace iodide (38.1–3810 ppb) using a near-real-time (on-line/off-line) PZ sensor in aqueous continuous flow conditions. This again demonstrates the potential application and long-term uses of such systems if applied to *in situ* aquatic monitoring or chemical laboratory instrument systems. Nomura *et al.* (1982) showed another potential system for a hyphenated approach (Figure 6.22) to using piezoelectric sensors, via solvent-extracted pre-concentration of lead 8-quinolinolate, for measurement of lead in the range 621–6210 ppb. Interferences of iron(III), nickel, cobalt(II), zinc, cadmium and silver(I) could be masked using L-ascorbic acid and cyanide. Another hyphenated approach that may be possible is a PZ cyclic voltammetry system (Figure 6.23).

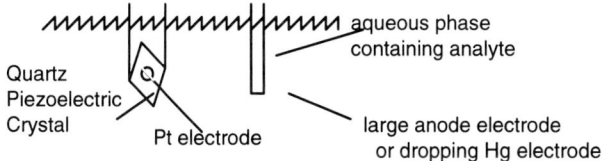

Quartz
Piezoelectric
Crystal

Pt electrode

aqueous phase
containing analyte

large anode electrode
or dropping Hg electrode

Figure 6.23 Piezoelectric cyclic voltage sweep electrochemical sensor.

The autosolvent extraction system (Figure 6.22) consists of a solvent mixing coil (S) (causes mixing of the organic and aqueous phase) followed by a sandwich phase separator (PS) (de Ruiter, 1989). The organic layer containing the trace species of interest (complexed) passes through the phase-separation membrane and the trace species measurement is made at the PZ detector surface. This system does demonstrate another approach that might be employed perhaps on route to the ultimate objective of simple continuous aqueous flow systems and reversible immobilised surface activated piezoelectric systems for aquatic environment.

These systems are capable of nanogram mass detection from a large volume of sample and have the ultimate trace analysis capability. Also they can sequentially detect and analyse a range of elemental ions from one stop flow volume. The reverse bias voltage sweep time governs the data update times (i.e. this may not be regarded as a rapid near-real-time system unless single species are being analysed).

Understanding the fundamental characteristics of integrated PZ oscillators for varying oscillator types, varying crystal type and cut, and varying solvent type hold the clue(s) to controlled systems development. In particular for natural waters analysis, the effect of solute type and loading (i.e. concentration) in water is important.

6.7 PZ sensor coatings operating in liquids

Any coating material acting as a stationary-phase detector in aquatic on-line analysis must possess certain basic properties prior to reactivity with specific analytes. These properties will include insolubility, inertness to water permeation, and dimensional stability, i.e. lack of swelling or shrinking. Also the coatings must be stable or at least have a characterised response in saline as well as in non-saline conditions. Coating materials should not be attractive sources of nutrients for microorganisms, to reduce the chance of sensor surface damage from these sources; it should be recognised that in flow cell systems this type of damage is not envisaged because we have continuous flow. However, in some circumstances, it may be necessary to use stop-flow procedures if chemical reaction rate with the immobilised active species on the sensor coating is slow. Also if piezoelectric sensors are to be used as aquatic biomonitors, then the stop-flow piezo-bio incubator with nutrient-coated crystals could be used as a special case. Such systems are likely to pose problems much more difficult to solve than those concerned with chemical speciation sensors. It is more likely that piezo-bio aquatic sensors will operate on specific bioproducts from a specific microorganism rather than via colony growths at a piezoelectric surface, i.e. using immobilised antibodies at the PZ surface to react with specified biomolecules from specific microorganisms.

Immobilised functional groups may have an affinity preference for certain cations, anions or organic molecules and the process of interaction at the piezoelectric surface may be a physical or chemical sorption process forming a loose bonding arrangement or it may be a chemical reaction forming a much stronger bond; all of which may be used with the piezoelectric micromass sensor.

Sometimes sorption reactions, and/or chemical reactions can be aided by the use of photo or thermal means. If we consider an analyte reacting with an immobilised functional group at the surface of a coating, and call the reaction forming the product the forward reaction (i.e. the reaction responsible for sensor response and directly relating to analyte concentration in the aqueous phase) then the backward or reverse reaction is that

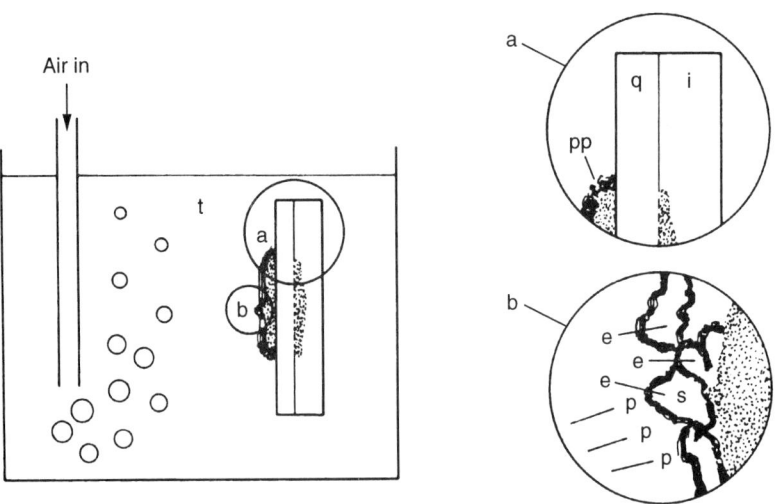

Figure 6.24 DMC substrate occluded in polypyrrole at a PZ gold electrode for real-time measurement of a protease enzyme. t, buffer solution liquid sorption trap; pp, polypyrrole; q, quartz crystal; i, insulator to prevent shorting of back electrode; s, occluded substrate within the polypyrrole coating; e, analyte enzyme; p, degradation products leaving coating surface.

which returns the surface of the coating to its original state. Such reversals may also be aided by photo and thermal means.

Emerging technologies that also have great potential with PZs include polymeric membranes (Schafer, 1976) and template polymers (Shea and Dougherty, 1986). In the enzyme dust research programme at Glasgow Caledonian University, polypyrrole has been used to entrap substrates at a PZ surface in a novel PZ enzyme dust monitor (Figure 6.24) (Hepher *et al.*, 1992); this polymer also has many other potential uses as a PZ coating material as it may be prepared in either a conducting or non-conducting form. Modification of PVDF and manufacturing procedures for PZ sensor elements from this material have also been studied at Glasgow Caledonian University (Hepher, 1989).

References

Alder, J.F. and Isaac, C.A. (1981) *Anal. Chim. Acta*, **129**, 163.
Beadle, D.G., Kitto, P.H. and Blignaut, P.J. (1954) *AMA Arch. Ind. Hyg. Occup. Med.*, **10** (5), 381–389.
Belcher, R. and Morris, S.J. (1981) *Talanta*, **28**, 599–600.
Bickerman, J.J. (1970) *Physical Chemistry/A Series of Monographs*, Vol. 20. Academic Press, New York.
Bond, R.L., Mullin, M.J. and Pinchin, F.J. (1963) *Chem. Ind.*, **Nov 30**, 1902–1903.
Brown, J.K., Hosey, A.D. and Jones, H.H. (1951) *AMA Arch. Ind. Hyg. Occup. Med.*, **3**, 198–203.
Cheney, J.L. and Homolya, J.B. (1975) *Anal. Lett.*, **8** (3), 173–193.
Cheney, J.L., Norwood, T. and Homolya, J.B. (1976) *Anal. Lett.*, **9**, 361.
Cheng, Y.S., Yeh, H.C. and Kanapilly, G.M. (1981) *Am. Ind. Hyg. Assoc. J.*, **42**:605–10.
Clemser, O. (1939) *Ber. dt. chem Ges.*, **72B**, 1879.
Cooke, S. (1979) PhD Thesis, University of Aberdeen, Scotland, UK.
Crawford, H.M. (1964) *Analysis and Instrumentation*, pp. 105–108. Plenum Press, New York.
Daley, P.S. and Lundgren, D.A., (1975) *Am. Ind. Hyg. Assoc. J.*, **36**, 518–532.
Durrant, P.J. (1952) *General and Inorganic Chemistry*, 2nd edn. Longmans, Green and Co, London.
Earp, R.B.W. (1966) PhD Thesis, University of Alabama, USA.
Edmonds, T.E. (1976) PhD Thesis, Imperial College, London University.
Edmonds, T.E., Hepher, M.J.and West, T.S. (1986) *Anal. Chim. Acta*, **187**, 293–399.
Edmonds, T.E., Hepher, M.J. and West T.S. (1988) *Anal. Chim. Acta*, **207**, 67–75.
Feigl, F. and Frankel, E. (1932) *Ber. dt. Chem. Ges.*, **65**, 545.
Frenchette, M.W. and Flashing, J.L. (1973) *Environ. Sci. Tech.*, **7**, 1765.
Frerking, M.E. (1978) *Crystal Oscillator Design and Temperature Compensation*. Van Norstrand, New York.
Geistmayer, J.W., Harteck, P. and Reeves, R.R. (1972) *J. Phys. Chem.*, **76**, 474.
Gilchrist, T.L. and Storr R.C. (1972) *Organic Reactions and Orbital Symmetry*. Cambridge University Press, Cambridge.
Gjessing, D.T., Holm, G. and Lanes, T. (1967) *Electronic Letters*, **3** (4).
Guiochon, G. and Pommier, C., (1973) *Gas chromatography in Inorganic and Organometallics*. Ann Arbor Scientific Publishers, Michigan.
Guilbault, G.G. and Lopez-Roman, A. (1971) *Environ. Lett.*, **2**, 35.
Hayward, G. (1984) A system feedback representation of piezoelectric transducer operational impedance, *Ultrasonics*, **July**, 153–162.
Hepher, M.J. (1984) PhD Thesis, University of Aberdeen, Scotland, UK.
Hepher, M.J. (1989) *RSC Small Grants Report*.

208   SENSOR SYSTEMS FOR ENVIRONMENTAL MONITORING

Hepher, M.J. (1994) *Proc. 2nd Int. Sym. Exhibition Environ, Contamination in Central and Eastern Europe*, Budapest, 20–31 September, pp. 27–30.
Hepher, M.J., Hamilton A., Wilson, L., and Teitje-Girault, J. (1993) *Aerosols Conf.*, Oxford, 6–11 September.
Hepher, M.J. and Wilson, L. (1995) *Proc. of Scotsense Symp. Envirotech '95*, Glasgow.
Hepher, M.J., Smith, C., Blundell, K. and Whitaker, R. (1995) *Proc. Scotsense Symp. Envirotech '95*, Glasgow, 9 March.
Hlavay, J. and Guilbault, G.G. (1977) *Anal. Chem.*, **49** (13), 1890–1898.
Holmyard, E.J. and Palmer, W.G. (1960) *Inorganic Chemistry*, p. 442, Dentl, London.
Janghorbani, M. and Freund, H. (1973) *Anal. Chem.*, **45**, 325.
Karmarker, K.H. and Guilbault, G.G. (1974) *Anal. Chim. Acta*, **71**, 419.
Karmarker, K.H. and Guilbault, G.G. (1975) *Anal. Chim.*, **75**, 111.
Karmarker, K.H., Webber, L.W. and Guilbault, G.G. (1975) *Environ. Lett.*, **6**, 345.
Karmarker, K.H., Webber, L.M. and Guilbault, G.G. (1976) *Anal. Chem. Acta*, **81**, 265.
Kasai, P.H. and Bishop, Jr, R.J. (1973) *J. Phys. Chem.*, **77**, 2308–2312.
King Jr, W.H. (1964) *Anal. Chem. Acta*, **36**, 1735.
King Jr, W.H. (1965) *US patent 3,164004, Jan 5.*
King Jr, W.H. (1969) *Res/Devel.*, **20** (4), 28.
Lowe, H.T. and Lucas, D.H. (1953) *Brit. J. Appl. Physics*, **24**, (Suppl. 2), 40–47.
Liu, B.Y.H. and Yeh, H.C. (1967) *Effect of Pressure and Electric Field on the Charging of Aerosol Particles*, Pub. No. 118. Particle Tech Lab., Mech. Eng. Dept., Univ. Minn., Minneapolis, MN.
Lu, C.S. and Lewis, O. (1972) Investigation of film-thickness determination by oscillating quartz resonators with large mass load, *J. Appl. Phys.*, **43** (11), 4385–4390.
McBain, J.W. (1932) *The Sorption of Gases and Vapours by Solids*, George Routledge & Sons Ltd, London.
Machala, O. (1965) In *Aerosols–Physical Chemistry and Applications*, (ed. K. Spurny), pp. **00-00**. Gordon and Breach, New York.
Mansfield, T.A. (1983) In *Pollution: Causes, Effects and Control*, (ed. R. M. Harrison), Publication 232. Royal Society of Chemistry, London.
Martin, R.W. and Sigelmann, R.A., (1975) *J. Acoust. Soc. Am.*, **58** (2) 475–489.
Mercer, T.T. (1957) *Atomic Energy Project Ur-475*, Univ. Rochester.
Morrison, R.T. and Boyd, R.N. (1959), *Organic Chemistry*. Allyn and Bacon, Boston, USA.
Nomura, T. and Hattori, O. (1980) *Anal. Chem. Acta*, **115**, 323–326.
Nomura, T. and Minatsu, T. (1982) *Anal. Chim. Acta*, **143**, 237–241.
Nomura, T. and Minemura, A. (1980) *Nippon Kagaku Kaishi*, 1621.
Nomura, T., Yamashita, T. and West, T.S. (1982) *Anal. Chim. Acta*, **143**, 243–247.
Olin, J.G. (1971) *Inst. Soc.Am. Int. Conf. and Ex.*, Chicago, Il, 4–7 Oct, Vol. 558, pp. 1–10..
Ostojic, N. (1974) *Anal. Chem.*, **46**, 1653.
Penney, G.W. (1962) Role of adhesion in electrostatic precipitation, *Am. Med. Assoc. Arch. Env. Health*, **4** (3), 301–305.
Rachkovskaya, M.M. and Kim, L.O. (1980) *Gazoustoich Rast*, (ed. Nikolaevskii), pp. 117–126. *Chem. Abs.* (1981) **94**, 151301r.
Ron, T. and Folman, M. (1971) *J. Phys. Chem.*, **75**, 2602.
de Ruiter, C. (1989) PhD Thesis, Vrije Universiteit, Amsterdam, pp. 37–48.
Saleh, J.M. (1973) *J. Phys. Chem.*, **77**, 1849.
Sauerbrey, G.Z. (1959) *Z. Physik*, **155**, 206–222.
Schafer, O.F. (1976) *Anal. Chim. Acta*, **87**, 495.
Shea, K.J. and Dougherty, T.K. (1986) *J. Am. Chem. Soc.*, **108**, 1091–1093.
Silverman, L. and Dennis, R. (1956) *Air. Cond. Heating and Vent.*, **12**, 75–80.
Spedding, D.J. and Stewart, G.M. (1980) *Analyst*, **105** (12), 1182.
Stow *et al.* (1981) *Comprehensive Analytical Chemistry*, Vol. XI, (ed. J. Grimshaw, P. Moritz and W.E. der Linden), p. 353. Elsevier, Amsterdam.
Street, D. (1976) PhD Thesis, Imperial College, University of London, UK.
Turk, S.D. and Cobb, R.L. (1967) *1,4-Cycloaddition Reactions*, (ed. J. Hamer), Academic Press, New York.
Vincent, J. (1989) *Aerosol Sampling*, J. Wiley and Son Ltd.
White, H.J. (1963) *Industrial Electrostatic Precipitation*, Addison-Wesley, Reading, MA.

Wilson, L. (1992) MPhil Thesis, Glasgow Caledonian University.
Wilson, L. (1996) PhD Thesis, Glasgow Caledonian University, UK.
Wilson, L.W., Hepher, M.J. and Reilly, D. (1994) *Scotsense Envirotech '94*, Glasgow, 17 February.
Yao, H.C. and Shelef, M. (1974) *J. Phys. Chem.*, **78**, 2490–2496.
Yoshio, O., Susiki, K. and Keii, T. (1974) *J. Phys. Chem.*, **78**, 218–220.

7 Biosensor devices

M. CARDOSI and B. HAGGETT

7.1 General introduction

Biosensors are analytical devices that transduce biological reactions into electrical signals. The unique feature of a biosensor is that the probe incorporates a *biological sensing element* close to the signal transducer, resulting in a device that is specific either for a particular chemical or for a group of related chemicals (Figure 7.1). Although the major thrust in biosensor development since the early 1970s has been for health care applications, a survey of the market potential (Hall, 1990) has identified the increasing importance of biosensors in environmental monitoring.

While the possible combinations of biocomponent and transducer are extensive (Table 7.1), it is the electrochemical probes, and in particular those based on amperometry, that represent the most advanced designs to date (Mascini and Palleschi, 1989; Hall, 1990; Scheller and Schubert, 1992). Consequently, the content of this chapter will reflect this point. (Optical biosensors are dealt with in Volume 1, Chapter 2.) In addition, the discussion of biological elements will focus on enzymes, microorganisms and anti-

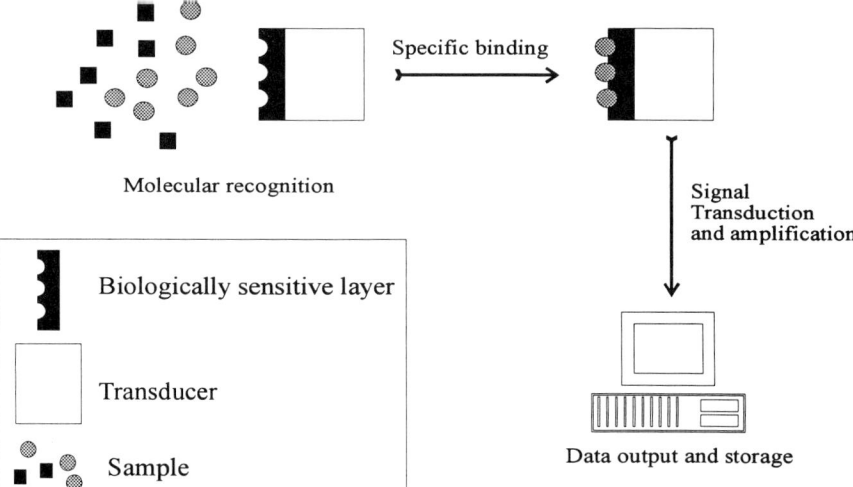

Figure 7.1 Schematic representation of a biosensor: a surface-modified transducer that is reactive toward either a chosen analyte or a group of analytes.

Table 7.1 Examples of possible sensing element–transducer combinations

Biological elements	Transducers
Tissues	Potentiometric
Cells	Amperometric
Organelles	Conductometric
Membranes	Impedimetric
Enzymes	Optical
Receptors	Calorimetric
Antibodies	Accoustic
Nucleic acids	Mechanical

bodies, since these represent the most practical approaches detailed to date for environmental applications. The interested reader is, however, referred to the following texts that cover other possible combinations of sensing element and transducer (Turner *et al.*, 1987; Cass, 1990; Hall, 1990).

7.2   The enzyme electrode

The enzyme electrode is a combination of any type of electrochemical probe with a thin layer (10–200 μm) of immobilised (insolubilised) enzyme. Typically, the progress of the enzymic reaction (which is related to the concentration of analyte) is measured by monitoring the rate of formation of a product or the disappearance of a reactant. If either the product or the reactant is electroactive, the reaction may be monitored directly using amperometry.

7.2.1   Enzyme kinetics

Enzymes are proteinaceous molecules (molecular masses typically 10^4–10^5 Da) that catalyse the myriad of biochemical reactions that occur within living cells. Like their chemical counterparts, enzymes accelerate the rate of chemical reactions without themselves being changed in the overall process. A fundamental difference between enzymes and industrial catalysts, however, is that enzymes function at physiological temperatures and near neutral pH.

An important feature of enzymes is that they possess specific three-dimensional (3-D) configurations that are fundamental to their biological function. This is because the overall shape of the molecule stabilises the precise geometric structure of the *active site*, the region in the enzyme where the substrate is converted into the product. The importance of the active site (which normally makes up only a small percentage of the entire molecule) is to stabilise the transition state between the substrate and its products, thus lowering the activation energy for the reaction. (Lowering this energy by about $34\,kJ\,mol^{-1}$ is calculated as bringing a million-fold increase in the rate of a reaction at 298K.) For this to occur, the substrate must fit precisely into the active site (*shape recognition*).

Clearly, enzymes cannot work at a distance. Kinetic modelling of enzyme-catalysed reactions must, therefore, allow for the formation of an *encounter complex* between the enzyme and its substrate. Furthermore, because the formation of the transition-state complex involves complexation with the enzyme, there will be a maximum in the concentration of substrate that can be processed at once since, at some point, all of the enzyme active sites will be occupied (*saturation*). When the enzyme is saturated, the reaction velocity is dependent only upon the turnover number (number of substrate molecules converted to product per second) of the enzyme.

The classical mechanism describing an enzyme-catalysed reaction is:

$$E + S \underset{k_{-1}}{\overset{k_1}{\rightleftharpoons}} ES \overset{k_2}{\rightarrow} E + P$$

where E is the enzyme, S is the substrate, P is the product, ES is the enzyme substrate encounter complex and k_1, k_{-1} and k_2 are reaction rate constants. If one assumes that any back reaction between the enzyme and the product can be ignored, then at steady state one can write:

$$\frac{d[ES]}{dt} = k_1[E][S] - k_{-1}[ES] - k_2[ES] = 0$$

Furthermore, if the reaction is characterised by the constant K_m (the *Michaelis constant*) where:

$$K_m = \frac{(k_{-1} + k_2)}{k_1}$$

and if the enzyme concentration is described as total $[E_\Sigma]$, rather than bound and unbound enzyme, where $[E_\Sigma] = [E] + [ES]$, then it follows that:

$$[ES] = \frac{[E_\Sigma][S]}{K_m + [S]}$$

The enzyme-catalysed reaction $(mol\,s^{-1})$ is then given by the *Michaelis–Menten equation* (Fersht, 1985):

$$v = \frac{-d[S]}{dt} = k_2[ES] = \frac{k_2[E_\Sigma][S]}{K_m + [S]}$$

If one considers the limiting case when $[S] \gg K_m$, a maximum rate V_{max} is reached such that:

$$V_{max} = k_2[E_\Sigma]$$

This implies that V_{max} depends upon the amount of enzyme and on the rate constant k_2 (also known as k_{cat}). When $[S] = K_m$, however, the velocity of the enzyme-catalysed reaction becomes:

$$v = \frac{V_{\text{max}}}{2}$$

At low substrate concentrations ($[S] \ll K_m$) the reaction rate is linearly related to the substrate concentration such that:

$$v = \alpha[S]$$

where:

$$\alpha = \frac{V_{\text{max}}}{K_m}$$

A plot of the Michaelis–Menten relationship is shown in Figure 7.2. It is to be stressed that the preceding treatment is the simplest available for enzyme-catalysed reactions and will not apply under every circumstance. More complex reaction schemes do exist; for example, the reverse reaction where breakdown of the enzyme–substrate complex cannot be ignored and reactions involving multiple substrates with additional reaction steps. The Michaelis–Menten scheme serves as an adequate

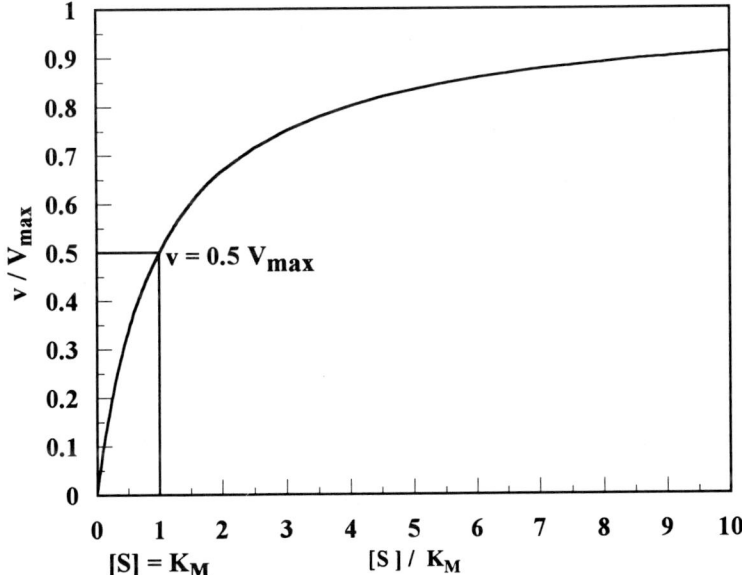

Figure 7.2 Plot of the Michaelis–Menten relationship. $[S] = K_m$ when the rate is equal to half the maximum velocity.

example for the current purposes. Fuller accounts of the reactions encountered in catalysis and the methods for their analysis can be found elsewhere (Laidler and Bunting, 1973; Walsh, 1977; Carr and Bowers, 1980).

7.2.2   Inhibition of catalysis

A particular feature of enzymes is that their catalytic action can be inhibited by a variety of substances. Inhibitors with close structural similarities to the substrate of the enzyme can bind to the enzyme active site thereby competing with the binding of the substrate (*competitive inhibition*). The affinity of the inhibitor for the enzyme is expressed quantitatively through the inhibition constant K_i:

$$K_i = \frac{[E][In]}{[EIn]}$$

Using the steady-state assumption (cf. above) and writing a mass balance equation that includes not only E but also ES and EIn, it is possible to obtain an equation relating rate to substrate concentration that is analogous to the Michaelis–Menten expression. Here however, K_m is replaced by an apparent constant K'_m where:

$$K'_m = K_m \left(1 + \frac{[In]}{K_i} \right)$$

The relationship between the velocity of the reaction (v) and the substrate concentration then becomes:

$$\frac{v}{[S]} = \frac{k_2[E_\Sigma]}{K'_m} - \frac{v}{K'_m}$$

In *non-competitive* inhibition, the inhibitor and substrate can bind simultaneously to the enzyme because the two binding sites do not overlap. A non-competitive inhibitor acts by decreasing the turnover of the enzyme. The maximal velocity in the presence of a non-competitive inhibitor (V'_{max}) is given by:

$$V'_{max} = \frac{V_{max}}{1 + \frac{[I]}{K_i}}$$

7.2.3   Kinetic and mass transport considerations for an enzyme electrode

Because enzyme electrodes generally have the enzyme immobilised in a thin layer of some finite thickness on the surface of the probe, any modelling of the system response must consider the diffusional and partitioning effects of

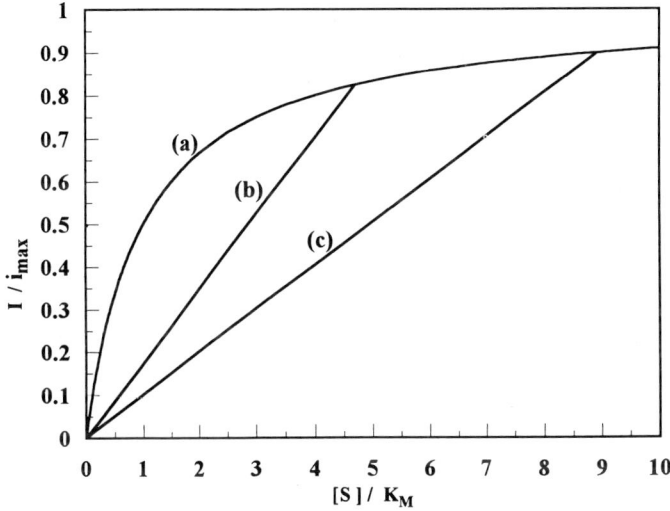

Figure 7.3 The effect that varying the permeability of the membrane has on the current response of an enzyme electrode to increasing concentration of substrate. In (a) the enzyme kinetics are rate limiting while (b) and (c) show the effect of using progressively less permeable membranes.

the enzyme layer. It is interesting to note that, unlike solution reactions where the rate is dependent upon a steady-state population of encounter complexes, with immobilised enzymes diffusion of the substrate to the enzyme is often, but not necessarily, the rate-determining step. Mass transfer by diffusion is a first-order reaction with respect to substrate concentration. Imposing a diffusional barrier, therefore, has the effect of extending the linear range of initial reaction velocity beyond the K_m value of the normal enzyme (cf. Figure 7.2). Because of this linear relationship, however, the observed rate of reaction and, therefore, the analytical signal is lower than it would have been in a kinetically controlled enzyme reaction conforming to the rectangular hyperbola of Michaelis–Menten kinetics (Albery and Cranston, 1987). This is illustrated schematically in Figure 7.3.

7.3 Preparation of the immobilised enzyme layer

The proper functioning of an enzyme-based sensor is, of course, heavily dependent on both the chemical and physical properties of the immobilised enzyme layer. To this end, there are many possible methods for immobilising an enzyme at the surface of an electrode. For convenience, these can be divided into *physical methods* and *chemical methods*. An important consideration when developing an immobilisation procedure is that the process should be applicable to a range of surfaces. This allows the choice of support to be as wide as possible and ensures that no re-fabrication of the support is required. Other advantages sought for in the immobilisation method include: (i) an ability of the biological component to operate at a wider pH range than in solution; (ii) attainment of greater stability resulting from the immobilisation; and (iii) the generation of a defined 'diffusion region' on the surface of the electrode.

7.3.1   Gel entrapment

Entrapment within a 3-D gel matrix is a common method of enzyme immobilisation. Numerous matrices have been employed, but the most favoured have been alginate cross-linked with linear chains of Ca^{2+} (Kuek and Armitage, 1985), gelatin (Romette et al., 1983) and polyacrylamide. With due attention to the degree of cross-linking and the nature of the gelling process, i.e. minimising the concentration of free radicals, gel entrapment can be applied to any enzyme. Unfortunately, this process does suffer from three major drawbacks: (i) large diffusional barriers to the transport of substrate or product resulting in reaction retardation and long response times; (ii) continuous loss of enzyme activity, since these materials generally do not have a narrow pore size distribution; and (iii) shrinkage and/or swelling of the polymer, depending upon the ionic strength of the milieu. A more straightforward method of enzyme entrapment is to retain the protein at the electrode surface behind a thin semipermeable membrane, such as dialysis tubing. Here, a thick paste of the enzyme in a suitable volume of buffer (1 or 2 mm³) is spread over the surface of the electrode. The layer is then covered with a 20–25 μm thick dialysis membrane, of about 10 000 Da molecular mass cut off, held in place by a suitably sized O-ring.

7.3.2   Adsorption

The great advantage of using adsorption as the immobilisation technique is that usually no additional reagents are required and only a minimum of activation or 'clean up' need be done. Unfortunately, however, only weak short-range interactions are involved, such as van der Waals forces, dipole–dipole interactions and hydrogen bonding. Because of this, the reversible nature of the binding equilibrium is highlighted by its suscepti-bility to changes in ambient conditions, i.e. pH, temperature, ionic strength, polarity, etc.

When a protein adsorbs to the surface of a support, the reaction will be influenced by non-covalent interactions between the surface of the protein and the support. Clearly the protein could interact with the surface in several different ways depending on the orientation with which it approaches the surface. In an unperturbed solution, this process, and the reverse process, is under mass transport control. If every molecule that encounters the surface is adsorbed, a concentration gradient rapidly develops at the surface. The rate of adsorption then becomes proportional to the rate of diffusion (Macritchie 1978) such that:

$$\frac{dn}{dt} = C_0 \sqrt{\frac{D}{\pi t}}$$

where n is the number of molecules, C_0 the bulk concentration of protein and D the diffusion coefficient. The model does not, however, account for the situation where the layer next to the surface is only partially saturated with protein (as happens during sensor use). Under these conditions, desorption becomes the dominant process and the biomolecule becomes lost from the electrode surface. Finally, Alvarez-Icaza and Scmid, (1994) have shown that the adsorption of glucose oxidase to the surface of a carbon electrode leads to the irreversible denaturation of the protein unless steps are taken to modify the interface. Indeed, recent results in this laboratory (unpublished) have shown that the adsorption of the enzymes glucose oxidase and horseradish peroxidase to carbon does not proceed via a simple Langmuir isotherm. Rather, three equilibrium steps are involved, suggesting that the protein exists in three different microcrystalline states on the surface, thereby making the immobilised protein difficult to quantify. Despite these apparent shortcomings, the ease of the technique and its general applicability make adsorption particularly attractive as a pilot method or for the production of devices not requiring long-term stability, e.g. one-shot disposable sensors.

7.3.3 Covalent immobilisation

Covalent binding of the enzyme to the surface of the electrode is generally the most irreversible immobilisation technique and, therefore, potentially the most stable. With the enzyme, chemical bonding to the surface must be effected by using nucleophilic groups present on the surface (carboxylic acid, amino acid, hydroxy, thiol, imidazole and phenolic groups) of the protein that are not involved in the biological function of the molecule. In general terms, the attachment of an enzyme to an electrode surface (e.g. a metal or carbon disc) is a two-step process. The first involves activating the surface of the electrode, i.e. imparting some useful chemical reactivity to the otherwise inert surface. The second step involves the binding of the enzyme to the chemically activated electrode surface.

7.3.4   Immobilisation of enzymes to metal electrodes

Although metal surfaces are in themselves unreactive, when they are covered by a thin surface oxide film they can be functionalised by reagents such as chloro- or alkylsilanes (Murray, 1980). By analogy with silica surfaces, Pt/PtO, Au/AuO and SnO_2 surfaces have many M−OH sites (where M is the metal) and when they are contacted with, for example, a solution of dichlorodimethylsilane $[Cl_2Si(CH_3)_2]$ under anhydrous conditions the organosilane reagent becomes immobilised by formation of chemically stable MO−Si bonds (Figure 7.4).

The silanisation reaction can be monitored using surface analysis techniques such as X-ray photoelectron spectroscopy (XPS). Here, the experimenter would detect the silanisation reaction of a 'clean' metal surface by noting the appearance of new XPS Si 2p, Si 2s and O 1s bands in the spectrum. Alternatively, other techniques such as reflectance FTIR may also prove useful if the modifying agent contains a suitable chromophore. Finally, it may also be possible to monitor the surface modification by measuring changes in the double layer capacitance resulting from the replacement of polar surface groups by non-polar methyl groups. Accurate measurements of the change in the double layer capacitance may allow quantification of the surface groupings (Moses *et al.*, 1975).

Although the reaction shown above produces a functionalised electrode surface, the resulting electrode is not suitable for further synthetic modification because of the chemical inertness of the methyl groups. To bond an enzyme to a silanised surface, it is important that the organosilane reagent itself bears chemical functionality, such as a primary amine group or a carboxylic acid. A particularly useful reagent in this context is propylaminosilane $[(CH_3CH_2O)_3Si(CH_2)_3NH_2]$, which will not only functionalise a metal electrode but will then allow coupling chemistry to take place with proteins through the attendant amine grouping. A typical synthetic scheme using this reagent is shown in Figure 7.5, by which an enzyme is coupled to a silanised metal electrode via the formation of an amide bond between the amino group of the silane and a surface carboxylic acid on the protein. Note, however, that to make the amide linkage, the

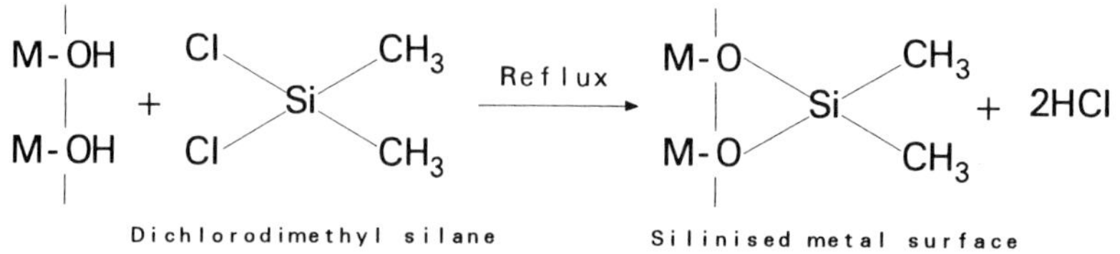

Figure 7.4 Schematic representation of the silinisation of a metal surface by dichlorodimethylsilane

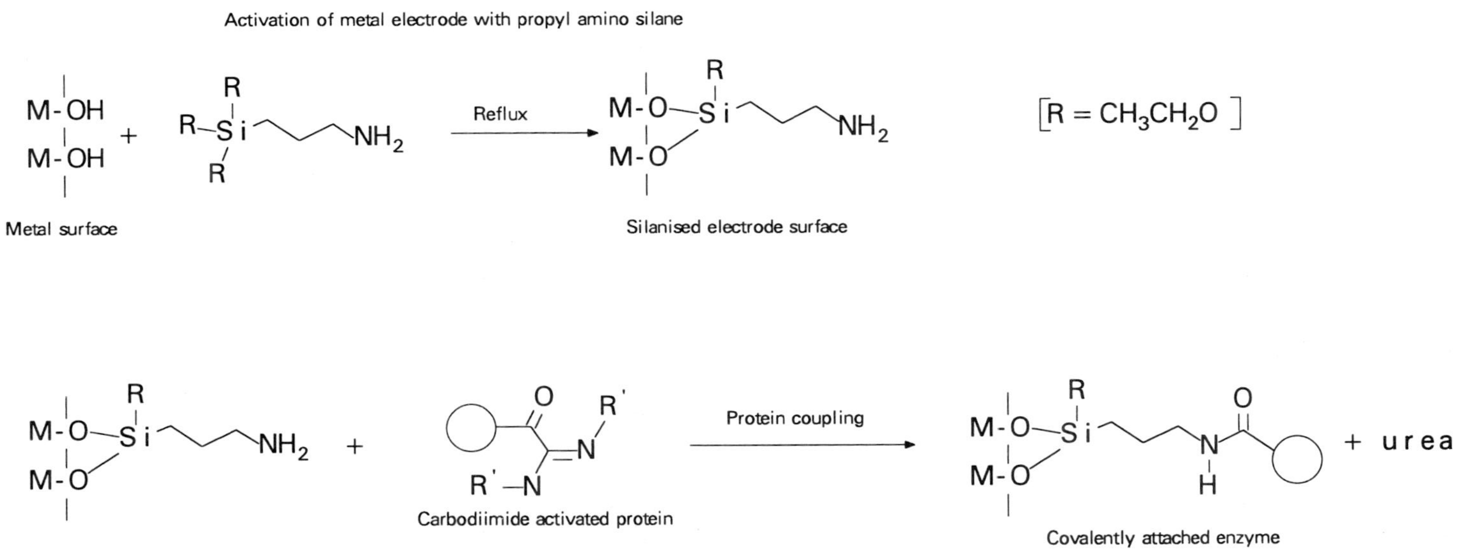

Figure 7.5 Organic synthetic schemes outlining the activation of a metal oxide electrode with propylaminosilane and the subsequent coupling of a carbodiimide-treated protein to the electrode surface.

carboxylic acid on the protein must itself be activated. This is normally achieved by using carbodiimide chemistry. The reaction uses attractively mild conditions and can draw upon a variety of diimide reagents developed largely for solid-phase peptide synthesis. The incorporation of the enzyme onto the electrode surface can be monitored by challenging the electrode with the substrate of the enzyme and detecting the production of an electrochemical product.

7.3.5   Immobilisation of enzymes to carbon electrodes

Carbon electrodes represent another popular material for the manufacture of enzyme electrodes. The attachment of enzymes to carbon surfaces is performed in an analogous manner in as far as the process requires both the activation of the carbon surface and the chemical bonding of the enzyme. Graphitic carbon consists of giant sheets of fused aromatic rings, stacked coplanarly as shown in Figure 7.6. An uninterrupted basal plane surface is non-ionic, of low polarity, hydrophobic and rich in π-electron density. Without alterations, the basal plane surface of carbon is somewhat barren to synthetic coupling reactions. (However, the high π-electron density is conducive to strong chemisorption interactions and could be used as a basis of adsorptive modification, particularly where unsaturated

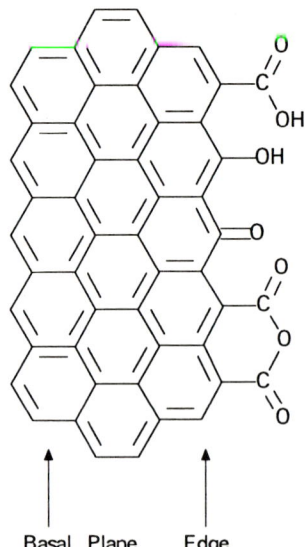

Basal Plane   Edge

Figure 7.6 Schematic representation of the chemical functionalities at the edge plane of pyrolytic graphite.

compounds and aromatic rings are involved.) The perimeter of the basal plane structure is terminated as a graphitic edge plane on which chemical functionalities such as carboxylic, hydroxy, phenolic, quinone, lactone and other carbonyl containing-groups abound. Any cleavage of graphite results in reactive edge plane dangling valencies that become satisfied by reaction with oxygen and water. Graphitic edge planes are, therefore, attractive surfaces for chemical coupling and modification procedures. Edge planes are also polar and so fairly hydrophilic, implying that aqueous coupling reactions can be used with a good degree of success.

Several chemical pretreatments have been proposed for enhancing synthetically useful carboxylic acid and hydroxylic groups on the edge planes. Carboxylic acid coverages on spectroscopic carbon rods, highly orientated pyrolytic graphite, glassy carbon and pyrolytic graphite can be enhanced by simply heating in air at 400–500°C (Elliot and Murray, 1976 (and references therein); Labuda, 1992 (and references therein)). Treatment in a radio frequency (RF) oxygen plasma is an equally effective procedure (Evans and Kuwana, 1977). To enhance hydroxylic groupings on pyrolytic graphite surfaces, Lin and co-workers (1977) first thoroughly oxidised the surface with an oxygen RF plasma and then reduced it with ethereal $LiAlH_4$. More recently, work in this laboratory has shown that simple oxidation of graphite particles with 1 M nitric acid at room temperature can also greatly increase the surface concentration of hydroxyl functionalities (Cardosi, 1994). Hydroxyl-containing surfaces are particularly attractive because they can react with polyfunctional cyanuric chloride (trichloro-sym-triazine) producing an ether linkage that is chemically and electrochemically stable in organic solvents and aqueous solutions (Lin et al., 1977). The attached cyanuric chloride (CC) can further react with a variety of substances including hydroxyl and amino compounds, alkyl and aryl Grignard reagents and organic hydrazine reagents. This approach has been used to attach the enzymes horseradish peroxidase and glucose oxidase to the surface of graphite particles (Cardosi and Birch, 1993; Cardosi, 1994). The synthetic scheme for the attachment of horseradish peroxidase to the graphite particles is shown in Figure 7.7. In an analogous fashion to metal oxide electrodes, the derivitisation of the carbon surface with CC can be easily followed by XPS. Here, use is made of the appearance of the Cl 1s peak in the XPS spectrum following reaction with CC.

It must be stressed, that the actual knowledge of the absolute population and chemical reactivities of the various groupings on carbon surfaces is still limited. Although one of the treatments discussed above may appear to enhance the efficacy of a chemical coupling procedure, e.g. by producing larger electrochemical waves for attached redox species, this in itself is an indirect form of evidence which explicitly pre-supposes that the coupling reaction proceeds as planned.

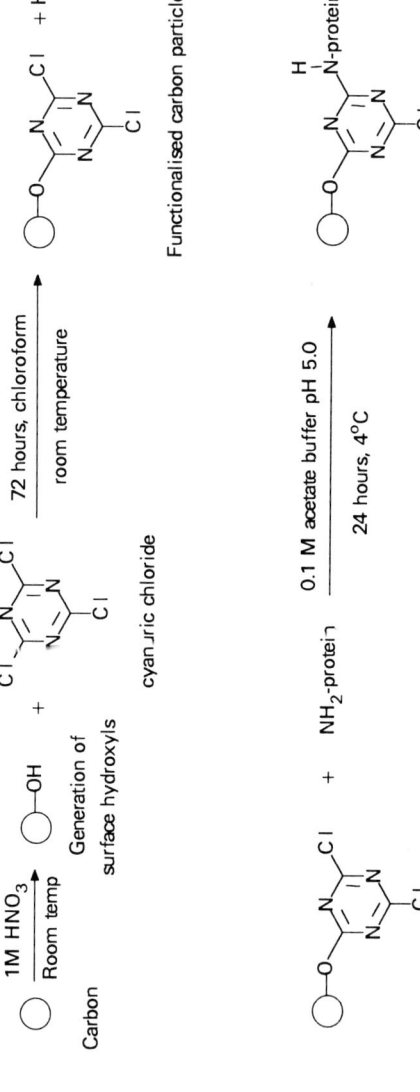

Figure 7.7 Organic synthetic scheme for the attachment of horseradish peroxidase to carbon using cyanuric chloride.

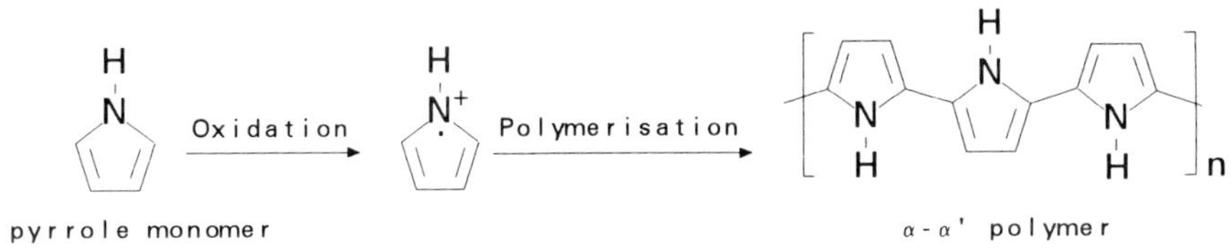

Figure 7.8 Oxidation of the pyrrole monomer and subsequent polymer growth.

7.3.6   *Immobilisation in an electrochemically grown organic polymer*

Direct *in situ* formation of a polymer film from a solution of monomers can be induced electrochemically. Through electrochemical initiation, a monomer such as pyrrole or thiophene is oxidised to a polymerisable radical, as illustrated in Figure 7.8.

This method is applicable to many electrode materials and also to many monomer compounds. The technique is often simple to carry out, requiring only basic electrochemical instrumentation. Furthermore, the growth of the polymer film can be controlled by stepping or cycling the electrode potential (Dicks *et al.*, 1993). The properties and the reproducibility of the surface polymer ultimately depend upon the nature of the polymerisation solution, the electrode material and, to some extent, the cell geometry and attendant hydrodynamic conditions. Another material that has proved particularly useful for the immobilisation of enzymes at electrode surfaces is polyaniline (Cooper and Hall, 1993). Figure 7.9 shows representative cyclic voltammograms obtained during polyaniline film growth from aqueous solution on a platinum disc electrode.

If an enzyme is present in the aqueous solution containing the monomer, molecules of the enzyme will become physically trapped within the growing matrix during film growth. This approach has several significant advantages as a general method for preparing enzyme electrodes. First, the method is flexible and can be readily controlled. Second, it is simple to carry out and usually results in enzyme loadings with high activity; third, it is possible to co-immobilise more than one enzyme either in the same film or by growing different layers one on top of each other. (N.B. this is only possible with conducting polymers.) Finally, the polymer deposition is localised at the surface of the electrode so that the method is suited to the spatially localised deposition of enzymes onto microelectrode arrays. An example of this approach is the entrapment of glucose oxidase within a conductive matrix of polypyrrole at the surface of a platinum electrode (Foulds and Lowe, 1988, Bartlett and Whitaker, 1987a,b). Film growth occurs because the oxidation of the monomer (pyrrole) results in the formation of a highly reactive radical that reacts with neighbouring pyrrole molecules to give a polymer that is predominantly α–α' coupled (Diaz, 1981), although some branching of the polymer chains is thought to take place through β-coupling reactions. The

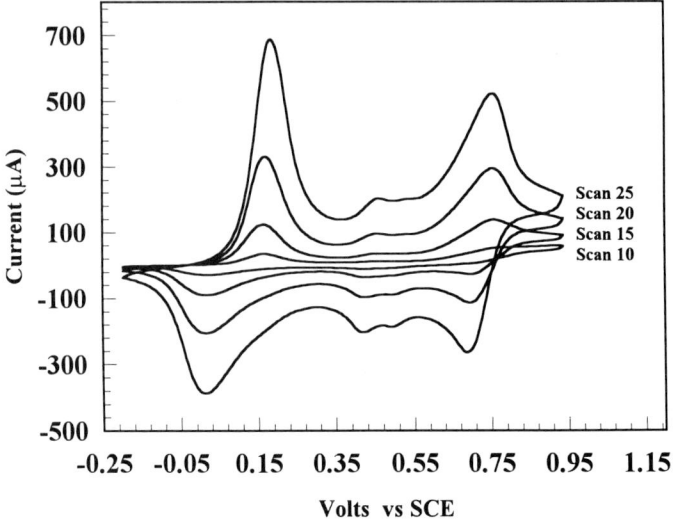

Figure 7.9 Cyclic voltammograms obtained from growth of a polyaniline film on the surface of a platinum disc electrode. The film was grown from a 0.1 M solution of sodium sulphate pH 1.0 M aniline. The voltage was cycled repetitively from −0.3 to 0.95 V versus SCE at a scan speed of 0.1 V s^{-1}. Scan numbers 10, 15, 20 and 25 are shown in the figure.

resulting polymer has a net positive charge and incorporates anions from the bulk solution to maintain charge neutrality (Bartlett and Cooper, 1993).

Although the above example makes use of conducting polypyrrole films, conductivity is not a prerequisite for a suitable immobilisation medium. Indeed, electrode films incorporating enzymes have been constructed from a wide range of materials, either conducting or insulating, including polyaniline (Shaolin *et al.*, 1991), polyphenols (Bartlett *et al.*, 1992), polypyridine (Gregg and Heller, 1990) and cobalt tetrakis (*o*-aminophenyl) porphyrin (Oyama *et al.*, 1988).

Clearly, the successful preparation of the enzyme layer is a primary concern in the design of an enzyme electrode. Unfortunately, this still requires a certain degree of empiricism and an appreciation of the fact that what is a suitable method for one enzyme may not work as well for another one. The relevant effects that immobilisation can have are summarised as follows:

- The apparent activity of an enzyme can be reduced after immobilisation. This in turn may be caused by several factors. First, the chosen immobilisation chemistry may not be optimal and may lead to modification of the active site. Even when the immobilisation does not interfere with the active site, the nature of the support may produce diffusional barriers. Second, the very act of immobilisation may lead to unfavourable conformational changes in the protein or reduce the conformational mobility of the enzyme. Finally, changes in the microenvironment resulting from immobilisation may also lead to apparent changes in activity.

- Although the immobilisation matrix is often seen as solely a support for the enzyme, it may nevertheless introduce partitioning effects. A positively charged matrix, for example, will exclude protons so an enzyme in this matrix will exhibit a lower optimal pH than usual. Similarly, a hydrophobic substrate will partition into a hydrophobic matrix and thus lower the apparent K_m for that substrate. This second consideration may be particularly significant when analysing for low levels of pollutants.

7.4 Theoretical considerations for amperometric enzyme electrodes

Amperometry is the branch of electrochemistry that deals with the addition (*reduction*) or the removal (*oxidation*) of electrons from a molecule or atom. In theory, any atom, molecule or assembly of atoms can be oxidised or reduced if sufficient energy can be provided. However, the range of energies that can be applied is limited by the experimental conditions. Molecules that can be oxidised or reduced in easily available energy ranges are said to be electroactive. The amount of energy required for a redox process is characteristic of the system under examination and is called the redox potential.

Oxidation or reduction processes cause a *Faradaic current* to flow in the electrochemical cell when a voltage is applied between the working and reference electrodes. This imposed potential encourages electron transfer reactions to occur at the working electrode, resulting in a current that is directly proportional to the concentration of the electroactive analyte. The transfer of charge occurring in an electrochemical cell can be described by:

$$O + ne^- \rightleftharpoons R$$

where n is the number of electrons (e^-) transferred between oxidant (O) and reductant (R). This equation describes a simple, chemically reversible charge-transfer process that takes place across the phase boundary between an electrode (which acts as a source or sink of electrons) and an ionically conducting medium that contains the electroactive analyte. The current (I) measured during the electrolysis is a direct measure of the rate of the electrochemical reaction at the electrode, as described by Faraday's law:

$$I = nF\frac{dN}{dt}$$

where dN/dt is the oxidation or reduction rate ($mol\,s^{-1}$) and F is the Faraday constant ($96\,485\,C\,mol^{-1}$). Because of the heterogeneous nature of the process, the reaction depends upon the rate of electron transfer at the surface and on the mass transport of analyte to the electrode.

An electrochemical experiment should be designed so that the mode of transport of the electroactive species to the electrode surface is well defined. The most important mass transport processes in an electrochemical measurement are *diffusion* and *forced convection*.

Diffusion is the movement of ions or molecules caused by a concentration (activity) gradient. The movement is from a region of high concentration to one of low concentration. A consideration of the above reaction shows how concentration gradients arise at stationary electrodes. Before the initiation of the electrolysis, the concentration of O is uniform throughout the solution. When current flows, the concentration of O at the electrode surface becomes less than in the bulk solution because of the electro-chemical conversion of O into R. Because diffusion is the only mode of transport, the flux of O to the electrode surface at time t at a distance x from the electrode surface is proportional to the steepness of the concentration gradient. The flux of O is given by Fick's first law of diffusion:

$$J(x, t) = D_o \frac{\partial[c_o(x, t)]}{\partial x}$$

where c_o is the concentration of O and D_o is the diffusion coefficient of O.

The current flowing in the cell is dependent upon the flux of material at the electrode surface, i.e. at $x = 0$. The current for an electrode with area A can, therefore, be defined as:

$$I = nFAJ(0, t)$$

The time dependence of the concentration of O is given by Fick's second law:

$$\frac{\partial[c_o(x, t)]}{\partial t} = D_o \frac{\partial^2[c_o(x, t)]}{\partial x^2}$$

All voltammetric equations at planar electrodes are derived from these basic equations using the proper initial and boundary conditions characteristic for the particular experimental situation. For example, when a constant potential is applied to an electrode in a quiescent solution so that O is immediately reduced upon its arrival at the electrode surface, the current shows a decrease with time (t) because of the slow spread of the *diffusion layer* out into the bulk solution. The current change under these conditions is characterised by the Cottrell equation, which predicts a $1/\sqrt{t}$ expansion of the diffusion layer (Cottrell, 1902):

$$I = nFAc_o \sqrt{\frac{D_o}{\pi t}}$$

In most practical situations, forced convection is provided by moving the electrode with respect to the solution (as with rotating electrodes) or vice versa. Under these conditions, a stagnant layer of thickness L (a concept

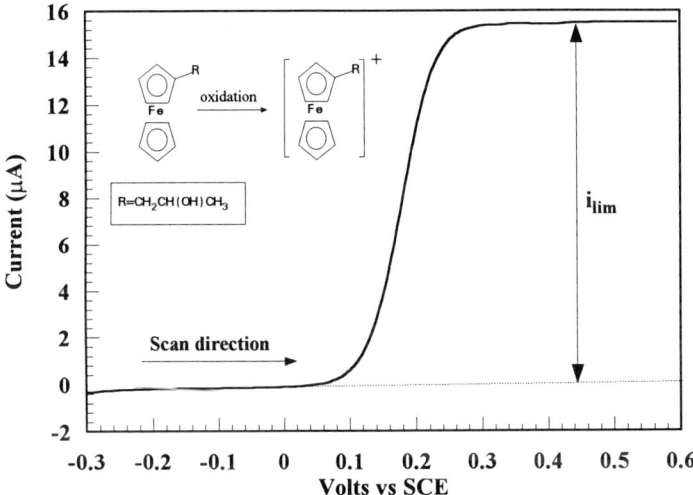

Figure 7.10 Hydrodynamic voltammogram obtained for the oxidation of 2-hydroxypropylferrocene in 0.1 M potassium phosphate pH 7.0 at a gold disc electrode rotated at 200 rpm.

introduced by Nernst in 1904) is formed at the surface of the electrode (Levich, 1962). Within this layer, the solution is considered motionless and the only mode of mass transport is diffusion. The forced convection, caused by the rotation of the electrode for example, serves to replenish the concentration of analyte at the boundary of the diffusion layer. The thickness of the layer depends upon the properties of the solution (kinematic viscosity, temperature, etc.) and on the relative movement of the electrode and the liquid. Consequently, mass transport to the electrode surface is controlled by diffusion through this layer. Because the thickness of the layer is independent of time (e.g. as with an electrode rotated at constant speed) the resultant current reaches a steady state that is also independent of the timescale of the experiment (Figure 7.10). This is described by the following equation, that can be obtained from Fick's laws under conditions of steady state:

$$I = \frac{nFAc_oD}{L}$$

The boundary conditions assume uniform analyte distribution in the bulk solution up to the stagnant layer and the condition of mass transport control, i.e. $c = 0$ at $x = 0$ (Figure 7.11).

With these basic ideas in mind, it is now possible to develop a simple model to explain the behaviour of enzyme electrodes. As has been stated above, the enzyme electrode has a layer of enzyme immobilised onto the surface of the probe. To analyse the situation within this region, the following model has been proposed in which diffusion and reaction processes are separated in space

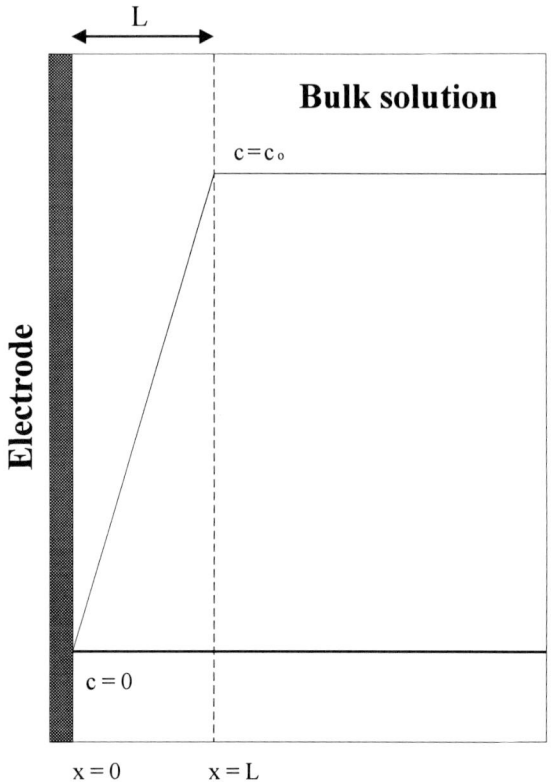

Figure 7.11 Schematic diagram of the concentration profile of an electrochemically active compound at an amperometric electrode.

(Figure 7.12(a)). The model consists of three regions: a convective region ($x > L$) where the analyte concentration is maintained constant ($c = S_{bulk}$); a diffusion-limiting region ($0 \leq x \leq L$) where a pure diffusion process takes place; and the region where the enzyme reaction takes place ($x < 0$). To keep the model simple, the following approximations are made: the partition coefficients for the three regions are considered to be unity, the supply of any cosubstrate for the enzyme is considered plentiful, the enzyme is distributed uniformly for $x < 0$; and the sensor signal depends on product generation and is, in turn, related to the flow of substrate into the sensor. Boundary conditions are established at the two interfaces ($c = S_1$) and very deep within the enzyme layer ($c = 0$). Alvarez-Icaza and Bilitewski (1993) have discussed the usefulness of the intermediate boundary condition ($x = 0, c = S_1$) in a recent article and shown that using S_1 and the appropriate boundary conditions it is possible to express the concentration profile within the enzyme region by a simple exponential:

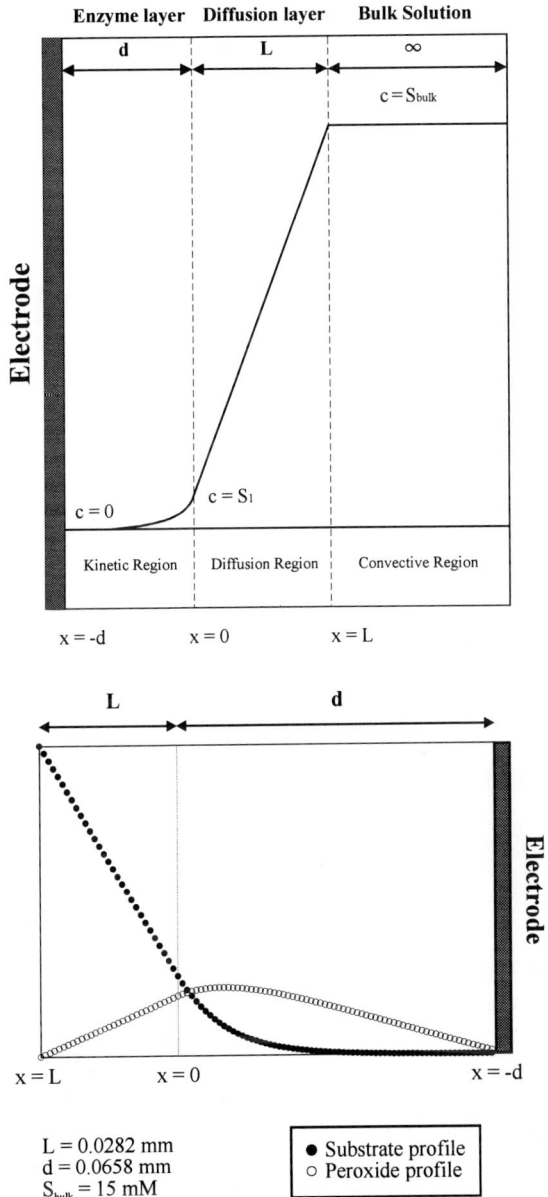

Figure 7.12 (a) Concentration profile of an enzyme substrate at an enzyme layer that includes a diffusion-limiting region of thickness L. The intermediate boundary condition $c = S_1$ is also shown. (b) Simulated concentration profiles for the transport of substrate through the diffusion barrier and into the immobilised enzyme layer and the simultaneous generation and diffusion of peroxide. Note that not all the peroxide generated by the enzymatic reaction diffuses to the electrode. (Data courtesy of Mr H. Anderson, 4th year Honours student, Dept of Physics, University of Paisley and Dr J. Nixon, Dept of Physics, University of Paisley.)

$$c(x) = S_1 \exp \sqrt{\frac{\alpha}{D_e}} x \, ; \, x < 0$$

where the term α is V_{max}/K_m and D_e is the diffusion coefficient of the substrate within the enzyme layer. The simulated response of such an electrode is shown in Figure 7.12(b).

The sensor response is kinetically controlled if the enzyme reaction is slower than the overall diffusion process. Under these conditions, and assuming a boundary condition where only half of the enzyme generated product diffuses to the electrode, the current can is expressed as:

$$I_k = \frac{nFAL\,k_2[E_\Sigma][S]}{2(K_m + [S])}$$

which gives a maximum current response I_{max} (when $[S] \gg K_m$) that is directly proportional to the enzyme loading and the thickness L:

$$I_{max} = \frac{nFAL\,k_2[E_\Sigma]}{2}$$

Combining the above equations gives a general expression for the kinetic current:

$$I_k = \frac{I_{max}[S]}{K_m + [S]}$$

If the enzymatic reaction in the enzyme layer of thickness d is faster than the transport process (i.e. the concentration $S_1 \rightarrow 0$), then the sensor response is under mass transport control. The concentration S_1 is negligible if the substrate concentration reaches zero inside the layer. This is possible when the dimensionless parameter σ^2 is greater than unity (Mell and Maloy, 1975) where:

$$\sigma^2 = \frac{\alpha d^2}{D_e} = \frac{V_{max} d^2}{D_e K_m}$$

This parameter essentially compares the rate of enzyme reaction (V_{max}/K_m) with diffusion through the enzyme layer (d^2/D_e). If $\sigma^2 < 1$, then enzyme kinetics predominate. If, however, $\sigma^2 > 1$, the response is under diffusion control.

In an analogous manner to the kinetic situation, one can derive a general expression for a sensor operating in the diffusion-limiting regime:

$$I_d = \frac{2I_{max}D_e[S]}{k_2[E_\Sigma]d^2}$$

Comparing the two limiting situations reveals that I_d is dependent on both d and D, whereas I_k is independent of D and increases with L.

The concentration profile in the diffusion control layer can also be written in terms of the intermediate variable S_1 (Alvarez-Icaza and Bilitewski, 1993) as:

$$c(x) = \frac{x}{L}(S_{bulk} - S_1) + S_1 \, ; \, 0 < x < L$$

The boundary conditions at $x = 0$ can be used to determine the value of S_1 as a function of S_{bulk}. The flux of material moving into the enzyme region can be calculated by applying Fick's first law to the concentration profile within the enzyme layer evaluated at $x = 0$. The flux out of the diffusion control layer is the product of $(S_{bulk} - S_1)$ and (D_S/L), where D_S is the diffusion coefficient of the substrate in that layer. The condition for both fluxes to be equal is defined when (Alvarez-Icaza and Bilitewski, 1993):

$$S_1 = \frac{S_{bulk}}{1 + \frac{L}{D_S}\sqrt{\alpha D_e}}$$

Since S_1 represents the maximum possible concentration in the enzyme layer, the linear range of the sensor will be determined by S_1 (i.e. when $S_1 \ll K_m$). The above expression implies that S_1 can be optimised by careful selection of the parameters D_S, $\sqrt{(\alpha D_e)}$ and L. In operational terms, however, increasing the thickness of the diffusion layer or decreasing the diffusivity may be of limited value because of the relationship between the characteristic diffusion time across the membrane (τ) and D_S and L:

$$\tau = \frac{L^2}{D_S}$$

By necessity, this section has dealt only with a specific model, which may not apply to all enzyme electrode configurations. It does, however, illustrate the theoretical problems, and their implications, which should be considered in the design of an enzyme electrode. The interested reader is referred to the following monographs which deal with the topic in greater detail: Blaedel and Jenkins (1972), Mell and Maloy (1975), Carr and Bowers (1980), Gough and Leypoldt (1981), Albery and Cranston (1987), Bartlett and Whitaker (1987a,b), Eddowes (1990) and Scheller and Schubert (1992).

7.5   Enzyme electrodes for substances of environmental interest

Substances that are generally considered pollutants can be:

- Synthetic organic compounds derived from industrial wastes, biocides and detergents, which may contain in their structures organic phosphates or carbamates or the inert aromatic rings that have to be prepared for cleavage and degradation by hydroxylation to phenols.
- Compounds that normally occur in nature such as nitrite, nitrate, sulphate, phosphate and even some heavy metals. When these substances accumulate in abnormal quantities, however, they upset the established ecological equilibria and become potential hazards.

7.5.1 Pesticides

The most commonly used enzyme for the detection of pesticides is the nerve enzyme *acetylcholinesterase,* which catalyses the hydrolysis of acetylcholine to choline and acetic acid. Indeed one of the largest classes of organic phosphate insecticides, which includes parathion, malathion, tabun and sarin, inhibit this enzyme by forming very stable covalent phosphoryl-enzyme complexes. Acetylcholine, a neurotransmitter, is released at many nerve synapses and is very toxic in excess; it must be destroyed rapidly by acetylcholinesterase to prepare the synapse for transmission of the next impulse. Inhibition of this reaction leads to respiratory paralysis and death.

Sensors that use immobilised acetylcholinesterase measure the inhibition of enzyme activity upon exposure to samples containing the pesticide. The first electrode based on this enzyme was described by Goodson and Jacobs (1974; 1976). In their system, the biocatalyst was immobilised on an open-pore polyurethane foam pad placed on the surface of a platinum electrode. This allowed for easy replacement of the biocatalyst when the activity fell below an acceptable level. For the detection of inhibitors, the synthetic substrate butyrylthiocholine was used. Hydrolysis of the thioester by cholinesterase results in an electroactive thiol that is readily oxidised to a disulphide at the platinum anode. This gives a steady-state signal that is at a maximum with the uninhibited enzyme and decreases with the degree of inhibition. The sequence of reactions is summarised below:

$$C_3H_7COS-R \xrightarrow{\text{cholinesterase}} C_3H_7COOH + R-SH$$

$$2\ \ R-SH \xrightarrow{\text{Pt anode}} RS-SR + 2H^+ + 2e^-$$

$$R = CH_2CH_2N^+(CH_3)_3$$

Using the above components, Goodson and Jacobs (1976) constructed a continuous aqueous monitor (CAM-1) designed for remote operation. The sensitivity of the monitor to various water-borne pesticides is shown in Table 7.2.

Since this original publication, other workers have described detection systems for pesticides based on the same enzyme. Kulys and D'Costa

Table 7.2 Detectable levels of various pesticides with the CAM-1 monitor

Substance	Detectable level (ppm)
Azordin	20
DDVP	1.0
Diazinon	1.2
Dursban	4.5
Dimetilan	10
Sevin	20
Temik	4.2

(1991), for example, have used acetylcholinesterase together with screen-printing technology to produce disposable electrodes for the determination of pesticide levels in water. Like the previous workers, their method monitors the decrease in acetylcholinesterase activity in the presence of the pesticide using butyrylthiocholine as the substrate. The thiocholine produced during the catalytic reaction was determined amperometrically at a 7,7,8,8-tetracyanoquinodimethane (TCNQ)-modified carbon electrode polarised at 0.1 V versus the Ag/AgCl half cell. The sensor was constructed by printing the TCNQ-modified carbon working electrode and the Ag/AgCl reference half cell onto a PVC support. The enzyme layer was prepared by covering the working electrode with a 10 μl solution of acetylcholinesterase containing 2.5% (v/v) glutaraldehyde. The authors characterised and optimised the behaviour of the immobilised enzyme for effects of pH, temperature, substrate concentration and incubation time with the organophosphate. In a more recent publication, Skladal and Mascini (1992) immobilised acetylcholinesterase onto the surface of a nylon net. This was then placed over the surface of a carbon electrode modified with cobalt phthalocyanine. Using the synthetic substrates acetyl- or butyrylthiocholine, 1.5 and 8.4 μg l^{-1} of paraoxon and heptenophos, respectively, were detected in less than 3 min using a similar detection scheme to that described above. A device based upon immobilised acetylcholinesterase is also currently in use by NATO armed forces. The sensor, known as NIAD (Nerve-Agent Immobilised-Enzyme Alarm Detector) is used in the field to give advance warning of a chemical attack involving nerve gas. The NIAD device was in use during the Gulf War (Allen, 1994).

7.5.2 Phenols

Two types of enzyme have been used to construct enzyme electrodes for phenol determination. The first belong to the family of copper-containing proteins designated phenol oxidases. They are obtained from fungi, potatoes and several other sources. The other type is the well-defined flavoprotein phenol 2-monoxygenase isolated from the soil yeast *Trichosporon cutaneum*. This enzyme does, however, have the disadvantage of requiring NADPH as an electron donor. The function of NADPH is to reduce the FAD prosthetic group of the enzyme to activate it towards oxygen. The cost of this requirement and the extra degree of complexity has made this enzyme less popular in the design of enzyme electrodes.

The catalytic reactions of phenol oxidases can be illustrated by the oxidation of L-dopa to L-dopaquinone by the enzyme tyrosinase, also known as polyphenol oxidase (Figure 7.13).

Because the enzyme uses dioxygen, the rate of the enzyme-catalysed reaction can be monitored directly with a Clark type oxygen sensor. The oxygen sensor is based on a pair of electrodes immersed in an electrolyte

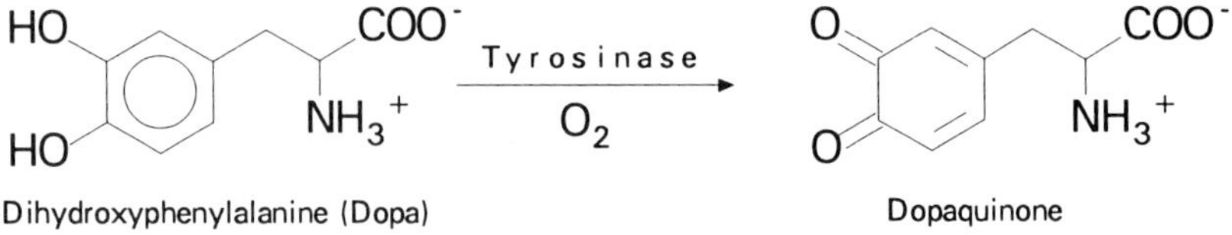

Figure 7.13 Pathway for the oxidation of L-dopa to L-dopaquinone by the enzyme tyrosinase.

solution and separated from the test solution by a gas-permeable hydrophobic membrane, usually made of Teflon, silicone or polyethylene. Oxygen diffuses through the membrane and is reduced at a platinum cathode (polarised at -0.8 V versus Ag/AgCl half cell) according to the following scheme:

$$2O_2 + 4e^- + 2H_2O \longrightarrow H_2O_2 + 2OH^-$$
$$H_2O_2 + 2e^- \longrightarrow 2OH^-$$
$$\text{Total: } O_2 + 4e^- + 2H_2O \longrightarrow 4OH^-$$

The resulting electrolytic current is proportional to the rate of diffusion of oxygen to the cathode and, therefore, to the partial pressure of oxygen in the sample:

$$I = nFA\alpha_m pP_s \frac{D_m}{L}$$

where α_m is the solubility of oxygen in the membrane phase, pP_s is the partial pressure of oxygen in the sample, D_m is the diffusion coefficient of oxygen in the membrane phase and L is the membrane thickness.

An enzyme electrode with immobilised tyrosinase was described by Macholan and Schanel (1977). In their system, a mixture of tyrosinase and bovine serum albumin (a neutral carrier protein) was reticulated onto the surface of a polyamide net with the bifunctional reagent glutaraldehyde. The enzyme-modified net was then stretched over the hydrophobic membrane of a Clark oxygen sensor. The electrode responded to phenol, p-cresol, pyrocatechol and pyrogallol in waste water. Calibration curves for these compounds, when tested individually in the laboratory, were linear in the range 6.6–66 µM although the relative activities ($-d[O_2]/dt$) towards the phenols were different. No attempts were made to quantify individual phenols in mixtures. Linear and stable responses corresponding to total phenol were obtained with graded amounts of coking water.

More recently, Hall and co-workers have described an amperometric enzyme electrode for the determinations of phenols in chloroform (1988a,b). The enzyme (tyrosinase) was immobilised by adsorption to a Hybond-N nylon membrane held securely over a carbon electrode. The

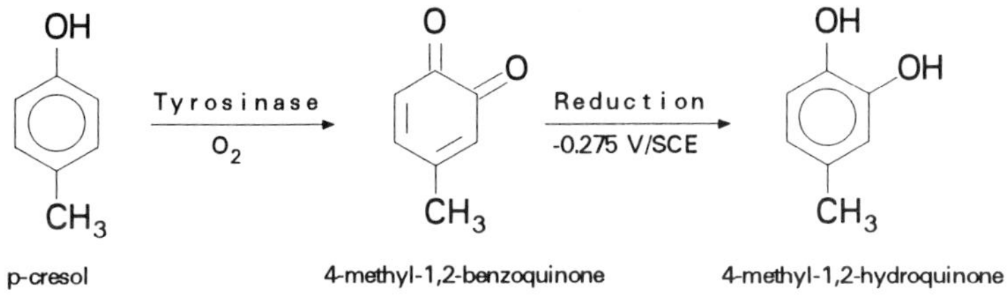

Figure 7.14 Detection scheme for *p*-cresol using the enzyme tyrosinase and a voltammetric indicator electrode.

detection scheme was based on the reduction of the product, 4-methyl-1,2-benzoquinone, at the basal electrode. The response of this electrode to *p*-cresol was linear in the concentration range 0–0.1 mM with a response time of approximately 3 min. The limit of detection for *p*-cresol was 1.0 μM. Other compounds that were detected using this electrode were *m*-cresol phenol, catechol, 4-methylcatechol, *p*-aminophenol and 4-chlorophenol. The advantages of using an organic-phase enzyme electrode in this work were high enzyme loadings, ease of immobilisation, extended substrate range, reduction of side reactions and lower detection limits due to the partitioning of the pesticides into the chloroform layer (Hall *et al.*, 1988). The scheme for the detection of *p*-cresol is shown in Figure 7.14. The area of organic phase enzyme electrodes has recently been reviewed by Saini and co-workers (1991).

7.5.3 Heavy metals

Heavy metals can, in principle, be detected as a group using any enzyme that carries a catalytically essential SH group. Complexation of this thiol with the heavy metal will tend to reduce the catalytic turnover of the enzyme. Despite this generic approach to heavy metal detection, few examples have been published in the literature. In a recent publication, Gayet and co-workers (1991) screened a number of oxidase enzymes for their suitability in detecting heavy metals. The enzymes were bound onto the surface of UltraBind US 450 affinity membranes with glutaraldehyde and placed over the surface of an oxygen sensor. The probes were then challenged with heavy metals and the resultant activity monitored. Mercuric and silver salts were found to be the most potent inhibitors. With pyruvate oxidase for example, 50% inhibition of the baseline response was achieved with a final concentration of $HgCl_2$ and $AgNO_3$ of 1.0 and 0.1 μM, respectively. An interesting feature of this work was that the sensor response could be recovered by treatment with 10 mM EDTA followed by 25 μM 1,4-dithio-DL-threitol, thus allowing multiple determinations to be carried out.

Finally, an interesting approach to the determination of Cu^{2+} was described by Mattiasson *et al.* (1979). These workers immobilised tyrosinase on the surface of a nylon net and then removed the enzyme-bound copper by treatment with 0.1 M NaCN. The resultant immobilised *apoenzyme* was mounted onto an oxygen sensor. This device has no tyrosinase activity unless Cu^{2+} is supplied to reconstitute the holoenzyme. The amount of reconstituted activity was, therefore, dependent on the amount of Cu^{2+} in the sample. For continuous use, the enzyme needed to be stripped of Cu^{2+} and then brought into contact with the sample. Thus, the system is operated in a semi-continuous cyclic fashion. Using this approach, the authors reported a useful response to Cu^{2+} of up to 50 µM with a linear range of approximately 25 µM.

7.5.4 Mass manufacture of enzyme electrodes

Despite the numerous publications that have appeared in the literature, very few enzyme electrodes have been successfully commercialised. In part, this is because the majority of enzyme electrodes have been developed to solve specific research problems in the laboratory. Additionally, not all experimental parameters affecting sensor response are completely understood. Consequently, this can result in large batch-to-batch deviations in the operational characteristics of the electrodes.

The most successful mass production technology that has been applied to the manufacture of enzyme electrodes to date is screen printing (Wring and Hart, 1992; Hart and Wring, 1994). In this technique, a squeegee is used to force a thick film paste, usually a preparation of carbon ink and enzyme, through a screen onto an inert support. The open pattern in the screen defines the final structures that are realised on the support. Additional screens can be used to print associated reference and auxiliary electrodes. The simplicity of this technique makes it suitable for low-cost and/or disposable sensor production even on a medium production scale. After the base electrode and enzyme layer have been printed, the production of the mass transfer control region (cf. above) remains of paramount concern. Possible existing technologies for membrane deposition are limited by the fact that the membranes have to be deposited on top of layers containing enzymes. High temperatures, ionising radiation, aggressive chemicals and high vacuums could lead to denaturation of the protein and, therefore, should be avoided. As a result researchers have been limited to spin coating, spraying or electrodeposition as the means of obtaining the surface membrane. Because the thickness of the layer and the viscosity of the film is critical to the sensor characteristics, much work remains to be done in optimising this aspect of enzyme electrode manufacture. If one assumes that the optimal design for an enzyme electrode is one where the probe has a fast response time and a linear output over a wide concentration range,

Table 7.3 Optimisation of theoretical parameters for an enzyme electrode

Parameter	Optimal value
D_s	Small
L	Large
α	Large
D_e	$D_e \gg D_s$
d	Large

See text for definition of parameters

which is insensitive to changes in the activity of the enzyme, then, in terms of the parameters described above, the optimal values would be those given in Table 7.3.

7.6 Whole cell sensors

Enzymes are perhaps the most widely used type of biological component in biosensors, but they are not necessarily isolated from the original biological source, e.g. they may be used as pastes or homogenates of the biological material (Sidwell and Rechnitz, 1985; Macholán and Boháčková, 1988). Another method is to maintain the enzymes in their native biological whole cells (Arnold and Rechnitz, 1980). This can confer advantages upon the resulting biosensor devices:

- greater stability of the biocatalyst since it is maintained in its native environment
- no need for cofactors
- possibility of multiple biochemical transformations using a single biocatalyst
- low cost of the biocatalyst.

There are also disadvantages:

- slow speed of response
- low selectivity
- complex cell types are often difficult to prepare for long-term storage
- lower quantitative reproducibility between sensors.

These potential disadvantages need to be borne in mind, but there are many applications where the advantages make such sensing devices attractive.
   Sensors that incorporate whole cells are of four broad types:

- tissue slices
- biological organelles
- isolated cells
- whole organisms.

Tissue slices have been widely used because of their low cost and ease of preparation. Biological organelles can demonstrate remarkable sensitivity to environmental stimuli, particularly when the assemblies include biological receptors. Isolated cells are of great importance in bioassays, particularly for toxicity testing, but have yet to make a significant impact in biosensor configurations for environmental monitoring. Sensing systems that incorporate whole organisms are of two types: those with higher organisms such as live fish, and those that incorporate microorganisms. The former type is perhaps more appropriately termed a 'biomonitor'; these are useful for environmental monitoring but suffer from high costs – both of initial outlay and of ownership. In addition, the monitors are most usually discontinuous since the fish are not always 'alert'. Potentially more practical and of widespread use are those devices that incorporate microorganisms, and the rest of this section focuses on this type of biosensor (Karube, 1987). Further details of the other types of sensor can be found in the literature (Arnold and Rechnitz, 1987; Wijesuriya and Rechnitz, 1993); (Table 7.4).

Table 7.4 Examples of sensors that incorporate different types of biological material

Sensor type	Biological component	Notes	Reference
Tissue slices	Potato	Hybrid biosensor with co-immobilised glucose oxidase; sensitive to phosphate and fluoride	Schubert et al. (1984)
	Walnut, mushroom	Sensitive to phenols	Macholán and Boháčková (1988)
Biological organelles	Crab antennae	Sensitive to trimethylamine oxide	Buch et al. (1991)
Isolated cells	Mouse leukaemia L1210	Antitumour drug screening	Liang et al. (1986)
	Chang liver cells	Toxicity assessment of formaline, surfactants and food additives	Karube et al. (1989)
	Human keratinocytes	Microphysiometer for toxicity assessment	Owicki and Parce (1990)
Cell fragments	Spheroplasts, membrane particles	Toxicity assessment	Gaisford et al. (1991)
Whole organisms	Trout	System developed commercially for protection of waste-water treatment works	Evans et al. (1986)
	Microorganisms	See Table 7.5	

7.6.1   Microbial biosensors

Microorganisms include a plethora of diverse species, strains and metabolic types:

- bacteria
- cyanobacteria
- microalgae
- microfungi
- protozoa.

Microorganisms are attractive sources of biological material since often they are cheap, easy and rapid to grow. Of particular relevance to environmental monitoring is that the use of microorganisms for toxicity assessment does not attract the same kind of public opprobrium as does the use of higher organisms.

Microorganisms may be used simply as 'bags' of enzymes rather than as 'living' cells. The term 'living' is used loosely since, in the biosensor literature, the existence or extent of any irreversible damage to biological function or integrity is not usually reported except where it is germane to the operation of the biosensor. However, this part of the chapter particularly concerns those devices that exploit metabolic processes such as respiration or photosynthesis that are an integral part of the living cells (Corcoran and Rechnitz, 1985; Bennetto *et al.*, 1987; Owicki and Parce, 1992)

7.6.2   Theoretical considerations

There are complex interactions between biological, chemical and physical processes within wholecells. Enzyme-catalysed processes are very important and cellular metabolism reflects some of the characteristic Michaelis–Menten behaviour of isolated enzymes (p. 212). For example:

- Stimulation of respiration by substrates, such as simple sugars, often can be characterised by the parameters K_m and V_{max} (Riedel *et al.*, 1988b; Riedel, 1991).
- Inhibition (reversible or irreversible) can be characterised in a similar manner to that of enzymes, e.g. Figure 7.15.

However, there are also significant differences between isolated enzymes and microbial whole cells. In particular, living cells frequently are able to adapt to environmental conditions. Other aspects of cell biology are also important. Biological membranes play a key role (Bitton *et al.,* 1988) since the biochemical centres of interest are most often separated from the 'outside world' by membranes. These membranes can be complex molecular assemblies that impart significant partitioning and mass

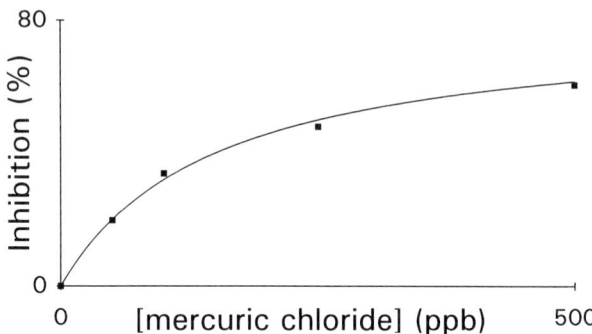

Figure 7.15 Inhibition of *Escherichia coli* respiration by mercuric chloride. The solid line shows the inhibition calculated according to an equation of Michaelis–Menten form, using $K_i = 150$ ppm and a maximum inhibition of 80%. (Original data from N. Richardson.)

transport effects on molecules that are required to traverse them. For example, the plasma membrane can present an almost insurmountable barrier to lipophobic molecules – such as small, polar or ionised molecules – that are unable to penetrate to the cytoplasm unless there are special transport mechanisms or ion pairing effects.

There is a large gap in understanding between the fundamentals of cell biology and the actual working and response of practical environmental monitoring devices. This gap is larger than with devices utilising isolated enzymes. Nevertheless, whole-cell-based biosensors are likely to play an important part in future environmental monitoring systems since the immobilised cells can act as surrogates for the more complex cells and organisms actually living in the environment.

7.6.3   Immobilisation of microbial whole cells

Microbial cells may be incorporated into sensing systems as isolated species, as collections of isolated species, as aggregated collections of species or as microbial consortia (Table 7.5). However, the immobilisation of living cells presents particular problems in the development of biosensor devices. Ideally, there should be intimate contact between the cells and the signal transducer, so there are few possibilities for the incorporation of cells into the types of support materials that are of widespread use in other areas of biotechnology. Immobilisation of cells by covalent attachment to the transducer surface is also of limited utility since the chemical treatments required usually have an adverse effect on the cells. For example, the entrapment of *Scenedesmus obliquus* within serum albumin, cross-linked with glutaraldehyde, disables the proper functioning of photosynthetic

Table 7.5 Examples of sensors incorporating microbial cells

Microorganism(s)	Origin	Notes	References
Synechococcus	Cyanobacterium	Sensitive to herbicides	Rawson *et al.* (1987; 1989)
Pseudomonas putida	Bacterium	Sensitive to ϵ-caprolactam	Riedel *et al.* (1989)
		Sensitive to 3-chloro-benzoate	Riedel *et al.* (1991)
		Sensitive to benzene	Tan *et al.* (1994)
Escherichia coli	Bacterium	Sensitive to phenols and heavy metals	Atkinson and Rawson (1994)
Chlorella vulgaris	Alga	Sensitive to herbicides	Pandard and Rawson (1993)
		Sensitive to herbicides	Pandard *et al.* (1993)
Trichosporon cutaneum	Yeast	Biological oxygen demand	Riedel *et al.* (1990b)
Bacillus subtilis + *Trichosporon cutaneum*	Bacterium + yeast	Biological oxygen demand	Riedel *et al.* (1988a)
Microbial consortium	Sewage sludge	Biological oxygen demand	Tsushima *et al.* (1992)
		Biological oxygen demand	Kong *et al.* (1993)

activity (Jeanfils, 1986). The most successful immobilisations are limited to physical methods of entrapment, e.g. behind a membrane or within a hydrogel or other polymer matrix.

7.6.3.1 Entrapment behind a membrane. A simple approach to cellular entrapment is to centrifuge cells and apply them as a paste (Walters *et al.*, 1980), or to filter a microbial suspension onto a membrane support such as cellulose acetate, nylon or alumina.

Tortuous membranes can be used to support the cells within the pores of the membrane. In these cases, the cells can be prevented from leaking out by the use of a semi-permeable membrane that still allows access of substrates and analytes to the cells from the external medium.

Alumina membranes (Anopore™, Whatman International) have an exceptionally high void volume of homogeneous capillary pores, so that there is a very small barrier to the diffusion of molecules across the membrane. Cells are easily trapped on the alumina surface since the pores have a narrow distribution of diameters. However, this membrane material is fragile and has to be handled with care. Other filters with capillary pores are available in less brittle materials (e.g. Nucleopore™ and Cyclopore™), but these generally have a smaller void volume and a wider distribution of pore sizes, so there is potentially a smaller flux of materials across the membrane and cell capture on the membrane surface is less rigorous than with the alumina type.

7.6.3.2 Gel entrapment. Acrylamide gels are of limited use for the immobilisation of living cells because of the toxicity of the monomers (Starostina *et al.,* 1983). However, microbial whole cells can be entrapped in gels such as calcium alginate or κ-carrageenin, in a similar manner to enzymes (q.v.). The loss of biological material, by diffusion out of the gels, is less than with enzymes, but cell-loaded gels still suffer from the large diffusion barriers presented to substrates and analytes, and also suffer from shrinkage and swelling of the gel as the external medium changes.

7.6.3.3 Polymer entrapment. Poly(vinyl alcohol) (PVA) is widely used for immobilisation of microbial cells since the material has a low toxicity and the conditions are mild. The hydrolysed polymer may be dissolved in a hot aqueous solution; cells are subsequently incorporated as a suspension at ambient temperatures and become entrapped by the air drying or freezing of thin films. The polymer can be further treated by, for example, freeze–thawing to produce a more rubbery polymer (Ariga *et al.,* 1987). Such procedures tend to reduce cell viability, but, without additional cross-linking, the polymer can re-dissolve over a period of time with a consequent loss of biological material. Martens and Hall (1994) immobilised *Synechococcus* in 2% PVA (average molecular weight 115 000); this was air-dried on a carbon-paste electrode and protected with a dialysis membrane. The resultant sensors were reported to be usable for 4–10 days when stored in fresh buffer (5°C) and used intermittently.

The *in situ* entrapment of enzymes by the electrochemical growth of organic polymers was discussed earlier (p. 223). Electropolymerisation also presents the possibility of using relatively thin, insoluble polymer films for the entrapment of biological whole cells. The successful immobilisation of viable microbial cells, using such a technique, requires a number of attributes:

- monomers or oligomers that are water soluble, since microbial cells have a limited tolerance to non-aqueous solvents
- concentrations of starting materials that are non-toxic
- maintenance of local reaction conditions (i.e. at the electrode surface) that are non-toxic and not detrimental to the integrity of the cellular membranes
- formation of conducting polymer films that can be laid down in sufficient thicknesses to entrap whole cells.

Deshpande and Hall (1990) immobilised viable banana cells in electropolymerised polypyrrole. They also reported a similar immobilisation of *Pseudomonas putida* and *Agaricus bisporus,* but, in these cases, the existence or extent of cellular damage was not reported since the cells were being used simply as a source of polyphenol oxidase for the determination of dopamine.

7.6.4   Methods of monitoring microbial whole cells

The activity of immobilised cells may be monitored in a variety of ways. Optical techniques can be used in sensor configurations, but electrochemical methods predominate (Table 7.6).

Oxygen respiration is a metabolic activity common to many microbial cells. Consequently, the membrane-covered oxygen sensor (Clark electrode) is perhaps the most widely used transducer for sensing microbial activity. It can be used either to follow oxygen consumption through respiration, or to follow oxygen evolution as a result of photosynthesis (Figure 7.16). Most usually, whole cells are immobilised directly onto the outside surface of an

Table 7.6 Examples of transduction techniques used with microbial biosensors

Transduction technique	Microorganism(s)	Immobilisation method(s)	References
Oxygen uptake or evolution: oxygen sensor	*Chloella vulgaris, Scenedesmus subspicatus*	Alumina membrane	Pandard *et al.* (1993)
	Bacillus subtilis, Pseudomonas aeruginosa, Trichosporon cutaneum	Poly(vinyl alcohol) or paper	Riedel *et al.* (1990a)
	Bacillus subtilis + Bacillus licheniformis	Polycarbonate membrane	Li and Tan (1994b)
	Pseudomonas aminovarans	Nylon membrane	Gamati *et al.* (1991)
Hydrogen evolution	*Chromatium* sp.	Polycarbonate membrane or alginate gel	Matsunaga *et al.* (1984)
Carbon dioxode uptake or evolution: potentiometric carbon dioxide sensor	*Escherichia coli*	Agar gel	Dorward and Barisas (1984)
	Saccharomyces cerevisiae	None	Campanella *et al.* (1987)
Ammonia flux: potentiometric ammonia sensor	*Streptococcus faecium*	Dialysis membrane	Rechnitz *et al.* (1977)
	Pseudomonas sp.	Dialysis membrane	Walters *et al.* (1980)
	Escherichia coli	Dialysis membrane	Corcoran and Kobos (1987)
Redox potential	*Saccharomyces cerevisiae*	None	Campanella *et al.* (1987)
Redox-mediated current	*Escherichia coli*	Alumina membrane	Richardson *et al.* (1991)
	Synechococcus	Poly(vinyl alcohol)	Martens and Hall (1994)
Amperometric determination of product	*Pseudomonas putida, Agaricus bisporus*	Polypyrrole	Deshpande and Hall (1990)
Resistance change	*Saccharomyces cerevisiae*	Agar gel	Palmquist *et al.* (1994)
UV spectroscopy	*Escherichia coli*	Agarose gel	Bains (1994)
Luminescence	*Photobacterium phosphoreum*	Cellulose nitrate membrane	Lee *et al.* (1992)

which is a glycoprotein (molecular weight $\sim 160\,000\,\text{Da}$) in which binding occurs at two identical sites. Antigens typically have a molecular weight greater than 1500. To elicit an efficient antibody response to smaller molecules (haptens), these are chemically attached to a carrier protein. The resulting conjugate will elicit the formation of antibodies, some of which will be directed against the hapten.

From an analytical point of view, an antibody can be viewed simply as a molecular recognition complex that binds selectively to the antigen in question in a reversible manner. It is estimated that an animal can synthesise up to 10^7–10^8 antibodies of different specificities. This variability is the source of the immense versatility in achieving high specificity of binding for a broad range of chemical features.

7.7.1 Kinetics of antibody–antigen binding

The reaction of antibody (Ab) with antigen (Ag) to give the complex Ab–Ag is an example of a simple binding equilibrium. The association reaction will be characterised by a second-order rate constant k_a and the dissociation reaction by the first-order rate constant k_d. The corresponding association and dissociation rates (v_a and v_d) are given by:

$$v_a = k_a[\text{Ab}][\text{Ag}]$$
$$v_d = k_d[\text{Ab–Ag}]$$

Equating v_a to v_d and rearranging gives an expression for the association constant K_a according to:

$$K_a = \frac{[\text{Ab–Ag}]}{[\text{Ab}][\text{Ag}]} = \frac{k_a}{k_d}$$

Values of K_a range from about 10^4 to $10^{12}\,\text{M}^{-1}$. Immunoglobulins with K_a values $< 10^4$ would be ineffective as antibodies against a particular antigen.

When making a measurement with immobilised antibody, the concentration of the biological recognition element is essentially fixed. The concentration of the antigen will be variable or unknown. Under these circumstances, the sum of the equilibrium concentrations of free antibody $\{[\text{Ab}]_{eq}\}$ and bound antibody $\{[\text{Ab–Ag}]_{eq}\}$ will be constant irrespective of the added concentration of antigen. The relative proportion of free and bound antibody will vary with the equilibrium concentration of antigen $\{[\text{Ag}]_{eq}\}$ and the nature of this variation will be of analytical significance. Given a starting concentration of antibody equal to $[\text{Ab}]_\Sigma$, the equilibrium concentration $[\text{Ab}]_{eq}$ will simply be $[\text{Ab}]_\Sigma - [\text{Ab–Ag}]_{eq}$, i.e. the starting concentration less the complexed concentration. Substituting this into the above expression for K_a gives, after some rearrangement, an expression for the concentration of complex as a function of antigen concentration:

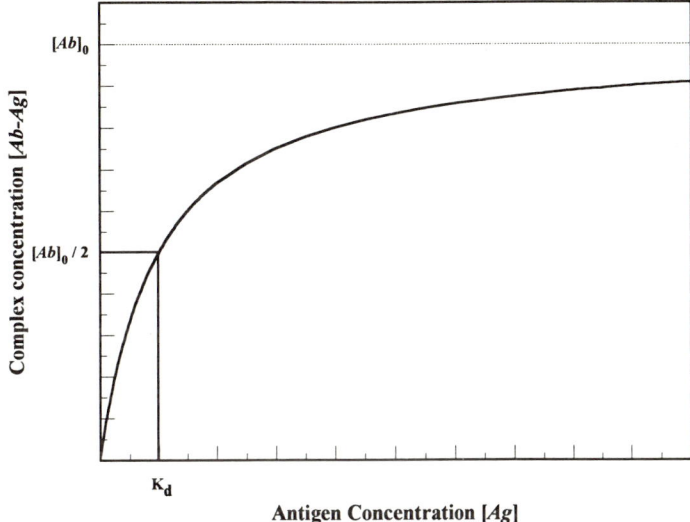

Figure 7.20 Variation of complex concentration, [Ab–Ag] as a function of antigen concentration.

$$[Ab-Ag]_{eq} = \frac{[Ag]_{eq}}{K_d + [Ag]_{eq}} \times [Ab]_{\Sigma}$$

where K_d is simply $1/K_a$ and has units of molarity. The form of this relationship is shown in Figure 7.20.

With immunosensors, it is often important to consider the binding interactions at the surface of the probe. If one considers a surface with a certain density of binding sites γ_{lim}, in units of moles per surface area $(mol\,m^{-2})$, this can be equated directly to the starting concentration of antibody $[Ab]_{\Sigma}$ in the description of complex formation in solution. The whole of the above treatment can then be applied to give an expression for the equilibrium surface coverage of the Ab–Ag complex γ_{eq} according to:

$$\gamma_{eq} = \frac{[Ag]_{eq}}{K_d + [Ag]_{eq}} \times \gamma_{lim}$$

Taking reciprocals of the above equation gives:

$$\frac{1}{\gamma_{eq}} = \frac{1}{\gamma_{lim}} + \frac{K_d}{\gamma_{lim}} \times \frac{1}{[Ag]_{eq}}$$

A double reciprocal plot of experimental data will give a straight line and the binding parameters γ_{lim} and K_d can be obtained from the intercept on the y-axis and the slope, respectively. It should be noted, however, that for the case described by the above equation the solution species [Ag] is assumed to be in considerable excess over the surface antibody. With immunoprobes, this assumption is often valid.

7.7.2   Immobilisation of antibodies

Because the reaction of an antibody with an antigen involves the formation of a bimolecular complex, the spatial orientation of the antibody on a surface is very important. Optimisation of both the surface coverage (obtaining a crystalline lattice) and the orientation of the antibody on the surface of an immunological probe is, therefore, a crucial design consideration. For example, in Figure 7.21 only antibody C is in the correct orientation to bind antigen. Both A and B cannot take part in the immunological reaction because of steric considerations.

To achieve the correct orientation, there are a number of techniques that can be used. In essence, these involve the use of suitable surface-modifying agents that will couple to the antibody in such a way that the antigen-binding sites are orientated away from the surface of the probe. One such example is the coupling of antibodies via the carbohydrate moieties in the hinge region. The first step in this technique involves the oxidation of the carbohydrate *cis*-diol groups to the corresponding reactive aldehydes. These aldehydes can then react with surface hydrazine groupings to form stable, leak-resistant hydrazone linkages. (N.B. It is necessary to modify the surface of the probe so that the hydrazine moieties are present to react with the oxidised antibody.) Alternatively, the immobilisation of antibody can be mediated by *protein G*, a 22 000 Da molecular weight protein that binds specifically to the Fc region of mammalian antibodies. These two schemes are shown diagrammatically in Figure 7.22.

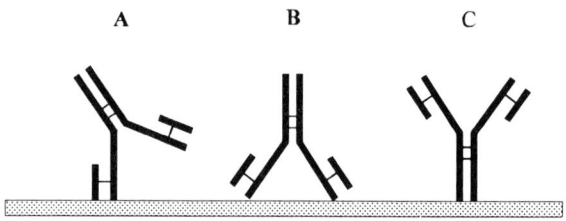

Surface of the immuno-probe

Figure 7.21 Possible steric complications resulting from the immobilisation of antibody molecules to a solid surface.

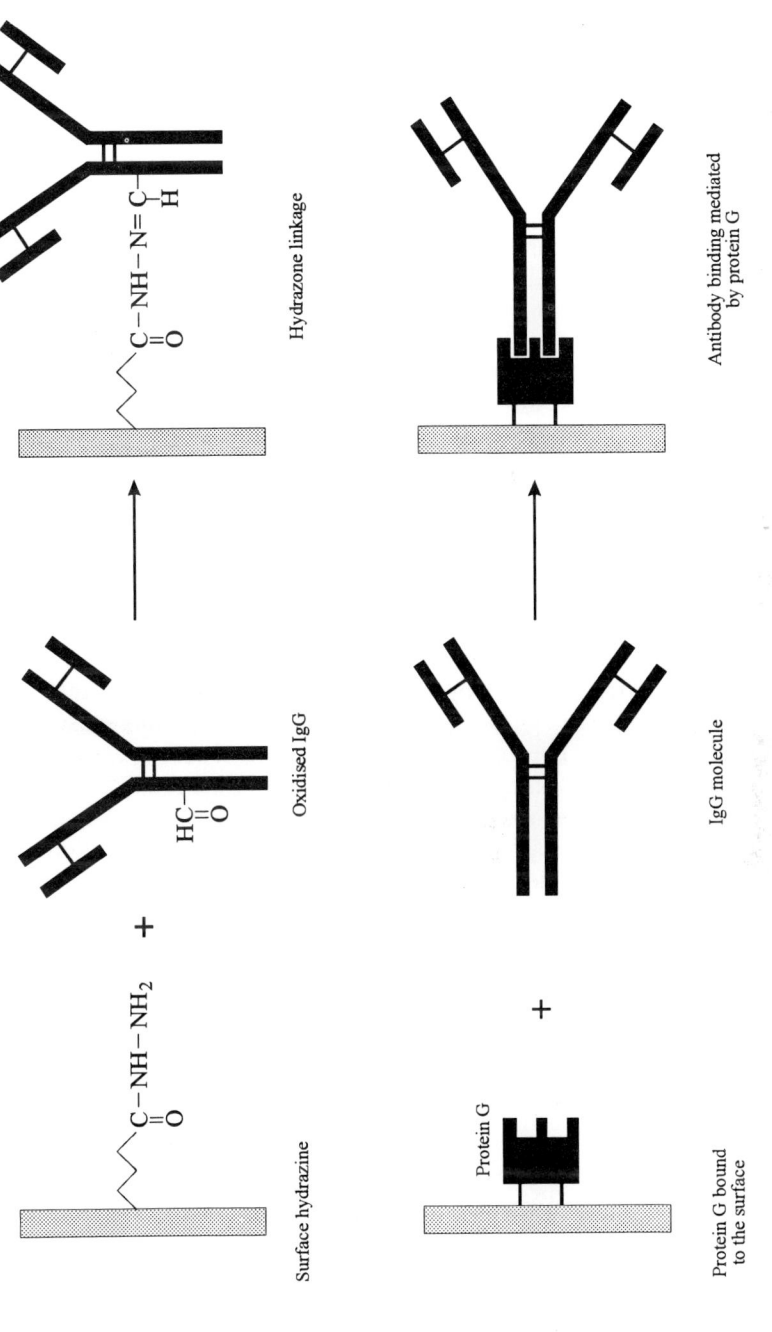

Figure 7.22 Two schemes for attaching antibody molecules to the surface of a transducer. (a) The attachment of antibody to the surface by a stable hydrazone linkage. (b) The protein G-mediated immobilisation of the antibody.

7.8   Schemes for electrochemical/electrical immunosensors

7.8.1   Reagentless immunoprobes

With the exception of catalytic antibodies, which are beyond the scope of this chapter, antibodies do not catalyse the conversion of a substrate into a product. No electroactive species are, therefore, generated or consumed during the antibody antigen–reaction. Consequently, techniques that measure Faradaic currents are not easily applicable to the direct determination of the Ab–Ag complex. One electrochemical approach that has shown promise, however, is based upon double-layer capacitance measurements.

When an electrode is placed into a solution of an electrolyte, the excess charge residing at the surface of the electrode must be exactly balanced by an equal but opposite charge on the side of the solution. In the simplest case, this charge is balanced by a layer of solvated ions held at the electrode surface by Coulombic attraction (e.g. Figure 7.23).

The line drawn through the centre of the cations at this distance of closest approach marks a boundary known as the *outer Helmholtz plane*. The region within this plane makes up the compact part of the double layer. Occasionally, specific adsorption of ions may occur at the electrode surface in which van der Waals and chemical forces participate. Most anions are specifically adsorbed, thereby losing most, if not all, of their inner hydration shell. These species can, therefore, approach more closely to the electrode surface. A line drawn through the centre of such species aligned at the electrode surface defines a further boundary within the Helmholtz layer called the *inner Helmholtz layer* (Figure 7.24). The extent to which specific adsorption occurs is controlled by the nature of the ions in solution, by the nature of the electrode material and by the magnitude and sign of the applied potential.

In summary, the interfacial region can be viewed as two layers of equal and opposite charge separated by a dielectric material, i.e. a capacitor.

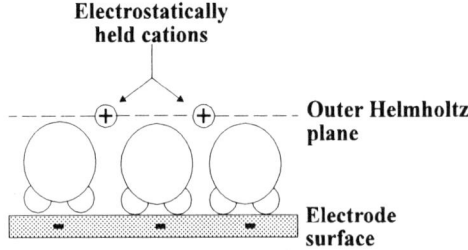

Figure 7.23 Microstructure at a charged electrode/electrolyte interface showing the outer Helmholtz plane.

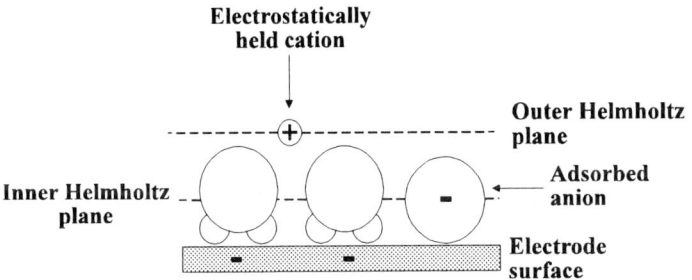

Figure 7.24 Microstructure at a charged electrode/electrolyte interface showing both the outer Helmholtz plane and the inner Helmholtz plane.

Indeed, the current response of an electrode to an applied potential in the absence of a Faradaic process can be modelled using a series combination of a capacitor of capacitance C_{dl} (the *double layer capacitance* of the interfacial region) and a resistor R_u (the *uncompensated resistance* of the electrochemical cell. If an oscillating potential is applied to the electrode, a 'base current' flows as a result of charging the electrical double layer (Arwin *et al.*, 1982). The base current is directly proportional to C_{dl} and consequently can be used to follow changes in the microstructure of the electrical interface under various conditions, such as the displacement of ions from the electrical double layer or the insertion of polyelectrolytes (Hertl, 1987). Biological macromolecules contain on their surface several positive and negative charges. Deposition of such molecules at the interface generates alterations in the electrical double layer.

Using this approach, Hertl (1987) described electrical immunoassays for IgG, anti-IgG, anti-ferritin and *Staphylococcus aureus* cells. In this work, antibodies were immobilised onto the surface of a glassy carbon electrode by adsorption from solution followed by cross-linking with glutaraldehyde. When a specific antibody is added to the system, it binds to the immobilised antibody thus perturbing the electrical double layer and causing a change in capacitance. This change is detected as a change in the current flow. The apparatus consisted of a potentiostat (a PAR model 174A polarographic analyser) used together with a model 124A lock-in amplifier. The applied DC potential was set at zero and the applied AC potential was 10 mV peak-to-peak with a frequency of 10 Hz. The applied potential was referenced to a platinum wire counter electrode and the current was monitored using a strip y–t chart recorder.

The author reported that IgG levels could be detected in the range 0.5–10 µg in a 4 ml volume of serum using an anti-IgG electrode IgG. The change in current (ΔI) over the linear portion of the curve (up to 1.5 µg) was in the order of 0.3 µA, giving a detection limit of 250 ng IgG. The response time of the electrode was, however, approximately 1 h.

Staphylococcus aureus contains a membrane protein (protein A) that agglutinates, i.e. sticks together, IgG. It can, therefore, be detected using the anti-IgG electrode. Hertl reported that in his system, up to 5×10^7 cells could be detected with an estimated detection limit of 1×10^6 cells. At saturating levels, ΔI; was in the order of $0.7\,\mu A$.

Although the examples quoted in this work are not directly applicable to environmental monitoring, they do illustrate the generic nature of the technique. Because antibodies can, in principle, be raised against any molecule, it is possible to develop 'environmentally relevant' probes using this approach. The technique does suffer from two major drawbacks. First, the current flowing in an RC circuit is dependent on both the capacitance and the resistance. Changes in the resistance of the test solution (ionic strength) could, therefore, result in misleading analytical information. Hertl (1987) studied this problem and concluded that, in his system, a change of 10% or less in the electrolyte concentration (which was KCl) could be tolerated. To overcome this problem, sample pre-treatment may have to be mandatory. Second, because the measurement depends on changes in the double layer capacitance upon binding of the target analyte, the technique is not very specific. Although antibodies do provide a degree of specificity, other proteins or macromolecules present in the solution could adsorb to the surface in a non-specific way, resulting in an erroneous analytical response.

7.8.2 Semiconductor systems

Another type of immunoprobe is based upon the antigen-induced potential shift of chemically modified semiconductor electrodes (Yamamoto *et al.*, 1983). Here, the surface of a titanium dioxide electrode is covered by a cyanogen bromide-activated polymer membrane and inserted into an antibody-containing solution. This results in the covalent attachment of antibody to the electrode surface. The antigen to be detected is then added and the potential difference between the sensor and a reference half cell measured. The potential change (ΔE) is a measure of the antigen content of the sample. As a practical example, an electrode was prepared with a monoclonal antibody selective to trinitrophenol group-containing substances. The sensor responded to trinitrophenol-γ-globulin in the concentration range 0.066–$1\,\mu mol\,l^{-1}$ with a response time of approximately 20 min. The authors speculated that the changes in potential were the result of perturbations of the double layer by the dipole of the bonding entity (Yamamoto *et al.*, 1983). Yamamoto and co-workers (1978) also proposed the use of antibody-coated titanium oxide to measure levels of human chorionic gonadotrophin (hCG) in the urine of pregnant women. However, the measured potential of such sensors was found to vary with buffer composition, pH and ionic strength. Moreover, the non-specific adsorption of proteins to the surface of the sensor falsified the analytical response. This implies that the accuracy of the measurement depends upon the ratio of specific and non-specific interactions, suggesting that a degree of sample preparation and pre-treatment must be carried out.

7.8.3 Piezoelectric systems

Immunosensors using piezoelectric mass balances coated with antigen or antibody were proposed as early as 1972 (Shons et al., 1972). The mass increase resulting from the immunological complex formation leads to a measurable change in the resonance frequency of the piezoelectric crystal, as predicted by the Sauerbrey equation (Sauerbrey,1959):

$$\Delta f = -2.3 \times 10^6 f^2 \frac{\Delta m}{A}$$

where f is the vibrational frequency of the crystal (Hz), Δm is the mass change (g) and A is the adsorbing or sensing area of the crystal (cm). The change in resonant frequency resulting from adsorption of the detected analyte can be calculated (Alder and McCallum, 1983; Guilbault and Jordan, 1988; Guilbault and Schmid, 1991) as providing extremely high sensitivity (approximately 500 to 2500 Hz µg^{-1}) with picogram detection limits for commercial piezoelectric crystals. In reality, however, there is usually little correlation between the predicted frequency shifts and those obtained experimentally with antibody-coated crystals. This is probably because of the inability to produce monolayer coatings on the crystal surface.

Shons and co-workers (1972) coated a piezoelectric crystal with bovine serum albumin (BSA) and measured the adsorption of BSA antiserum. The sensor responded to anti-BSA antibody over a concentration range of three orders of magnitude. Ngeh-Ngwainbi et al. (1986) proposed an antibody-coated piezoelectric crystal for the measurement of parathion and other organophosphorous pesticides in the gas phase. In this work, anti-parathion antibodies were layered onto piezoelectric crystals. The devices showed fast response times, typically 2–3 min, and a reproducibility of better than 4%. The calibration curve for parathion was linear in the parts per billion to parts per million range. In addition, no interference from moisture was observed if the relative humidity remained constant. The response of this device to various pesticides is shown in Table 7.8.

Finally, Muramatsu et al. (1986) reported a piezoelectric immunosensor for the detection of the potentially pathogenic organism Candida albicans. Here, the affinity binding of the organism resulted in a decrease in the resonance frequency in the range 0.5–1.4 kHz for 10^6 to 5×10^8 cells. (Readers should also refer to Volume 1, Chapter 6.)

Table 7.8 Detection of various pesticides with a piezoelectric immunosensor

Pesticide	Amount in gas phase (ppb)[a]
Parathion	36
Malathion	106
Methyl parathion	158
Disulfoton	560

[a]Amount of pesticide required to produce a frequency change of 400 Hz

7.8.4 Voltammetric immunoassay schemes

Although the direct measurement of the antigen–antibody complex is difficult in practice, it is possible to label either component with an electrochemical tag. If the electrochemistry of the tag is then modulated because of complex formation, the reaction can be monitored using sensitive measuring techniques such as *differential pulse voltammetry* (DPV). An example of this approach was described for the detection of *oestriol* in a competitive homogeneous assay (Wehmeyer *et al.*, 1982). In this work, the antigen was rendered electroactive (Ag*) by the introduction of two nitro groups ($-NO_2$) into the skeleton at positions 2 and 4. A DPV of Ag* gave a distinctive reduction peak (-0.45 V versus SCE) over a potential range in which unlabelled oestriol was electroinactive. The addition of antibody to the solution caused a decrease in the peak height to occur as Ag* became complexed with antibody. Consequently, peak height was used to monitor the degree of binding of Ag*. If any unlabelled oestriol (the analyte) was present in the sample, it competed for the antibody-binding sites, resulting in a decrease in the equilibrium concentration of the Ab–Ag* complex. Peak height could then be related to oestriol concentration. In an analogous fashion, this strategy has been used with other antigens, such as morphine labelled with ferrocene (Weber and Purdy, 1979); ovalbumin labelled with diazotised 4-aminobenzoic acid (Breyer and Radcliffe, 1951); and oestriol labelled with mercuric acetate (Doyle *et al.*, 1982).

Other useful tags to use in electrochemical immunoassay are enzymes (Blake and Gould, 1984). An enzyme that is particularly useful in this context is *alkaline phosphatase*, which catalyses the hydrolysis of phosphoric acid esters. The rationale of such assays is to measure the complex formation by electrochemically monitoring changes in the activity of the enzyme using a suitable 'electrochemical substrate'. The substrates should show either different electrochemical behaviour from the product, such as a positive or negative shift in E^0, or no electrochemical activity at all. Typical substrates reported include: phenyl phosphate (Rosen and

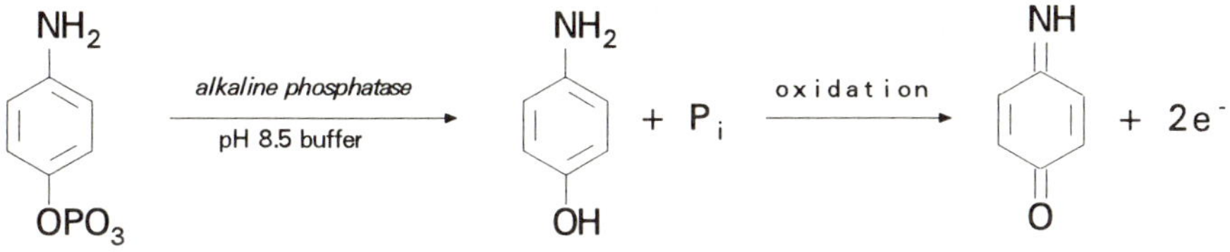

Figure 7.25 Electrochemical detection method for the enzyme alkaline phosphatase using an 'electroactive' substrate.

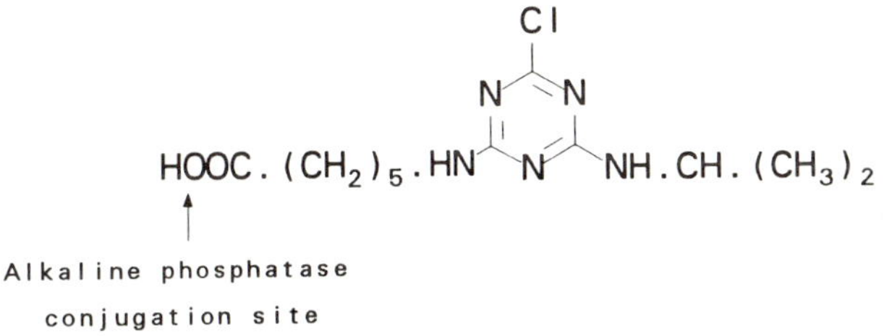

HOOC . $(CH_2)_5$. HN $\overset{N}{\underset{N}{\bigvee}}$ NH . CH . $(CH_3)_2$

Alkaline phosphatase
conjugation site

Figure 7.26 Atrazine analogue showing conjugation site for alkaline phosphatase.

Rishpon, 1989); 4-aminophenyl phosphate (Cardosi *et al.*, 1989) and [*N*-ferrocenoyl]-4-aminophenyl phosphate (McNeil *et al.*, 1988). The electrochemical scheme for the detection of alkaline phosphatase using 4-aminophenyl phosphate is shown in Figure 7.25.

Using alkaline phosphatase as the label, a competitive homogeneous immunoassay was developed for the commonly used photosystem-II herbicide atrazine (Cardosi and Turner, 1991). For the assay, an atrazine analogue was conjugated to alkaline phosphatase through the carboxylic acid residue, as illustrated in Figure 7.26. The activity of the alkaline phosphatase was detected using 4-aminophenyl phosphate as the substrate. The liberated 4-amino phenol was oxidised at a glassy carbon electrode. Because of the competitive nature of the assay, the peak current was inversely proportional to the concentration of atrazine in the sample. The response curve was linear over the range 1×10^{-7} to 1×10^{-9} g ml^{-1} (Figure 7.27).

Although the above are examples of homogeneous assays, enzyme labels can also be used in a 'probe configuration'. In each case, the activity of the enzyme label is related to the antigen concentration and the electrode serves as both immobilisation support and transducer for detecting the enzyme-generated product. A popular choice of enzyme for use in these configurations has been glucose oxidase (glucox), which catalyses the

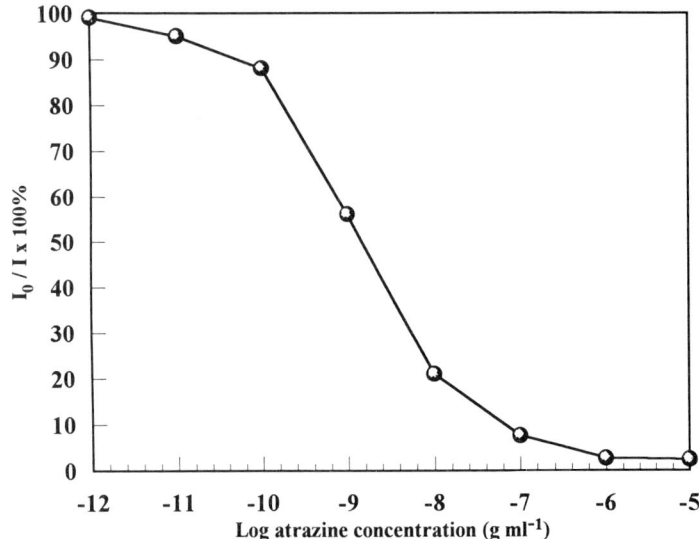

Figure 7.27 Calibration curve for the competitive electrochemical immunoassay for atrazine.

following reaction:

$$\text{Glucose} + O_2 \xrightarrow{\text{Glucose oxidase}} \text{Gluconic acid} + H_2O_2$$

The activity of the enzyme can be determined amperometrically either by monitoring the production of hydrogen peroxide or by measuring the decrease in the partial pressure of dioxygen.

As an example of this approach, Romette and Boitieux (1984) described an assay for hepatitis B surface antigen (HB_S). Here, anti-HB_S was immobilised onto a nylon net that was then placed over the surface of an oxygen electrode. For measurement, the membrane was contacted with the

Table 7.9 Examples of immunoprobes directed against various antigens

Antigen	Sensor	Label	Sensitivity	Reference
Theophylline	O_2	Catalase	1×10^{-3} to 5×10^{-3} M	Haga et al. (1984)
Insulin	O_2	Catalase	10^{-6} M	Mattiasson et al. (1977)
Alpha-fetoprotein	O_2	Catalase	5×10^{-11} to 10^{-8} g ml^{-1}	Aizawa et al. (1979a)
Human chorionic gonadotrophin	O_2	Catalase	3×10^{-9} to 1.5×10^{-5} g ml^{-1}	Aizawa et al. (1979b)
Albumin	O_2	Glucox	1×10^{-8} M	Mattiasson et al. (1977)
HB_S	O_2	Glucox	10^{-10} to 10^{-7} g ml^{-1}	Romette and Boitieux (1984)
IgG	O_2	Catalase	6×10^{-11} to 6×10^{-9} M	Aizawa et al. (1978)

sample in the presence of a known amount of glucox-labelled antigen. After washing the membrane to remove any adsorbed label, the activity of the 'bound' glucox was measured by adding glucose to the electrolyte solution and monitoring the rate of oxygen consumption. This approach has also been applied to the determination of other antigens (Table 7.9).

7.9 Conclusions

This chapter has given an insight into the applications of biosensors for environmental monitoring. Although the biosensor is perhaps unique in that it incorporates a biological element, the ultimate requirements of the device are no different to those of a more conventional sensor.

At present, more research is required to overcome the problems still associated with biosensors, the most serious of which is poor operational stability. When all the problems have been overcome, the applications of biosensors to environmental monitoring are undoubtably manyfold. It is important to state, however, that even after a successful period of research and development the ultimate routine adoption of a biosensor (or any sensor technology) lies in the acceptability of the technology by the user. This is an area that certainly requires a closer look.

References

Aizawa, M., Morioka, A. and Suzuki, S. (1978) Electrochemical determination of IgG, *Journal of Membrane Science*, 4, 221–226.

Aizawa, M., Suzuki, S., Nagamura, Y., Shinohara, R. and Ishiguro, J. (1979a) Enzyme immunoassay for human chorionic gonadotropin with electrochemical detection. *Journal of Solid-phase Biochemistry*, 4, 25–30.

Aizawa, M., Morioka, A., Suzuki, S. and Nagamura, Y. (1979b) Enzyme immunosensor III. Amperometric determination of human chorionic gonadotropin by membrane bound antibody. *Analytical Biochemistry*, 94, 22–28.

Albery, W.J., and Cranston, D.H. (1987) Amperometric enzyme electrodes: theory and experiment. In *Biosensors: Fundamentals and Applications*, (ed. A.P.F. Turner, I. Karube and G.S. Wilson), pp. 180–210. Oxford University Press, Oxford.

Alder, J.F. and McCallum, J.J. (1983) Piezoelectric crystals for mass and chemical analysis. *Analyst*, 108, 1169–1189.

Allen, C. (1994) *Thunder and Lightning: The RAF in the Gulf.* Warner Books, London.

Alvarez-Icaza, M. and Bilitewski, U. (1993) Mass production of biosensors. *Analytical Chemistry*, 65, 525A–533A.

Alvarez-Icaza, M. and Scmid, R.D. (1994) Observation of direct electron transfer from the active centre of glucose oxidase to a graphite electrode achieved through the use of mild immobilisation. *Bioelectrochemistry and Bioenergetics*, 33, 191–199.

Ariga, O., Takagi, H., Nishizawa, H. and Sano, Y. (1987) Immobilization of microorganisms with PVA hardened by iterative freezing and thawing. *Journal of Fermentation Technology*, 65, 651–658.

Arnold, M.A. and Rechnitz, G.A. (1980) Comparison of bacterial, mitochondrial, tissue and enzyme biocatalysts for glutamine selective membrane electrodes. *Analytical Chemistry*, 52, 1170–1174.

Arnold, M.A. and Rechnitz, G.A. (1987) Biosensors based on plant and animal tissue. In *Biosensors: Fundamentals and Applications*, (ed. A.P.F. Turner, I. Karube and G.S. Wilson), Oxford University Press, Oxford, pp. 35–78.

Arwin, H., Lundström, I. and Palmqvist, A. (1982) Electrode adsorption method for the determination of enzymatic activity. *Medical and Biological Engineering Computation*, **20**, 362–374.

Atkinson, A.L. and Rawson, D.M. (1994) Biosensors for pollution monitoring and toxicity assessment. In *Ecotoxicology of Soil Organisms*. (ed. M.H. Donker, H. Eijsackers and F. Heimbach), Lewis, Boca Raton, FL, pp. 68–92.

Bains, W. (1994) A spectroscopically interrogated flow-through type toxicity biosensor. *Biosensors and Bioelectronics*, **9**, 111–117.

Bartlett, P.N. and Cooper, J.M. (1993) A review of the immobilisation of enzymes in electropolymerised films. *Journal of Electroanalytical Chemistry*, **362**, 1–12.

Bartlett, P.N. and Whitaker, R.G. (1987a) Electrochemical immobilisation of enzymes, Part II. Glucose oxidase immobilised in poly-N-methylpyrrole. *Journal of Electroanalytical Chemistry*, **224**, 37–48.

Bartlett, P.N. and Whitaker, R.G. (1987b) Electrochemical immobilisation of enzymes, Part I. Theory. *Journal of Electroanalytical Chemistry*, **224**, 27–35.

Bartlett, P.N., Tebbutt, P. and Tyrrell, C.H. (1992) Electrochemical immobilisation of enzymes, Part III. Immobilisation of glucose oxidase in thin films of electrochemically polymerised phenols. *Analytical Chemistry*, **64**, 138–142.

Bennetto, H.P., Box, J., Delaney, G.M., Mason, J.R., Roller, S.D., Stirling, J.L. and Thurston, C.F. (1987) Redox-mediated electrochemistry of whole micro-organisms; from fuel cells to biosensors. In *Biosensors: Fundamentals and Applications*, (ed. A.P.F. Turner, I. Karube and G. S. Wilson), Oxford University Press, Oxford, pp. 291–314.

Bitton, G., Dutton, R.J. and Koopman, B. (1988) Cell permeability to toxicants: an important parameter in toxicity test using bacteria. *CRC Critical Reviews in Environmental Control*, **18**, 177–188.

Blaedel, W.J. and Jenkins, R.A. (1972) Study of a reagentless lactate electrode. *Analytical Chemistry*, **48**, 1240–1247.

Blake, C. and Gould, B.J. (1984) Use of enzymes in immunoassay techniques: a review. *Analyst*, **109**, 533–547.

Breyer, B. and Radcliffe, F.J. (1951) Polarographic investigations of the antigen antibody reaction. *Nature (London)*, **167**, 79.

Buch, R.M., Barker, T.Q. and Rechnitz, G.A. (1991) Intact chemoreceptors based on Hawaiian aquatic species. *Analytica Chimica Acta*, **243**, 157–166.

Campanella, L., Paoletti, A.M. and Tranchida, G. (1987) Biosensors of total toxicity. *Chimicaoggi*, **March**, 61–63.

Cardosi, M.F. (1994) Hydrogen peroxide-sensitive electrode based on horseradish peroxidase-modified platinised carbon. *Electroanalysis*, **6**, 89–96.

Cardosi, M.F. and Birch, S.W. (1993) Screen printed glucose electrodes based on platinised carbon particles and glucose oxidase. *Analytica Chimica Acta*, **276**, 69–74.

Cardosi, M.F. and Turner, A.P.F. (1991) Mediated electrochemistry: a practical solution to biosensing. In *Advances in Biosensors*, Vol. 1, (ed. A.P.F. Turner), pp. 125–169. JAI Press, London.

Cardosi, M.F., Birch, S.W., Talbot, J. and Phillips, A. (1991) An electrochemical immunoassay for *Clostridium perfringens* phospholipase C. *Electroanalysis*, **3**, 169–176.

Carr, P.W. and Bowers, L.D. (1980) Immobilised enzymes in analytical and clinical chemistry. In *Advances in Biochemical Engineering*, Vol. 15, (ed. A. Fiechter), pp. 89–129. Springer Verlag, Berlin.

Cass, A.E.G. (ed.) (1990) *Biosensors: A Practical Approach*. IRL Press, Oxford.

Cooper, J. and Hall, E.A.H. (1993) Catalytic reduction of benzoquinone at polyaniline and polyaniline/enzyme films. *Electroanalysis*, **5**, 385–397.

Corcoran, C.A. and Kobos, R.K. (1987) Selectivity enhancement of an *Escherichia coli* bacterial electrode using enzyme and transport inhibitors. *Biotechnology and Bioengineering*, **30**, 565–570.

Corcoran, C.A. and Rechnitz, G.A. (1985) Cell-based biosensors. *Trends in Biotechnology*, **3**, 92–96.

Cottrell, F.G. (1902) Der Restrom bei galvanischer Polarisation, betratchet als ein Diffusionproblem. *Zeitschrift für physik Chemie*, **XLII**, 385–431.

Deshpande, M.V. and Hall, E.A.H. (1990) An electrochemically grown polymer as an immobilisation matrix for whole cells: application in an amperometric dopamine sensor. *Biosensors and Bioelectronics*, **5**, 431–448.

Diaz, A. (1981) Electrochemical preparation and characterisation of conducting polymers. *Chemica Scripta*, **17**, 145–148.

Dicks, J.M., Cardosi, M.F., Turner, A.P.F. and Karube, I. (1993) The application of ferrocene-modified n-type silicon in glucose biosensors. *Electroanalysis*, **5**, 1–9.

Dorward, E.J. and Barisas, B.G. (1984) Acute toxicity screening of water pollutants using a bacteria electrode. *Environmental Science and Technology*, **18**, 967–972.

Doyle, M.J., Halsall, H.B. and Heinemann, W.R. (1982) A heterogeneous immunoassay for serum proteins by differential pulse anodic stripping voltammetry. *Analytical Chemistry*, **54**, 2318–2322.

Eddowes, M. (1990) Theoretical methods for analysing biosensor performance. In *Biosensors: A Practical Approach*, (ed. A.E.G. Cass), pp. 211–262. IRL Press, Oxford.

Elliot, C.M. and Murray, R.W. (1976) Chemically modified carbon electrodes. *Analytical Chemistry*, **48**, 1247–1254.

Evans, G.P., Johnson, D. and Withell, C. (1986) Development of the WRc Mk III fish monitor: description of the system and its response to some commonly encountered pollutants. *WRc Environmental Report TR233*, WRc Medmenham, UK.

Evans, J.F. and Kuwana, T. (1977) Introduction of functional groups onto carbon electrodes via treatment with radio-frequency plasma. *Analytical Chemistry*, **51**, 358–365.

Fersht, A. (1985) *Enzyme Structure and Mechanism*, 2nd edn, Freeman, New York.

Foulds, N.C. and Lowe, C.R. (1988) Immobilisation of glucose oxidase in ferrocene-modified pyrrole polymers. *Analytical Chemistry*, **60**, 2473–2478.

Gaisford, W.C., Richardson, N.J., Haggett, B.G.D. and Rawson, D.M. (1991) Microbial biosensors for environmental monitoring. *Biochemical Society Transactions*, **19**, 15–18.

Gamati, S., Luong, J.H.T. and Mulchandani, A. (1991) A microbial biosensor for trimethylamine using *Pseudomonas aminovarans* cells. *Biosensors and Bioelectronics*, **6**, 125–131.

Gayet, J.-C., Haouz, A., Geloso-Meyer, A. and Burstein, C. (1991) Detection of heavy metal salts with biosensors built with an oxygen electrode coupled to various immobilised oxidases and dehydrogenases. *Biosensors and Bioelectronics*, **6**, 55–72.

Goodson, L.H. and Jacobs, W.B. (1974) Application of immobilised enzymes to detection and monitoring. In *Enzyme Engineering*, Vol. 2, (ed. K.E. Pye and L.B. Wingard, Jr), pp. 393–400. Plenum Press, New York.

Goodson, L.H. and Jacobs, W.B. (1976) Monitoring of air and water for enzyme inhibitors. In *Methods in Enzymology*, Vol. 44, (ed. S.P. Colwick and N.O. Kaplan), pp. 647–658. Academic Press, New York.

Gough, D.A. and Leypoldt, J.K. (1981) Theoretical aspects of enzyme electrode design. *Applied Biochemistry and Bioengineering*, **3**, 175–200.

Gregg, B.A. and Heller, A. (1990) Cross-linked redox gels containing glucose oxidase for amperometric biosensor applications. *Analytical Chemistry*, **62**, 258–263.

Guilbault, G.G. and Jordan, J. (1988) Analytical uses of piezoelectric crystals. *CRC Reviews*, **19**, 28–60.

Guilbault, G.G. and Schmid, R. (1991) Electrochemical, piezoelectric and fibre-optic biosensors. In *Advances in Biosensors* (ed. A.P.F. Turner), JAI Press Ltd, London, pp. 257–289.

Hall, D.O. and Rao, K.K. (1994) *Photosynthesis*. Cambridge University Press, UK.

Hall, E. (1990) *Biosensors*. Open University Press, Milton Keynes.

Hall, G.F., Best, D.J. and Turner, A.P.F. (1988a) The Determination of *p*-cresol in chloroform with an enzyme electrode used in the organic phase. *Analytica Chimica Acta*, **213**, 113–119.

Hall, G.F., Best, D.J. and Turner, A.P.F. (1988b) Amperometric enzyme electrode for the determination of phenols in chloroform. *Enzyme and Microbial Technology*, **10**, 543–546.

Hart, J.P. and Wring, S.A. (1994) Screen-printed voltammetric and amperometric electrochemical sensors for decentralised testing. *Electroanalysis*, **6**, 617–624.

Hertl, W. (1987) Amperometric immunoassays. *Bioelectrochemistry and Bioenergetics*, **17**, 89–100.

Jeanfils, J. (1986) Immobilization of whole cells of green algae or cyanobacteria in insoluble matrices. Morphological observations and nitrate reductase activity of immobilized cells. *Archives of Biology (Bruxelles)*, **97**, 209–222.

Karube, I. (1987) Micro-organism based sensors. In *Biosensors: Fundamentals and Applications*, (ed. A.P.F. Turner, I. Karube and G.S. Wilson), pp. 13–29. Oxford University Press, Oxford, UK.

Karube, I. and Tamiya, T. (1987) Biosensors for environmental control. *Pure and Applied Chemistry*, **59**, 545–554.

Karube, I., Hiramoto, K., Kawarai, M. and Sode, K. (1989) Biosensor for toxic compounds using immobilized animal cell membrane. *Membrane*, **14**, 311–318.

Kong, Z., Vanrolleghem, P.A. and Verstraete, W. (1993) An activated sludge-based biosensor for rapid IC_{50} estimation and on-line toxicity monitoring. *Biosensors and Bioelectronics*, **8**, 49–58.

Kuek, C. and Armitage, T.M. (1985) Scanning electronmicroscopic examination of calcium alginate beads immobilising growing mycelia. *Enzyme and Microbial Technology*, **7**, 121–125.

Kulys, L. and D'Costa, E.J. (1991) Printed electrochemical sensor for ascorbic acid determination. *Analytica Chimica Acta*, **243**, 173–178.

Labuda, J. (1992) Chemically modified electrodes as sensors in analytical chemistry. *Selective Electrode Reviews*, **14**, 33–86.

Laidler, K.J. and Bunting, P.S. (1973) *The Chemical Kinetics of Enzyme Action*, 2nd edn, Clarendon Press, Oxford.

Lee, S., Sode, K., Nakanishi, K., Marty, J.-L., Tamiya, E. and Karube, I. (1992) A novel microbial sensor using luminous bacteria. *Biosensors and Bioelectronics*, **7**, 273–277.

Levich, V.G. (1962) *Physiochemical Hydrodynamics*. Prentice-Hall, Englewood Cliffs, NJ.

Li, F., Tan T.C. and Lee, Y.K. (1994) Effects of pre-conditioning and microbial composition on the screening efficacy of a BOD biosensor. *Biosensors and Bioelectronics*, **9**, 197–205.

Li, F. and Tan T.C. (1994b) Monitoring BOD in the presence of heavy metal ions using a poly(4–vinylpyridine)-coated microbial sensor. *Biosensors and Bioelectronics*, **9**, 445–455.

Li, F. and Tan T.C. (1994a) Effect of heavy metal ions on the efficacy of a mixed *Bacilli* BOD sensor. *Biosensors and Bioelectronics*, **9**, 315–324.

Li, Y.-R. and Chu, J. (1991) Study of BOD microbial sensors for waste water treatment control. *Applied Biochemistry and Biotechnology*, **28/29**, 855–863.

Liang, B.S., Li, X.-M. and Wang, H Y (1986) Cellular electrode for antitumor drug screening. *Biotechnology Progress*, **2**, 187–191.

Lin, A.W.C., Yeh, P., Yacynych, A.M. and Kuwana, T. (1977) Cyanuric chloride as a general linking agent for the attachment of redox groups to pyrolytic graphite and metal oxide electrodes. *Journal of Electroanalytical Chemistry*, **84**, 411–419.

Macholán, L. and Boháckoá, I. (1988) Non-traditional membrane biocatalysts for amperometric enzyme electrodes sensing phenolic substances. *Biologia (Bratislava)*, **43**, 1121–1130.

Macholán, L. and Schanel, L. (1977) Enzyme electrode with immobilised polyphenol oxidase for determination of phenolic substrates. *Collection of Czechoslovak Chemical Communications*, **42**, 3667–3675.

Macritchie, F. (1978) Dynamics of protein adsorption. *Advances in Protein Chemistry*, **32**, 283–289.

Martens, N. and Hall, E.A.H. (1994) Diaminodurene as a mediator of a photocurrent using intact cells of cyanobacteria. *Photochemistry and Photobiology*, **59**, 91–98.

Mascini, M. and Palleschi, G. (1989) Design and applications of enzyme electrode probes. *Selective Electrode Review*, **11**, 191–264.

Matsunaga, T., Tomoda, R. and Matsuda, H. (1984) Photomicrobial electrode for the selective determination of sulphide. *Applied Microbiology and Biotechnology*, **19**, 404–408.

Mattiasson, B. and Nilsson, H. (1977) Competitive immunoelectrode for the determination of albumin. *Febs Lett.* **78**, 251–256.

Mattiasson, B., Nilsson, H. and Olsson, B. (1979) An apoenzyme electrode. *Journal of Applied Biochemistry*, **1**, 377–384.

McNeil, C.J. Bannister, J.V. and Higgins, I.J. (1988) Amperometric determination of alkaline phosphatase activity: application to immunoassays. *Biosensors*, **3**, 199–209.

Mell, L.D. and Maloy, J.T. (1975) A model for the amperometric enzyme electrode obtained through digital simulation and applied to the glucose oxidase system. *Analytical Chemistry*, **47**, 299–307.

Moses, P.R., Wier, L. and Murray, R.W. (1975) Chemically modified tin oxide electrode. *Analytical Chemistry*, **47**, 1882–1886.

Muramatsu, H., Kajiwara, K., Tamiya, E. and Karube, I. (1986) Piezoelectric immunosensor for the detection of *Candida albicans* microbes. *Analytica Chimica Acta*, **188**, 257–261.

Murray, R.W. (1980) Chemically modified electrodes. *Accounts of Chemical Research*, **13**, 135–141.

Ngeh-Ngwainbi, J., Foley, P.H., Kuan, S.S. and Guilbault, G.G. (1986) Parathion antibodies on piezoelectric crystals. *Journal of the American Chemical Society*, **108**, 5444–5450.

Nishikawa, S., Sakai, S., Karube, I., Matsunaga, T. and Suzuki, S. (1982) Dye-coupled electrode system for the rapid determination of cell populations in polluted water. *Applied and Environmental Microbiology*, **43**, 814–818.

Owicki, J.C. and Parce, J.W. (1990) Bioassays with a microphysiometer. *Nature*, **344**, 271–272.

Owicki, J.C. and Parce, J.W. (1992) Biosensors based on the energy metabolism of living cells: the physical chemistry and cell biology of extracellular acidification. *Biosensors and Bioelectronics*, **7**, 255–272.

Oyama, N., Ohsaka, T., Mizunuma, M. and Kobayashi, M. (1988) Electropolymerised cobalt tetrakis (*o*-aminophenyl) porphyrin film mediated enzyme electrode for amperometric determination of glucose. *Analytical Chemistry*, **60**, 2534–2536.

Palmquist, E., Kriz, C.B., Khayyami, M., Danielson, B., Larsson, P.-O., Mosbach, K. and Kriz, D. (1994) Development of a simple detector for microbial metabolism, based on a polypyrrole DC resistometric device. *Biosensors and Bioelectronics*, **9**, 551–556.

Pandard, P. and Rawson, D.M. (1993) An amperometric algal biosensor for herbicide detection employing a carbon cathode oxygen electrode. *Environmental Toxicology and Water Quality: An International Journal*, **8**, 323–333.

Pandard, P., Vasseur, P. and Rawson, D.M. (1993) Comparison of two types of sensors using eukaryotic algae to monitor pollution of aquatic systems. *Water Research*, **27**, 427–431.

Rawson, D.M., Willmer, A.J. and Cardosi, M.F. (1987) The development of whole cell biosensors for on-line screening of herbicide pollution of surface waters. *Toxicity Assessment*, **2**, 325–340.

Rawson, D.M., Willmer, A.J. and Turner, A.P.F. (1989) Whole-cell biosensors for environmental monitoring. *Biosensors*, **4**, 299–311.

Rechnitz, G.A., Kobos, R.K., Riechel, S.J. and Gebauer, C.R. (1977) A bio-selective membrane electrode prepared with living bacterial cells. *Analytica Chimica Acta*, **94**, 357–365.

Richardon, N.J., Gardner, S. and Rawson, D.M. (1991) A chemically mediated amperometric biosensor for monitoring eubacterial respiration. *Journal of Applied Bacteriology*, **70**, 422–426.

Riedel, K. (1991) Biochemical fundamentals and improvement of the selectivity of microbial biosensors – a minireview. *Bioelectrochemistry and Bioenergetics*, **25**, 19–30.

Riedel, K. (1994) The alternative to BOD₅: ARAS-SensorBOD. *Application Report Bio Nr.202*. Dr Bruno Lange, GmbH, Berlin.

Riedel, K., Renneberg, R., Kühn, M. and Scheller, F. (1988a) A fast estimation of biological oxygen demand using microbial sensors. *Applied Microbiology and Biotechnology*, **28**, 316–318.

Riedel, K., Renneberg, R. and Liebs, P. (1988b) Biochemical basis of a kinetically controlled microbial sensor. *Bioelectrochemistry and Bioenergetics*, **19**, 137–144.

Riedel, K., Naumov, A.V., Grishenkov, V.G., Boronin, A.M., Stein, H.J., Scheller, F. and Mueller, H.-G. (1989) Plasmid-containing microbial sensor for δ-caprolactam. *Applied Microbiology and Biotechnology*, **31**, 502–504.

Riedel, K., Huth, J., Kuehn, M. and Liebs, P. (1990a) Amperometric determination of ammonium ions with a microbial sensor. *Journal of Chemical Technology and Biotechnology*, **47**, 109–116.

Riedel, K., Lange, K.-P., Stein, H.-J., Kühn, M., Ott, P. and Scheller, F. (1990b) A microbial sensor for BOD. *Water Research*, **24**, 883–887.

Riedel, K., Naumov, A.V., Boronin, A.M., Golovleva, L.A., Stein, H.J. and Scheller, F. (1991) Microbial sensors for determination of aromatics and their chloroderivatives. I. Determination of 3–chlorobenzoate using a *Pseudomonas*-containing biosensor. *Applied Microbiology and Biotechnology*, **35**, 559–562.

Roit, I. (1980) *Essential Immunology*, 4th edn., Blackwell Scientific, Boston.

Romette, J.L. and Boitieux, J.L. (1984) Oxidase enzyme: enzyme and immunoenzyme sensor. *Annals of the New York Academy of Sciences*, **434**, 533–535.

Romette, J.L., Yang, J.S., Kusakabe, H. and Thomas, D. (1983) Enzyme electrode for the specific determination of L-lysine. *Biotechnology and Bioengineering*, **25**, 2557–2566.

Rosen, I. and Rishpon, J. (1989) Alkaline phosphatase as a label for heterogeneous immunochemical sensors: an electrochemical study. *Journal of Electroanalytical Chemistry*, **258**, 27–39.

Saini, S., Hall, G.F., Downs, M.E.A. and Turner, A.P.F. (1991) Organic phase enzyme electrodes. *Analytica Chimica Acta*, **249**, 1–15.

Sauerbrey, G.Z. (1959) Use of a quartz vibrator for weighing thin layers on a microbalance. *Zeitschrift Physik*, **155**, 206–210.

Scheller, F. and Schubert, F. (1992) *Biosensors*. Elsevier, Amsterdam.

Schubert, F., Renneberg, R., Scheller, F.W. and Kirstein, L. (1984) Plant tissue hybrid electrode for determination of phosphate and fluoride. *Analytical Chemistry*, **56**, 1677–1682.

Shaolin, M., Huaiguo, X. and Biding, Q. (1991) Bioelectrochemical response of the polyaniline glucose oxidase electrode. *Journal of Electroanalytical Chemistry*, **302**, 7–16.

Shons, A., Dorman, F. and Najarian, J. (1972) Immunospecific microbalance. *Journal of Biomedical and Material Research*, **6**, 565–670.

Sidwell, J.S. and Rechnitz, G.A (1985) 'Bananatrode' – an electrochemical sensor for dopamine. *Biotechnology Letters*, **7**, 419–422.

Skladal, P. and Mascini, M. (1992) Sensitive detection of pesticides using amperometric sensors based on cobalt phthalocynaine-modified composite electrodes and immobilised cholinesterases. *Biosensors and Bioelectronics*, **7**, 335–343.

Starostina, N.G., Lusta, K.A. and Fikhte, B.A. (1983) Prediction of microbial resistance to immobilization in polyacrylamide gel. Translated from *Prikladnaya Biokhimiya i Mikrobiologiya*, **19**, 369–371.

Tan, H.-M., Cheong, S.-P. and Tan, T.-C. (1994) An amperometric benzene sensor using whole cell *Pseudomonas putida* ML2. *Biosensors and Bioelectronics*, **9**, 1–8.

Tan, T.C., Li, F., Neoh, K G and Lee, Y.K. (1992) Microbial membrane-modified dissolved oxygen probe for rapid biochemical oxygen demand measurement. *Sensors and Actuators B*, **8**, 167–172.

Tan, T.C., Li, F. and Neoh, K.G. (1993) Measurement of BOD by initial rate of response of a microbial sensor. *Sensors and Actuators B*, **10**, 137–142.

Tsushima, R., Kondo, A., Sakata, M. and Kawabata, N. (1992) Preparation of bacteria-adsorption polymer and its application to biosensor. *Polymeric Materials Science and Engineering*, **66**, 437–438.

Turner, A.P.F., Karube. I., and Wilson, G.S. (ed.) (1987) *Biosensors: Fundamentals and Applications*, Oxford University Press, Oxford.

Walsh, C.T. (1977) *Enzymatic Reaction Mechanisms*. Freeman, New York.

Walters, R.R., Moriarty, B.E. and Buck, R.P. (1980) *Pseudomonas* bacterial electrode for determination of L-histidine. *Analytical Chemistry*, **52**, 1680–1684.

Weber, S.G. and Purdy, W.C. (1979) Homogeneous voltammetric immunoassay. *Analytical Letters*, **12**, 1–9.

Wehmeyer, K.R., Halsall, H.B. and Heinemann, W.R. (1982) Electrochemical investigation of hapten–antibody interactions by differential pulse polarography. *Clinical Chemistry*, **28**, 1968–1972.

Wijesuriya, D.C. and Rechnitz, G.A. (1993) Biosensors based on plant and animal tissues. *Biosensors and Bioelectronics*, **8**, 155–160.

Wring, S.A. and Hart, J.P. (1992) Chemically modified, screen printed carbon electrodes. *Analyst*, **117**, 1281–1286.

Yamamoto, N., Nagasawa, Y., Sawai, M., Sudo, T. and Tsubomura, H. (1978) Potentiometric investigations of antigen–antibody and enzyme–enzyme inhibitor reactions using chemically modified metal electrodes. *Journal of Immunological Methods*, **22**, 309–317.

Yamamoto, N., Nagaoka, S., Tanaka, T., Shiro, T., Honma, K. and Tsubomura, H. (1983) Potentiometric detection of biological substances by using chemically modified electrodes. *Analytical Chemistry Symposium Series, No. 17 (Chemical Sensors)*, pp. 699–704.

8 Automated measurement

E.A. KNIGHT and J.R. PUGH

8.1 Introduction

This chapter is intended to provide an awareness of the current techniques in the use of on-line computers for measurement. This will be of interest to both the developers and users of environmental sensors, for example:

- those developing and evaluating sensors in the laboratory will find a wide range of on-line techniques of interest (A/D modules, GPIB, RS 232, CAMAC, Rack systems)
- those involved in the use of sensors for environmental measurement will be interested in techniques of portable data logging and in methods of transmitting data over long distances
- those developing sensors for the market place should be aware of the techniques that are being incorporated into the sensor unit and the systems that may be employed by future customers using the sensors (smart devices, distributed systems, fieldbus).

An in-depth treatment is not possible in a review of this type; however, the topics chosen are those that have proved important from our experience in employing on-line measurement in a variety of fields.

An on-line computer is able to acquire data in a variety of ways. Perhaps the most direct way is to convert a voltage from a transducer to digital form via an analogue/digital (A/D) converter mounted in the computer. If this approach is adopted the user must ensure that the signal from the transducer spans the optimum voltage range and has been adequately filtered. Another technique is to measure the transducer output with an instrument such as a digital meter, an oscilloscope, or a frequency meter and pass a digital result to the computer.

Fundamentally there is no difference between the two methods described, the only difference being where the A/D conversion takes place. In practice, however, there are advantages and disadvantages associated with each of the approaches.

The instrument-based approach has advantages in that an instrument designed for a particular purpose (e.g. microwave power measurement, extremely low-noise measurement and high-frequency measurement) may be utilised directly in the measurement system and additional electronic processing prior to conversion does not have to be undertaken. In addition, the instrument set-up, via the front panel (in the case of a conventional instrument) or via software (for a programmable instrument), is a familiar activity. The user does not need to be concerned with the matching of the signal to the A/D converter. The advantage of the direct analogue interfacing approach is that it is a more versatile and generally less expensive method of acquiring data; however, the user has to be more aware of the details of sampling speed and amplitude resolution.

For signals of a moderate frequency and amplitude, the difference between the two techniques is more artificial with the increasing use of virtual instruments. This is a technique whereby the computer VDU is employed to simulate the front panel of an instrument and the user is able to set up the front panel in a similar manner to the real instrument. The A/D converter parameters are then configured to reflect these settings. It must be remembered with such systems that the signal is fed directly to the analogue input board, which will, at best, have only limited signal processing capability. The virtual instrument approach may, therefore, be viewed as a trade-off between versatility and the performance that would be obtained with a specialist instrument. However, it is also true that the performance associated with specialist instruments may not be required in many measurements.

8.2 General principles

8.2.1 Sampling

Whenever an analogue signal is to be processed by a digital system, it has to be converted to a digital form. The major concerns in converting an analogue signal into digital form are (i) ensuring that the level is sampled with sufficient resolution; and (ii) ensuring that the signal is sampled at a sufficiently high frequency.

The relationship between the frequency of sampling (measurement) and the time scale of changes in the signal is encompassed by the *sampling theorem*, which states: 'If a signal is sampled at a rate that is at least twice the highest frequency present in the signal, the sampled values exactly describe the original signal'.

If the above condition is not met, although the values of the signal at the instant of measurement are known, the values in the intervening times are not. In fact the sampled measurements can lead to confusion, as Figure 8.1 indicates. This source of error is termed aliasing and, in order to counteract it, a low-pass filter which cuts out all frequencies above half the sampling frequency should be incorporated prior to sampling. This low-pass filter is termed an anti-aliasing filter.

In summary, a continuous measurement is one that is made often enough to satisfy the sampling theorem, namely at a rate of twice the highest

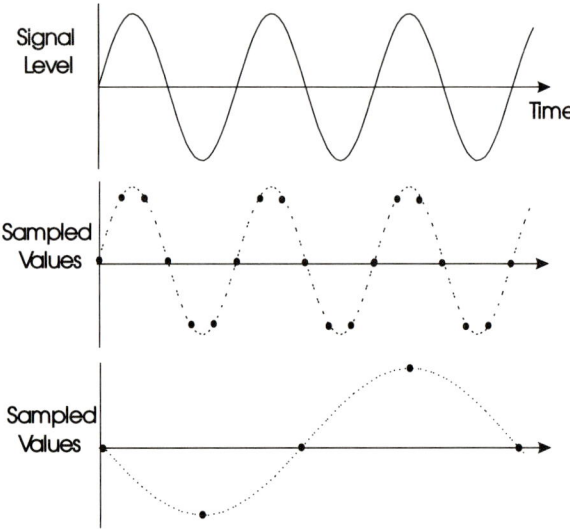

Figure 8.1 Signal aliasing.

frequency present in the signal. In practice, since filters do not have an infinitely sharp cut-off frequency, sampling frequencies of around five times the filter cut-off frequency provide a suitable margin of safety.

8.2.2 Resolution

The resolution of a measurement system is defined as the smallest change in the measurand that may be detected. In an A/D converter, the resolution of the signal level is determined by the number of binary digits to which the signal level is converted. For example, an '8 bit' converter will convert the analogue signal into one of 256 levels whereas a '10 bit' converter will have 1024 possible levels for the signal. This fact, coupled with a knowledge of the input range of the converter, will enable the designer to calculate the resolution of the converter in millivolts. This is summarised in Table 8.1 for a converter with an input range 0–5 V.

Table 8.1 The resolution of a converter with an input range 0–5 V

Number of bits	Number of levels	Resolution (mV)
8	256	19.5
10	1,024	4.89
12	4,096	1.22
16	65,536	0.076

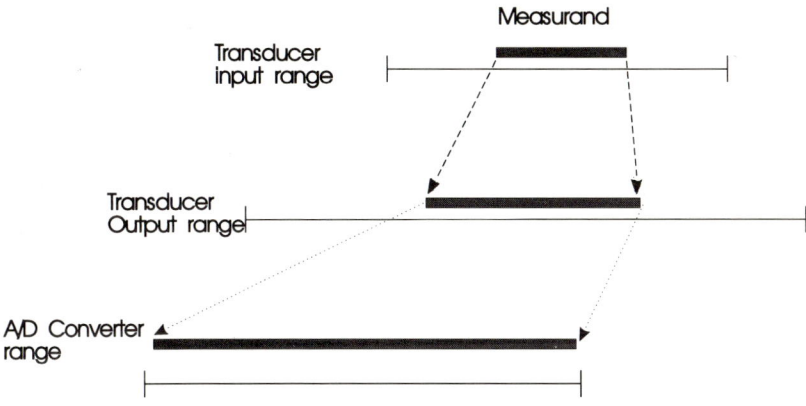

Figure 8.2 Conditions for optimum resolution.

The upper limit of resolution can be obtained by mapping the range of the measurand (not the transducer) to the A/D converter input range. This process may involve amplification/attenuation and DC shifting and is illustrated in Figure 8.2.

Assuming optimum range matching, one is able to calculate the minimum number of bits required for the A/D converter to give a particular resolution of the measurand. However, it should be remembered that the calculation gives the minimum number of bits required and it may be useful to build in some factor of safety by mapping a range larger than the measurand range to the A/D converter range. However, any of the system elements may limit the resolution in a measurement system, and the A/D converter may not be the determining factor.

In an instrument-based approach, the matching of range and filtering functions tends to be incorporated into the instrument. For example, digital multimeters may be autoranging, the range may be set manually or, indeed, may be programmable over an appropriate interface.

8.3 Analogue interfacing

The analogue signal from the transducer or from an instrument with an analogue output may be either a voltage or a current over a variety of ranges that depend on the application. For example, a range of a few millivolts is typical from a thermocouple while many instruments have either a 0–1 V or a 0–10 V analogue output.

The main technique used for the transmission of analogue signals in a distributed instrumentation system is by means of a current rather than a

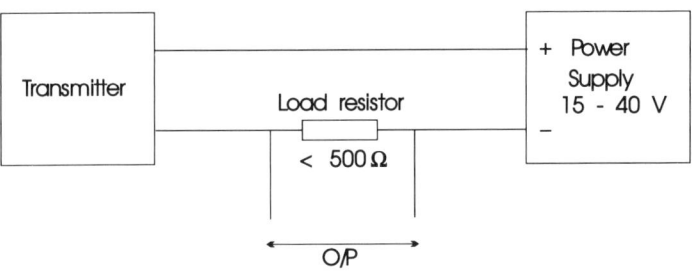

Figure 8.3 The 4–20 mA current loop.

voltage. This choice arises primarily from the fact that the resistances are low and electrical interference is less likely to be a problem. The most commonly used system is the 4–20 mA current loop, incorporating the power supply, the signal transmitter and the receiver or receivers (Figure 8.3). Transmission over distances up to a few kilometres is possible. The advantages of such systems is that impedance changes in the loop (the addition of extra cable or an additional receiver) do not affect the signal level, provided the additional impedance does not exceed the drive capacity of the transmitter. Transmitters are normally sold integral with the transducer, with the input range mapping to the entire 4–20 mA output current range. A zero current indicates a fault condition.

Most A/D converters can accept a voltage input in a number of voltage ranges for example ±5 V, 0–1 V, 0–10 V. These ranges are either set in hardware or are selectable in software. In order to provide the mapping between ranges, and to ensure that the requirements of the sampling theory are met, signal processing is usually required. Some of the more commonly required functions are outlined below.

8.3.1   Buffer amplifier

To ensure impedance matching between the transducer and the A/D board, a buffer stage may be required in some applications. The circuit has a high input impedance, which avoids loading and feeds the signal to the A/D board from a low impedance source.

8.3.2   Voltage attenuator

Attenuation is required if the output range is larger than that desired. A simple circuit, equation of operation, and input/output impedance conditions are presented in Figure 8.4(a).

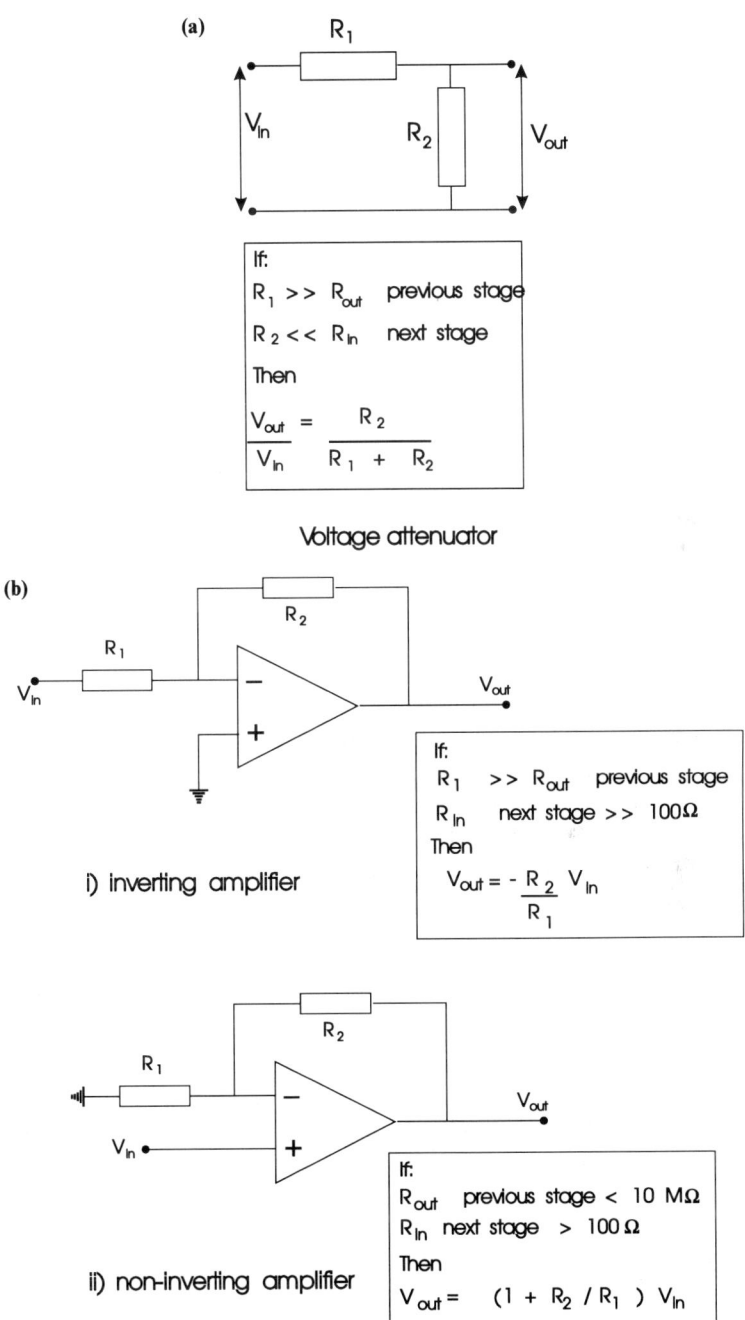

Figure 8.4 Signal conditioning circuits. (a) Voltage attenuator. (b) Single-stage amplifier circuits: (i) the inverting amplifier, (ii) the non-inverting amplifier.

(c)

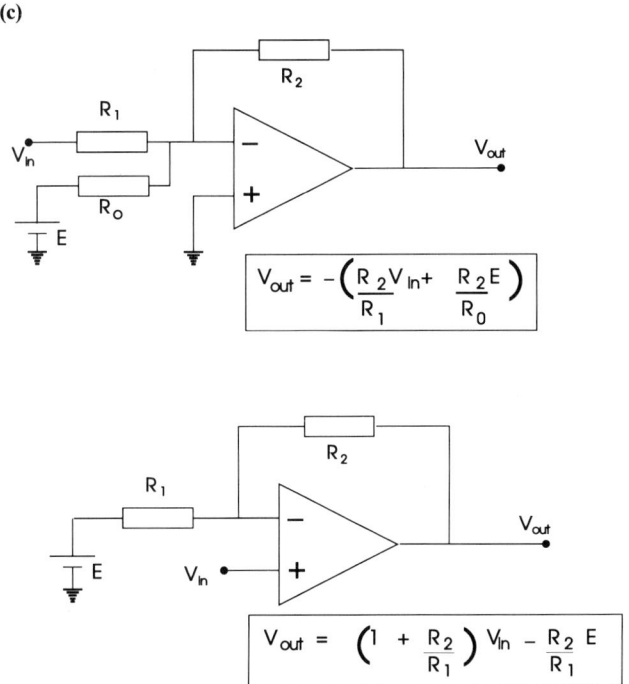

$$V_{out} = -\left(\frac{R_2 V_{in}}{R_1} + \frac{R_2 E}{R_0}\right)$$

$$V_{out} = \left(1 + \frac{R_2}{R_1}\right) V_{in} - \frac{R_2 E}{R_1}$$

Figure 8.4 (c) DC shifting circuits.

8.3.3 Amplification

8.3.3.1 Single-stage amplifiers. Single-stage amplifier circuits are presented in Figure 8.4(b). The inverting amplifier (Figure 8.4(i)) is particularly simple; however, care has to be taken because of its rather low input impedance (equal to R_1).

8.3.3.2 DC shifting circuits. The circuits shown in Figure 8.4(c) are able to provide amplification in addition to a DC shift and are often used prior to A/D conversion.

8.3.3.3 Difference amplifiers. These amplifiers ideally only the amplify the *difference* between two input signals. They are used extensively in instrumentation systems to amplify the output of bridge circuits. They are also used in other applications where neither of the inputs should be connected to ground. Instrumentation amplifiers are high-quality

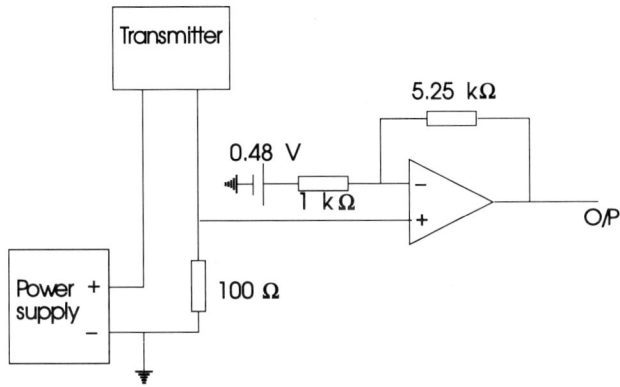

Figure 8.5 Example of a processing system.

differential amplifiers with a high-input impedance at both inputs and improved CMRR over the single-stage differential amplifier.

8.3.4 Current-to-voltage converters

The current output from a 4–20 mA system can be easily matched to the required voltage input range using the circuit shown in Figure 8.5 . The input range of the converter is 0–10 V in this example.

8.3.5 Filters

The most commonly used filters in instrumentation tend to be low-pass filters. They are used to enable the signal frequencies of interest to pass (below the cut-off frequency) whilst attenuating noise or interference at higher frequencies. They are also used in A/D conversion to ensure that signals at a frequency greater than twice the sampling frequency are attenuated and do not cause aliasing. Two examples of low-pass filters are presented in Figure 8.6: (a) passive and (b) active filters.

8.3.6 Sample and hold circuits

The conversion time of the A/D converter may be such that there is a possibility of the signal level changing appreciably while the conversion is being undertaken. If this is likely, the idealised circuit shown in Figure 8.7 may be used before the converter.

The approach here is either to custom-build such circuits or to use commercially available signal-conditioning boards or modules.

It should be noted that many analogue boards are capable of performing

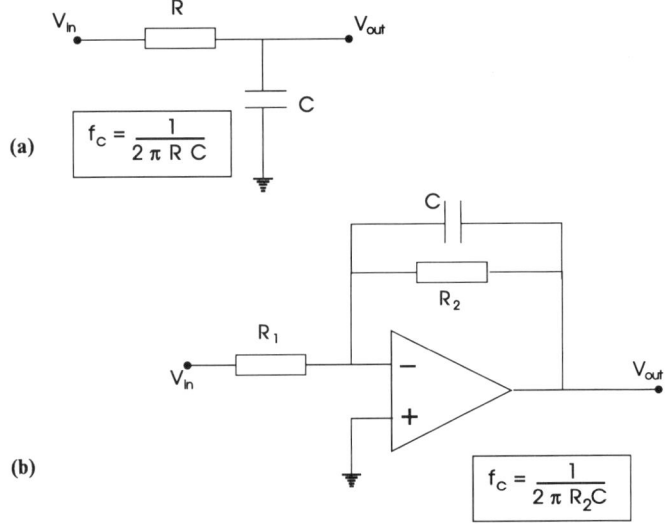

Figure 8.6 Low-pass filters (a) passive, (b) active.

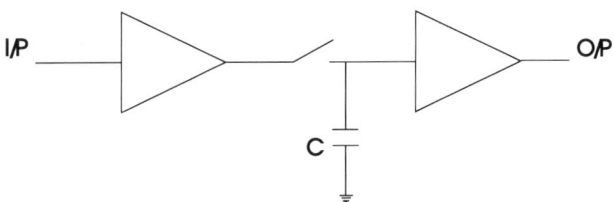

Figure 8.7 Sample and hold circuit.

some of these functions. For example, a board with software selectable gain may well dispense with the need for amplification prior to input to the board; nevertheless the principles of setting the gain are still required.

8.4 Digital interfacing

8.4.1 Introduction

Rather than transmit an analogue signal (voltage or current), there are several digital transmission systems used between transducers and the processing/recording electronics or between the processing units themselves.

There is a range of methods for encoding information in digital form. The function of an interface is to translate these codes into a form that is acceptable to a computer system. The use of standard interfaces in the development of measurement systems allows for maximum flexibility and interoperability.

Digital data can be transferred either in serial or parallel form. In the simplest form of serial transmission, only two wires are required, a data line and a reference line. Data-bits are sent in succession on the single-signal line. In 'bit parallel form', all bits of the data word are transmitted simultaneously and in the case of an 8 bit system, this requires 8 data lines plus a reference line. Parallel data transmission is a more rapid method than serial transmission, but it requires a greater number of data lines.

A handshaking process may be used to control data transfers. The handshaking sequence may, for instance, enable data to be transmitted only when the receiver (for example a printer or some other external device) is ready to receive data. Once data are valid, the transmitter signals that data are available and waits for the receiver to signal that data have been received before sending more. This process tends to slow down data transmission, but it does ensure that data are accurately transmitted and received. Such digital information transmitters/receivers in instrumentation can be:

8.4.1.1 Programmable instruments. These devices are conventional instruments such as multimeters and oscilloscopes, which can be both controlled locally via the front panel and programmed via a standard interface. Such interfaces include RS 232C and IEEE 488.

8.4.1.2 'Instruments on a card'. These instruments are built on a PCB and may be mounted in a rack or directly into a PC using one of the standard expansion slots. The rack-mounted instruments communicate via a backplane bus such as the VXIbus. Depending on the configuration, the computer may also be a module in the rack.

8.4.1.3 'Smart' sensors. Transducers traditionally had an analogue voltage or current (4–20 mA) output. Increasingly, these devices are fitted with a microprocessor that allows for digital output, and two-way communication. Communication protocols such as HART and Fieldbus are being developed for such applications.

8.4.2 The general purpose interface bus: IEEE.488.2

The general purpose interface bus IEEE.488.2 was initially based on the Hewlett-Packard interface bus (HPIB) and the first standard appeared in 1975. It is also known as the General Purpose Instrumentation Bus (GPIB).

The present standard ANSI.IEEE.488.2 was adopted in 1987 and strengthened the original standard by defining precisely communication between instruments and controller. In 1990, *Standard Commands for Programmable Instruments* (SCPI) was introduced. This provides a single command set for programmable instruments.

The IEEE.488 interface system permits bi-directional, asynchronous communication with up to 14 compatible instruments connected by a total of 20 m of cable. Data are transmitted via 8 bi-directional lines. In addition, there are three handshake and five interface management lines. There are also eight ground lines, five of which are usually twisted as pairs with the five interface management lines. Provided the devices connected into the bus at that time can handle the rate, approximately 1 megabyte of information can be transmitted per second using a fast handshake routine. Some manufacturers have developed proprietary solutions that are capable of a transfer rate up to 8 Mbytes s^{-1}, however, these rates are dependent on the computer architecture and the system configuration.

Various categories of instrumentation, e.g. measurement, stimulus, display, processor and storage devices, can all be connected and the data produced by each can be transmitted over the bus. The functions of the individual instruments are of no consequence to the bus since all instruments adhere to the standard.

An IEEE.488 interface contains a microprocessor and some ROM which handles all the standards' protocols. This approach reduces the amount of software required to control the interface and normally a two or three line BASIC program is sufficient to transmit or receive data. The problem of writing machine code programs is thus avoided and in the majority of cases data transfers can proceed at an adequate speed. In addition, there are a variety of software packages that support the IEEE.488 interface.

The IEEE.488 standard was established when most microprocessors were 8-bit devices. Now that 16-bit and 32-bit microprocessors have been produced, a new interface standard based upon a wider data bus highway would now be advantageous. However, the large number of instruments fitted with the IEEE.488 interface that are already available will guarantee its continuance for the foreseeable future.

8.4.2.1 The bus and connector.   All interface lines are common to all instruments on the bus. The standard allows up to 15 devices to be connected to the bus simultaneously.

A universal connector is used by all IEEE.488 standard compatible devices. This enables the devices to be quickly connected to the bus using either a star or party line connection. In both these modes of connection, several cables may enter or leave one particular device. This is achieved by using a special IEEE-type plug, which contains both male and female sockets so that the plugs may be piggy-backed together.

The IEEE.488 standard stipulates the number of interfaces and the maximum length of cables that may be used in a system. However, bus extenders can be used to increase this length if necessary, for example a high-speed fibre optic GPIB extender extends the maximum distance to 1 km. Additionally, with the appropriate hardware, it is possible to control GPIB devices anywhere on an ethernet network.

8.4.2.2 Devices on the bus. The devices connected to the interface bus, at any one time fall into certain distinct categories.

talker: this is a device that transmits information onto the bus; there can only be one talker active at any one time on the bus

listener: this is a device that receives information from the bus; there can be any number of listeners on the bus at any one time

controller: this is the device that uses the control lines for sending or transmitting messages and data; there can be only one controller on the bus at a particular time, known as the 'active' controller. The system controller is the device that assumes control of the bus when the system is switched on. Inactive controllers may be present on the bus.

All devices on the bus have a talk/listen address, which is used by the controller to assign the state of devices on the bus.

8.4.2.3 SCPI. The IEEE 488.2 standard does not define the device-dependent commands used to control an instrument. This allows instrument developers flexibility in designing new instruments or adding additional features to existing devices; however, when programming such devices, the device-dependent commands are required. Applications software written for one type of oscilloscope cannot be used with another type of oscilloscope. The introduction of SCPI in 1991 means that instruments which comply with SCPI use the standard command set, hence developed software can be used with other similar devices without extensive re-programming. SCPI is a language that is not hardware dependent. SCPI messages and responses may be communicated over RS-232, VXIbus and other communication methods. As new instruments develop, additional commands can be submitted to the SCPI consortium for inclusion in the standard.

8.4.3 Serial interface standards RS 232, RS 422, RS 423, RS 449 and RS 485

RS 232C is probably one of the most commonly implemented serial data transmission techniques. RS 232D is the most recent revision of the original RS 232 standard; RS standing for recommended standard. The standard was established in the late 1960s when it became desirable for computers in separate locations to communicate with each other. It was originally designed to provide an interface between Data Terminal Equipment (DTE), for example a computer, and Data Communications Equipment (DCE) for example, a modem. It was subsequently adopted for instrument communication. The RS232C standard is still widely used in micro-computer systems in situations where a data rate limit of 20 000 bits s^{-1} and maximum distance of 15 m are acceptable.

The standard defines 22 lines with 3 lines unassigned. No connector is defined but normally a 25 pin D connector is used with a male connector with data terminal equipment and a female connector with data communications equipment.

A logic 1 (Mark) is a voltage level between -3 and -25 V and a logic 0 (Space) is a voltage between $+3$ and $+25$ V. Data may be transmitted at a variety of different rates and these are referred to as baud rates (i.e. number of bits per second). The standard data rates may vary from 50 up to 20 000 baud with various discrete rates in between.

Information is transmitted asynchronously using ASCII characters. Asynchronous transmission means that the process is not synchronised with a clock waveform and, therefore, the receiver must be informed when each byte of data begins and ends, using start and stop bits. This means that the time between each byte of data is not fixed. However the process is often referred to as 'local synchronous' since the bits between the start and stop bits must be timed. A START bit begins the sequence, followed by 5-, 6-, 7- or 8-bit data format with 1 or 2 STOP bits. To ensure that the information has been received correctly, a PARITY bit may be included. For a parity check, the number of 1s in the word are counted. If the parity is even then the total number of 1s, including the parity bit must be an even number. If the parity is odd, then the total number of 1s must odd. Figure 8.8 shows the waveforms associated with the transmission of the ASCII codes for 'A' (ASCII code 65, 01000001) with odd parity and 'B' (ASCII code 66, 01000010) with even parity.

A feature of RS 232C is that, in many circumstances, many of the 22 lines of the standard may be dispensed with. Although 25 lines were specified in the original standard, serial data transfer may be achieved with just 3 lines, i.e. transmit line, receive line and the signal ground. Since the introduction of the IBM AT PC, all RS232 C serial ports on IBM PCs and compatibles have a 9 pin connector.

The RS 449, RS 422 and RS 423 standards are developments of the RS 232C that allow data to be transmitted over longer distances at higher speeds. It should be noted that the RS 232D is the latest version of the RS 232C standard but it does not allow for faster transfer rates or longer transmission distances. The RS 422 and RS 432 standards have been designed to take advantage of modern integrated circuits and are intended to replace the RS 232C. These standards specify the electrical character-

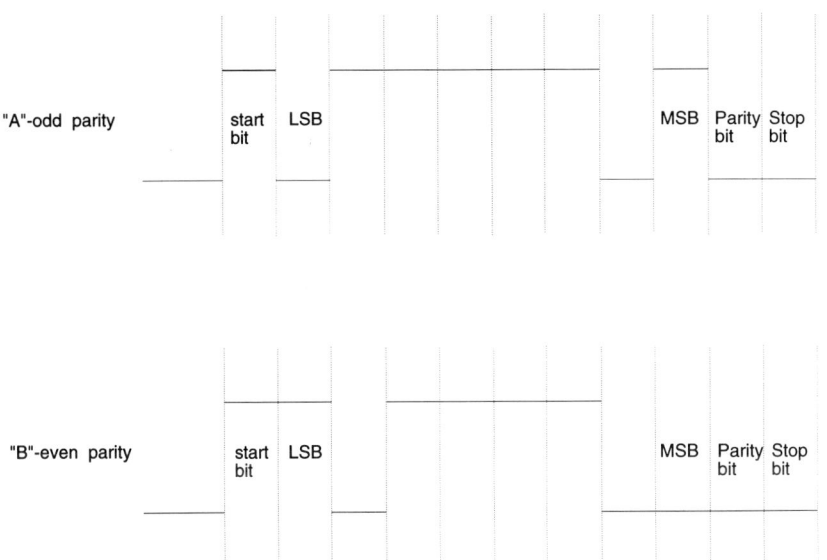

Figure 8.8 RS 232C Data transmission of the ASCII codes for 'A' and 'B'.

istics for balanced and unbalanced voltage–digital interface circuits, respectively. For each of these standards, the maximum data-transfer rate is dependent on the transmission distance, with the maximum transfer rate falling off as the distance increases. For example for the RS 422 at transfer rates of 10 Mbit s^{-1}, the maximum distance is 3 m. A comparison of their performance is presented in Table 8.2. The RS 499 standard is able to handle either single-ended or balanced voltages and was designed to replace the RS 232 standard. Unlike earlier standards, the RS 499 standard specifies the connector. A 37 pin connector is required for the signal, control and timing interface circuits; a separate 9 pin connector is required if a secondary channel circuit is used. The use of these connectors is possibly why the microcomputer manufacturers have failed to incorporate this standard despite its improved data-transfer rates and communication distances.

Current signals may be used instead of voltage levels particularly in applications requiring transmission over longer distances or in noisy environments. The current loop system offers considerable advantages in terms of distance (Table 8.2).

The standards previously described are essentially point-to-point systems. The RS 485 serial interface standard allows for multiple transmitters and receivers on the same wire pair. It is, therefore, a bus arrangement with addressing being dealt with in software. Up to 32 transmitters or receivers are allowed.

Table 8.2 Comparison of the performance of RC data transmission techniques and the current loop system

	RS 232C	RS 422	RS 432	Current loop
Communication	Point-to point	Point-to point	Point-to point	Point-to point
Signal type	Single-ended voltage	Balanced voltage	Single-ended voltage	
Signal level	± 3 V to ± 25 V	± 2 V to ± 6 V	± 2 V to ± 6 V	0–20 mA
Distance (m)	15	1000	90	1000
Speed (kbit s^{-1})	20	100	100	20

8.4.4   VME/VXIbus

The VMEbus (Versabus Modular European) standard defines two board sizes, connectors, pins and a backplane carrying up to a 32 bit wide data-transfer bus. It is capable of high-speed data transfer at up to 40 Mbyte s^{-1} The standard was accepted in 1986 as IEEE 1014 and in 1985 as IEC 821. The VXIbus is an extension of the VMEbus standard designed to overcome some of the deficiencies of the VMEbus with regard to instruments. The VXIbus essentially combines the VMEbus backplane with IEEE.488 protocols and provides for additional card sizes, instrument addressing, communications protocols, triggering lines and additional power. VXI card-based instruments communicate either with a computer external to the rack via, for example, an IEEE 488 interface or over the VXIbus to a computer module mounted in the rack.

8.4.5   CAMAC

CAMAC is a modular data-handling system used extensively in specialist applications, for example in nuclear physics research laboratories. The CAMAC standard (IEEE 583) covers the electrical and physical specifications for the modules, crates or racks and the backplane.

Individual crates have controllers tied together and ending in a branch driver that is then interfaced directly to the data-acquisition computer.

Crate controllers that can interface directly with the IEEE 488 bus are available, allowing for the convenient set-up of relatively small-scale applications.

The range of instruments available include scalars, counters and timers, multiple input A/D converters, gate and delay generators and programmable discriminators. These systems tend to be the system of choice in applications requiring fast counting.

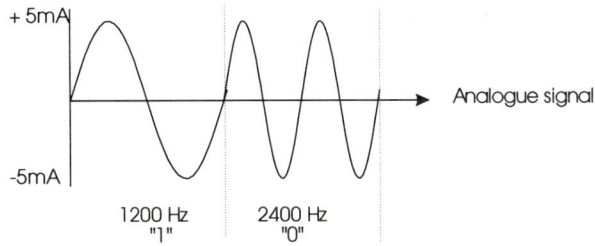

+ 5mA

Analogue signal

-5mA

1200 Hz   2400 Hz
  "1"   "0"

Figure 8.9 Frequency shift keying.

8.4.6 HART field communication protocol

The 4–20 mA analogue current signal is the method of transmission most
widely utilised in transducers measuring process variables in industry.
However, in recent years, 'smart' devices have been introduced. These
devices, in addition to measuring the process variable, are capable of
two-way communication, allowing for additional features such as remote
set-up.

The HART (Highway Addressable Remote Transducer) field communications protocol is a digital communications protocol for smart
transducers that is fast gaining acceptance in industry. This protocol retains
the widely utilised 4–20 mA current loop and provides an additional digital
signal simultaneously using the existing cabling.

The protocol operates using the frequency shift keying (FSK) principle,
which is based on the Bell 202 communication standard. The digital signal
is made up of two frequencies, 1200 Hz and 2400 Hz, which represent '1'
and '0', respectively. The digital signal, composed of sine waves of these
frequencies, is superimposed on the 4–20 mA DC analogue signal. Since
the average value of the AC signal is zero, the analogue signal is unaffected
(Figure 8.9).

The connection of a HART device to a data-acquisition or control
system can be done in one of two ways: point-to-point or multi-drop.

8.4.6.1 Point-to-point operation.

In this mode of operation (Figure 8.10)
the 4–20 mA signal can still be used for process variable measurement and
control purposes, while the digital data are available for measurement,
adjustment and set-up.

Two-way communication with the HART device can be achieved either
through the master computer or via a hand-held terminal or laptop
computer, which can be connected anywhere in the loop.

The protocol allows for poll/response transmission where the device
responds to a request from the master, or burst transmission where the

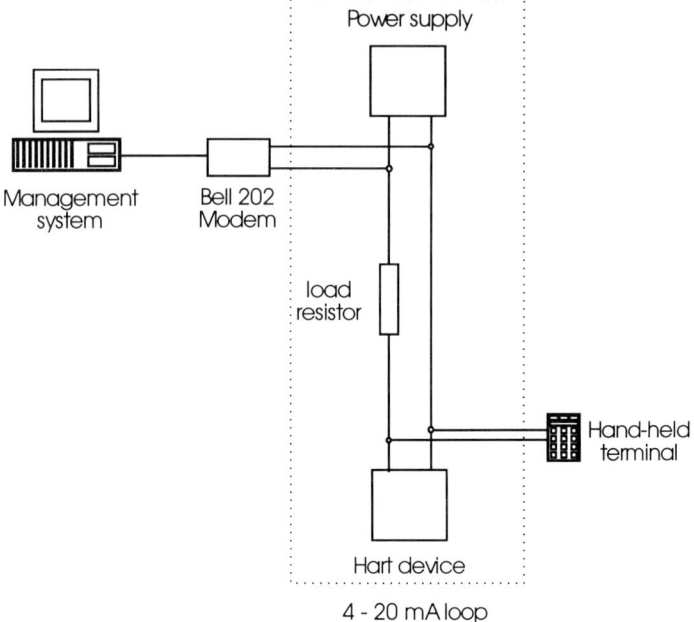

Figure 8.10 Point-to-point operation.

device continuously transmits process data. The single digital process variable rate is 2 Hz for poll/response mode and 3.7 Hz for burst mode.

Analogue meters, data loggers, data-acquisition boards and controllers can be connected across the load resistor as before.

This mode of operation allows for the gradual implementation of digital capabilities without modifying field devices or changing wiring.

8.4.6.2 Multi-drop operation.   In multi-drop operation (Figure 8.11) several 'smart' devices can be networked to a single communication line. In this mode, each HART device must have an address greater than zero, which causes it to place its analogue output at 4 mA, and all communication is digital.

The maximum number of loop-powered devices operating from one power source is 15, and in multi-drop networks only the poll/response mode of communication is possible.

8.4.7   Networks and fieldbus

As a result of the complexity of measurement and control in an industrial environment, a large number of computers may have to communicate with

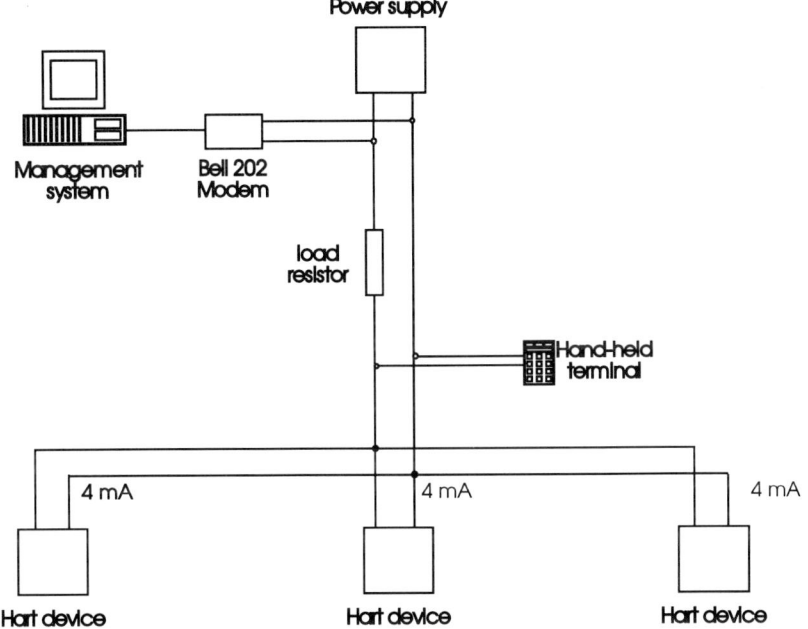

Figure 8.11 Multi-drop operation.

each other, with users and with other devices. This task can be complicated because of the variation in the type of computer and equipment that have to be interfaced. There is a need for a single standard connection for two-way communication for field-mounted devices such as transducers, PLCs and PCs. The advent of smart sensors led manufacturers to find individual solutions. The HART system developed by Rosemount in the early 1980s as an interim solution has been adopted by a group of manufacturers and major users. Such groupings allow for interchangeability between devices, which is clearly one of the major incentives to define a standard fieldbus.

Work has been going on over a number of years to develop a standard solution. Such a system requires a protocol that defines the rules under which communication can be achieved. The protocol consists of three elements:

- the structure of commands and responses in either field formatted or character string (usually ASCII) format, known as 'syntax'.
- the 'semantics', which are the requests and responses issued by either party
- the timing of events within the network.

The basis of any up-to-date protocol is the ISO standard protocol, incorporating the following seven layers:

1. the physical layer
2. the data link layer
3. the network layer
4. the transport layer
5. the session layer
6. the presentation layer
7. the application layer.

A number of major companies have developed local area networks (LANs) implementing the above. These include: DEC's Data-way, Hewlett-Packard's LAN, using coaxial cable and an ethernet compatible protocol with a transmission rate of $10\,\text{Mbits}\,\text{s}^{-1}$; Allen-Bradley's Data Highway, Gould's Modbus and Modway supporting programmable controller networks; General Electric's GEnet, using coaxial cable transmitting at $1\text{--}5\,\text{Mbits}\,\text{s}^{-1}$.

These systems do have differences, in terms of transfer rate for example. Groupings of companies and users have been working towards standardisation. WorldFIP and ISP have now emerged as the two major fieldbus groups, both having working standards. The International Electrotechnical Commission (IEC) has almost completed the specification of an international fieldbus standard. Both WorldFIP and ISP have now adopted the IEC physical layer and are committed to migrating towards the IEC standard.

The HART user group have recently introduced a device description language that will allow any device to communicate its description. This will allow for interoperability as the controller can immediately recognise not only the new device but also the menus, special functions, etc. This idea has been adopted by ISP in their latest fieldbus specification and may well appear in the final IEC standard.

8.5   Hardware aspects

8.5.1   Computers

The most commonly used computer in the small- to medium-sized application is the IBM PC standard, and its many clones. The Apple Macintosh computer may also be used in these applications. However, the range and versatility of both hardware boards and supporting data-acquisition software is greatest for the PC platform.

For larger applications, UNIX-based systems may be appropriate, while computers designed to withstand harsh industrial environments will be ideally suited to some environmental monitoring applications. These PCs have rugged waterproof and dust-proof enclosures and allow for the connection of specially designed I/O devices such as keyboards and monitors.

8.5.1.1 Expansion slots. Many computers contain expansion slots that allow the operator to install user-oriented boards, e.g. Ethernet cards, A/D cards, GPIB cards, etc. Compatibility between the computer bus and card must be ensured.

8.5.1.2 Serial port. Most computers are fitted with a standard serial interface. This may be used to connect any communications link (modem), serial printer or instrument using that particular interface. The IBM PC, for example, has an RS 232C interface while Macintosh use the RS 422 interface.

8.5.1.3 Parallel. IBM PCs and clones have a standard Centronics parallel interface primarily for use with a parallel printer. It can also, of course, be used with any compatible instrument. An enhanced version of this interface, the enhanced parallel port (EPP), offers improved communication rates. The standard interface typically operates at between 100 and 200 kbytes s^{-1} while the enhanced port is capable of up to 800 kbytes s^{-1}. EPP has been incorporated into INTELs 386SL chip sets and is available on a number of PCs and laptops.

8.5.1.4 SCSI. The small computer system interface (SCSI) is an interface originally designed to connect small computer systems with peripherals such as disk drives. SCSI ports on computers without expansion slots are used to provide a means of communicating with a data-acquisition system. For example, the SCSI port can be used with a SCSI-to-IEEE.488 converter to control IEEE instruments. Sun SPARC station has a SCSI port and an Sbus slot so GPIB instruments may be controlled through a card or an external box connected to the SCSI port.

8.5.1.5 PCMCIA slots. The development of laptop and notebook computers has led to the introduction of a small format expansion slot taking a card the size of a credit card. These slots were originally developed to take communications hardware, but recent developments have included data-acquisition cards.

8.5.2 Rack-based systems

These systems consist of a rack supplying mains power and a backplane with spaces for the system modules required to build a customised data-acquisition system. Such modules include A/D converters, multiplexers, analogue input and output, digital input and output, high-speed scanning units, counters and timers, and communication modules. Data are usually acquired into internal memory and are then off-loaded to the computer. For high acquisition rates, this memory may fill very quickly, requiring suspension of the data-acquisition while the data are off-loaded.

Communication between the computer and the data-acquisition system is dependent on the particular system. Some systems use a standard back-plane such as VME or VXI, while others use a proprietary design. In many cases, communication with the computer is over an RS 232 or IEEE 488 interface. In these cases, the data-acquisition rack appears as a single programmable instrument.

For all but the smallest applications, rack systems are required to provide the appropriate number of analogue input and output channels.

Standard VXI racks and modules are available from a number of the major companies such as Hewlett-Packard, National Instruments and Arcom. The extensive range of VXI racks and modules available allow the user to develop an extremely flexible data-acquisition solution no matter what the requirements are in terms of size, speed and processing.

The Series 500 Data Acquisition system from Keithley Instruments is a rack-based system that can be connected to a PC through a standard GPIB interface or a specialised parallel interface. Both of the solutions require a half card slot in the PC.

8.5.3   Interface boards/modules

Interface and data-acquisition boards have all the necessary interface or data-acquisition functions built on a single board that fits directly into an expansion slot on a computer or into an instrument rack. Boards are available, for example, for IBM PC/XT/AT, EISA, IBM PS/2, Power Macintosh, NuBus, and Sun SPARC Sbus, VXIbus and CAMAC systems.

8.5.3.1   Analogue boards.

These boards can range in price from £200 to £2000, depending on the resolution, conversion speed (and hence maximum sampling rate) and number of input channels. In choosing the number of channels, it is useful to bear in mind the possibility of future expansion. These boards can have up to a maximum of 64 analogue input channels but typically have 16 single-ended or 8 differential analogue inputs on a single board; they are relatively inexpensive and suitable for small applications.

Many of the boards have on-board amplification, which may or may not be software controllable. An additional option is the choice between differential or single-ended inputs. The differential inputs can be useful in situations that require a floating measurement; however, a board with differential inputs will usually have less channels than the equivalent single-ended version.

Boards are available with resolutions of 8–16 bits and maximum sampling rates up to 1 MHz depending on the computer platform. Some boards feature simultaneous sampling, on-board sample-and-hold, analogue or digital triggering, timing functions and digital I/O.

Data-acquisition boards acquire data and execute transfer to and from memory by programmed I/O, interrupts or direct memory access (DMA). DMA provides a direct link to the memory without CPU interruption, resulting in enhanced performance in terms of data-transfer rates. Boards with single channel DMA are capable of up to 100 kHz data through-put, while double-channel DMA boards increase the maximum through-put to 250 kHz.

An important factor in the selection of a board is whether it is compatible with both the platform and the software that is to be used.

8.5.3.2 Digital boards and devices. A range of boards and devices implementing the standards previously defined are available. These include:

- GPIB cards for a variety of computer platforms and rack-based applications
- serial interface cards, including RS 232, RS 422, RS 485
- Ethernet-to-GPIB devices, which allow networked computers to control GPIB devices anywhere on the network; platforms include Macintosh, PCs and workstations
- RS-232/GPIB, GPIB/RS 422 controllers: these devices allow GPIB instruments to be controlled via the serial port or allow a serial device to be integrated into an GPIB system
- GPIB controller for Macintosh SCSI port or serial port
- PCSMART from Arcom, which is capable of communicating with HART devices
- Digital signal-processing boards.

8.5.3.3 Card based instruments. Expansion card instruments fit into the standard PC expansion slot and have a cost advantage over programmable instruments or rack-based devices. The range of instruments is fairly limited but includes function generators, oscilloscopes, multimeters, counter timers and programmable power supplies. Software provides the user interface that mimics the front panel of a conventional instrument. Such systems are portable, easily configured and simple to operate since they require no knowledge of, for example, GPIB interfacing or commands.

8.5.4 Data loggers

These are stand-alone dedicated data-acquisition instruments accepting analogue signals from transducers. These usually incorporate such features as conditional triggering, data conversion, scaling, storage and display.

Most are capable of being interfaced to a computer via an IEEE.488 or RS 232C interface to allow programming and downloading of data.

For example, the Datataker from Data Electronics is a compact device that can be used as a stand-alone data logger using a PCMCIA-type card for data storage. The device is capable of communication via an RS 232 interface.

8.5.5   Portable data-acquisition systems

Portable data-acquisition systems are frequently required in environmental monitoring applications. The introduction of notebook and laptop computers has increased substantially the scope and versatility of such remote monitoring systems. Some are capable of all the features of a PC-based system while others are designed for small size and ease of use and may, therefore, have limitations. As with all acquisition hardware, a number of software options are available. System selection will, of course, be dependent on the application. It is not possible to cover the complete range of solutions available; however, the types of system and some examples are outlined below.

8.5.5.1   PCMCIA-based systems.   A number of companies have introduced PCMCIA data-acquisition cards. These cards require little power and are, therefore, suitable for battery operation. The DAQCard 700 from National Instruments has a 12 bit ADC, will accept up to 16 single-ended analogue inputs and can sample at up to 100 kHz. The Cardaqs range from IOTECH are available with 12 or 16 bit resolution and have software selectable gain on each channel. An expansion chassis capable of taking signal-conditioning units and of battery operation is available for use with these systems.

National Instruments have a GPIB interface on a type II PCMCIA card. This card has complete IEEE 488.2 capability.

All PCMCIA cards are available with a range of software solutions.

8.5.5.2   PC parallel port based systems.   A number of systems link to the Centronics port of a PC:

- the Data shuttle from Strawberry Tree is a compact data-acquisition unit weighing less than a kilogram, with 12 or 16 bit resolution, eight differential analogue inputs and sampling rates of up to 6 kHz; a number of options are available including thermocouple compensation and two analogue outputs
- the DAQPad from National Instruments is a low-cost data-acquisition system that communicates with IBM PC XT/AT and compatible computers through the standard Centronics parallel port; a second port on the unit allows the connection of a printer

- the DaqBook from IOTECH offers a number of options including the ability to expand the standard 16 channel analogue input to 256 channels using upto 16 expansion cards housed in three-slot expansion enclosures; either the standard Centronics port or the EPP with its higher data-transfer rates can be used for communication, depending on the computer platform

8.5.5.3 Expansion systems.   Some laptop PCs have an ISA docking port for the connection of an external expansion box.

The DacPac from Keithley Instruments is an expansion system operating via a notebook or laptop PC with an expansion box connector. The DacPac, which is approximately the same size as the laptop PC, takes up to two full sized cards and can be battery powered. (One of these cards can be an embedded PC allowing the complete system to be located within the DacPac.) This system is more flexible than some dedicated data-acquisition devices since any PC-based data-acquisition board or instrument may be used. PCs with docking ports are available from Toshiba, Compaq, AST and Dual.

8.6   Software

The function of the software in a data-acquisition application is to set-up boards or instruments, gather, analyse, store and present the data.

In developing an instrumentation system, it should be borne in mind that the software provides the interface between the user and the hardware. At the instrument-development stage, the requirement is for a system that is easy to use, flexible and highly interactive.

At a later stage, particularly if developing a commercial instrument, the requirement may be for a executable stand-alone programme capable of configuration, data-acquisition and presentation. At this stage a user-friendly system is essential.

Traditionally, software was developed using high-level languages such as FORTRAN and BASIC in combination with machine code subroutines, which were either self-written or supplied in the form of device drivers by the manufacturer of the data-acquisition boards. These systems had limited graphics capability, were difficult to modify (depending on the quality of the documentation) and required a specialist to implement or modify them.

Perhaps the greatest advance in the development of on-line data acquisition has been in the development of software packages. These packages range from low cost menu-driven or mouse-driven packages providing interactive control of board configuration and data-acquisition to extremely sophisticated programming environments allowing the development of turnkey software solutions for large-scale measurement systems.

Most data-acquisition hardware comes packaged with sufficient software either to drive the device in simple applications or to allow the development of more sophisticated or more user-friendly software solutions. This software may include device drivers or dynamic link libraries (DLL) for use with Windows applications.

DLL are the Windows equivalent of the DOS device driver. A DLL for an A/D board will be capable of providing all the necessary control of the board, including initialisation, setting sampling rates, channels, etc. National Instruments, for example, provide a data-acquisition DLL that will control more than 30 of their boards.

GPIB instruments are also packaged with device drivers that can substantially reduce the software development time. In Testpoint (Table 8.3), for example, a GPIB instrument may be controlled using SCPI commands or by using the device driver for the instrument and choosing the appropriate features from the panel. In either case, the code is generated and the final test routine can be used as a stand alone application.

Most data-acquisition software packages have the facility to export and import data to other packages, for example ASCII or text files. Microsoft Windows-based packages have the additional advantage that data transfers between the acquisition package and, for example, a spreadsheet can be easily achieved through Dynamic Data Exchange (DDE), which is the standard message-passing system used to import and export information between Windows applications.

A number of packages providing programming environments for high-level languages such as C, C++ and BASIC are available. Visual BASIC, for example, has become a popular programming environment for the development of data-acquisition software. It has no inherent data-acquisition capabilities; however, these are provided through DLLs and instrument drivers. Most of the major manufacturers will supply data-acquisition visual programming tools supporting their data-acquisition hardware and programmable instruments. Some of these tools will include additional features such as FFT analysis. Data Translation, for example, provide VB–EZ: a complete set of visual programming tools to simplify the operation of a supported board and the display of the resulting data.

Graphical programming environments using icons to build test routines allow sophisticated solutions to be developed without the need to write code. Packages utilising these intuitive programming methods tend to be user friendly and come with extensive tutorial support and on-line help.

Software selection will be based on a number of factors including: compatibility with data-acquisition hardware and the computer platform, requirements (run-time version, interactive, data analysis), user skills and experience (programming language, icon/menu based) and cost.

The packages in Table 8.3 have been selected from the wide range of commercial software to illustrate some of the features and approaches available.

Table 8.3 Packages selected from the wide range of commercial software to illustrate some of the features and approaches available in different systems

	Description
Asyst: Keithley Instruments	
Application	GPIB, RS 232, plug-in data-acquisition boards
Platform	IBM AT or compatible, DOS 3.0 or better
Features	A programming language-based package with extensive data analysis and graphics capability (including 3-D surface and contour plots). Interactive mode with on-line help. Applications may be distributed under a run-time licence
LabWindows CVI: National Instruments	
Application	GPIB, VXI, RS 232, CAMAC and plug-in data-acquisition boards
Platform	Windows for the PC, Solaris for Sun SPARC
Features	Interactive C programming environment for the development of a PC-based virtual instrument. Provides the tools to develop a graphical user interface. Extensive analysis library. Programmes can be complied for stand-alone execution
Lab Windows: National Instruments	
Application	GPIB, VXI, RS 232 and plug-in data-acquisition boards
Platform	IBM PC AT/EISA and PS/2,80286, co-processor, 2 Mbytes memory min, MS-DOS 3.0 or higher + version C or Basic as required
Features	A medium-cost package for IBM PC computers supporting A/D, IEEE 488 and RS 232. This mouse-driven package generates code for a background C or Basic programme and uses dedicated instrument drivers for particular instruments and converters. Programmes can be complied for stand-alone execution
DADisp: DSP Development Corporation (IOTech)	
Application	GPIB, data-acquisition boards
Platform	Windows, UNIX
Features	Graphical worksheet-based package capable of complex data-analysis and extensive graphics. Modules to support GPIB, data acquisition, statistics, etc.
Workbench: Strawberry Tree (Adept Scientific)	
Application	Strawberry Tree data-acquisition boards, GPIB, RS 232
Platform	PC versions for DOS 3.0, or Windows 3.1 or higher. Macintosh 68020 or SE
Features	Icon-based software for set-up of data acquisition and control tasks. Icons are linked on a worksheet that can be saved and run. For GPIB instruments device-dependent commands must be programmed as instrument drivers are not utilised
	No executable code is generated, though a runtime version is available for running developed worksheets remote from the base version
Testpoint: Keithley Instruments	
Application	Data acquisition, GPIB, RS 232, RS 485
Platform	PC compatible 386 or higher, Windows 3.1, 4 Mbytes RAM
Features	Object-oriented programming environment. Extensive analysis and graphics capability. Device drivers can be used with some GPIB instruments. Development system and runtime application builder. Runtime version can be distributed at no extra cost

Table 8.3 *Continued*

	Description
Easyst LX & AG: Keithley ASYST	
Application	Data-acquisition boards
Platform	PC 286 min, DOS 3.0 or later
Features	A low-cost mouse-driven system supporting A/D conversion. LX and AG includes basic data acquisition, storage and presentation, while the LX version includes data analysis (maths functions, curve fits, FFT, etc.) The system is extremely easy to use and is useful in basic data-logging applications
HPVee: Hewlett-Packard	
Application	GPIB, VXI, RS 232
Platform	PC Windows 3.1 or later, 12 Mbytes RAM; HP-UX, Sun workstations
Features	HPVEE (Visual Engineering Environment) is a graphical programming language for the development of automatic test routines. Integrates with standard languages including C, C++, Basic and Pascal. Run-time version available.
Lab View: National Instruments	
Application	Data acquisition, GPIB, Serial, VXI and CAMAC
Platform	Windows, Windows NT, HP-UX, Sun Solaris
Features	Provides an object-orientated graphical programming environment using NI Virtual Instrument libraries for a wide range of instruments, and plug-in boards. Extensive data analysis and presentation libraries are available. Using these VI libraries, the application is built through the medium of simulating instrument front panels on the computer screen (virtual instruments). An Application Builder is available for the development of runtime versions.

8.7 Summary

Recently, the most significant and relevant developments in data acquisition have been in the areas of portable data-acquisition systems and in the software applications packages. Predicting the future is an extremely hazardous process; however, the authors feel that the future of on-line computing will involve developments in all of the following areas:

- an expansion in the range and number of transducers incorporating digital communication (SMART devices)
- vastly improved communication and standardisation, particularly with the development of a standard fieldbus
- increase in PCMCIA board-based applications
- much improved MMI (man–machine–interface) and graphics capability
- further advances in signal processing, data storage and retrieval.

Further reading

Thompson, L.M. (1991) *Industrial Data Communications, Fundamentals and Applications.* Instrument Society of America.

Driscoll, F. (1992) *Data Communications*. Saunders College Publishing.

Howath, M. (1994) HART – Standard for 4–20 mA digital communication. *Measurement and Control*, **27**, 5–7.

Allen, C. (1994) The Interoperable Systems Project (ISP). *Measurement and Control*, **27**, 38–41.

Desjardins, M. (1994) Worldfip. *Measurement and Control*, **27**, 42–46.

Application notes and data sheets from Adept Scientific, Arcom, Biodata, Data Translation, Hewlett-Packard, IOtech, Keithley Instruments, Lecroy, National Instruments, Strawberry Tree.

Index

Volume denoted by **bold** Roman numerals.